AF443047

# THE DETERMINATION OF GEOPHYSICAL PARAMETERS FROM SPACE

# THE DETERMINATION OF GEOPHYSICAL PARAMETERS FROM SPACE

Proceedings of the Forty-Third Scottish
Universities Summer School in Physics,
Dundee, August–September 1994.

A NATO Advanced Study Institute.

Edited by

**N E Fancey** — University of Edinburgh
**I D Gardiner** — University of Dundee
**R A Vaughan** — University of Dundee

Series Editor

**P Osborne** — University of Edinburgh

Copublished by
Scottish Universities Summer School in Physics &
Institute of Physics Publishing, Bristol and Philadelphia

British Library Cataloguing-in-Publication Data:

*A catalogue record for this book is available
from the British Library*

*ISBN 0-7503-0-350-6*

Library of Congress Cataloging-in-Publication Data are available.

Copublished by

**SUSSP Publications**
The Department of Physics, Edinburgh University,
The King's Buildings, Mayfield Road, Edinburgh EH9 3JZ, Scotland.

and

**Institute of Physics Publishing**, wholly owned by
The Institute of Physics, London.

Institute of Physics Publishing, Techno House, Redcliffe Way, Bristol BS1 6NX, UK.
US Editorial Office: Institute of Physics Publishing, The Public Ledger Building,
Suite 1035, 150 South Independence Mall West, Philadelphia, PA 19106, USA

Printed in Great Britain by J W Arrowsmith Ltd, Bristol

# SUSSP Proceedings



/continued

# SUSSP Proceedings (continued)

# Lecturers

| | |
|---|---|
| Professor J Askne | Chalmers University of Technology |
| Dr L Bannehr | Freie Universität Berlin |
| Mr I D Downey | Natural Resources Institute |
| Professor A F G Fiuza | University of Lisbon |
| Dr T H Guymer | Chilworth Research Centre |
| Dr C G Helmis | University of Athens |
| Dr J McGregor | Victoria University of Wellington |
| Dr D Mantripp | Mullard Space Science Laboratory |
| Professor G E Peckham | Heriot-Watt University |
| Professor A C B Roberts | Simon Fraser University |
| Mr M Rohn | Freie Universität Berlin |
| Mr D Sloggett | Earth Observation Science Ltd |
| Dr C A Varotsos | University of Athens |
| Professor G Warnecke | Freie Universität Berlin |
| Dr A I Watson | University of Stirling |
| Mr Y Xue | University of Dundee |

# Executive Committee

| | | |
|---|---|---|
| Prof. A P Cracknell | University of Dundee | *Director* |
| Dr N E Fancey | University of Edinburgh | *Treasurer and Co-Editor* |
| Dr R A Vaughan | University of Dundee | *Secretary and Co-Editor* |
| Dr A O Tooke | University of Dundee | *Social Secretary* |
| Dr I D Gardiner | University of Dundee | *Co-Editor* |

# International Organising Committee

| | |
|---|---|
| Professor G Warnecke | Freie Universität Berlin |
| Professor A C B Roberts | Simon Fraser University |
| Professor M Ph Stoll | University of Strasbourg |
| Professor C A Varotsos | University of Athens |
| Professor G E Peckham | Heriot-Watt University |
| Dr C Duncan | University of Edinburgh |

# Preface

It is always a pleasure to write the Director's Preface to a set of our Summer School proceedings because it means that the developments in knowledge that were presented at the Summer School are at the final stage in the process of finding their way into the permanent scientific literature. This particular Summer School was concerned with the use of space-borne, and to a lesser extent airborne, techniques for studying various geophysical parameters. As will be seen from the Table of Contents, the topics convered included the study of the land, the sea and the atmosphere as well as air-sea interactions. The advantage of using remote sensing techniques is the rapid and frequent gathering of synoptic data, while the disadvantage is that the information collected does not necessarily constitute a direct measurement of a geophysical parameter; the values of geophysical parameters usually have to be deduced from the remotely-sensed data by some sort of inversion procedure. The effects of the intervening atmosphere between the instrument and the surface or cloud being observed and the errors in the values of the geophysical parameters retrieved from the remotely-sensed data are very important and very considerable effort has to be devoted to their study. Thus a large amount of the material presented at this Summer School was concerned with the problems of the various inversion procedures and the accuracy of the results.

I am, as always, very grateful to everyone who worked so hard to make the Summer School a success, both on the organisational side and also the lecturers and seminar speakers who contributed the scientific material which formed the academic content of the Summer School. As always, an important component of the Summer School was the informal contacts established among the participating students and between the students and the lecturers.

I should like to acknowledge the support that we received from our various sponsors and without which the Summer School could not have been held. Apart from the major support received from NATO, we also received support from the Commission of the European Communities (the Euro-conferences programme), Earth Observation Sciences Ltd, the Institute of Physics, the Natural Environment Research Council and the Governing Committee of the Scottish Universities Summer Schools in Physics. I am also very grateful to Dr Peter Osborne for his invaluable assistance in the later stages of the preparation of these Proceedings.

Arthur P Cracknell
Dundee, June 1996

# Contents



/continued

# Overview of Remote Sensing and Data Availability

Ian D Downey

Environmental Science Group
Natural Resources Institute, UK

## 1    Introduction

This chapter provides a brief overview of satellite remote sensing in the 1990s. The material is aimed at giving an appreciation of the (ever increasing) wealth of remote sensing satellite systems and measurement technology available for observing the Earth and its atmosphere to determine geophysical parameters. The presentation is not intended as an in depth treatment of the science of remote sensing, rather a brief reminder to researchers, teachers, scientists and other workers of the observations that are possible from current and future systems. The material is presented simply as a backcloth for the detailed technical materials presented elsewhere during the summer school. The reader is directed to any of the excellent texts (Kramer 1994, Lillesand and Kieffer 1994, Cracknell and Hayes 1993, Barrett and Curtis 1982) now available for this subject for further elaboration if required.

## 2    Definition

Remote sensing is the collection of information about an object from a distance without being physically in contact with it. The information is usually some kind of (geo)-physical measurement and most often transformed into some form of imagery or other visualisation.

Remote sensing is still a relatively new science, aerial photographic techniques having been used since the early 1900s and satellite imagery since the 1960s. Image processing techniques developed manually at first and then digitally by taking advantage of the advent of modern computing technology. The field is growing quickly and ever stronger as time progresses. More satellites carrying instruments with increased sensitivities mean that realising the potential for numerous, high precision geophysical measurements has been greatly advanced in the last years. Increasingly, therefore, a shift towards more quantitative information is being achieved (Cracknell 1994). This data explosion is expected to continue into the next century.

# 3  Mechanisms

## 3.1  Orbits

Measurements can be made from two types of satellite orbit; geostationary and polar orbiting. The former ensures that a satellite (some 35,000km above the Earth) orbits at the same rate as the Earth rotates thus appearing to be stationary over the same position. Satellites in these orbits view an entire hemisphere of the Earth at one time but with a coarse spatial resolution (2.5–5km). They are predominantly used for weather forecasting as many measurements can be made of the same position on the Earth in a given time period allowing rapid updates to models and predictions. Geostationary satellites also provide an ideal vehicle for relaying telecommunications so that information (images, charts, forecasts, etc.) from around the globe can be passed on and distributed routinely.

Polar orbiting satellites are in close proximity to the Earth (usually 700–800km) and make regular passes of the globe whilst it rotates beneath them. In this way observations are made in relatively narrow strips (100–2500km) but with higher spatial resolution than from geostationary satellites (*e.g.* 10–100m). These satellites tend to be used for higher definition weather forecasting or natural resources monitoring and mapping. Because of their orbit, the same point is not seen all the time but is revisited after some days.

In both cases, the satellite orbit enables a synoptic view of the Earth (and its atmosphere) over large areas, and at regular intervals. The instruments on board these satellites are designed to capitalise on this revisit capability by making observations at (unrivalled) high spatial sampling frequencies of the sub satellite region.

## 3.2  Instruments

Remote sensing satellites carry instruments to make measurements of atmospheric, terrestrial or oceanographic parameters. These instruments measure fluxes of electromagnetic energy impinging on and received from these different surfaces. That energy could either be reflected solar radiation (*e.g.* sunlight), absorbed and re-emitted radiation (*e.g.* terrestrial temperature) or energy directed at the Earth by the instrument itself (*e.g.* Radar). Remote sensing systems are designed to detect and quantify this returning radiation. Thus, there needs to be a physical relationship between electromagnetic energy and matter for these measurements to be made and have relevance. These can be summarised as follows:

- All matter above zero degrees Kelvin emits electromagnetic radiation.

- All matter absorbs and reflects radiation (depending on wavelength).

- As surface temperature increases, more energy is emitted (at shorter wavelengths).

- Measurements of the emittance, absorption and reflectance characterise objects and phenomena.

- Therefore, materials to be characterised must be spectrally separable.

Two types of instrument are used to measure these characteristics; passive and active.

**Passive sensor systems**

These systems rely entirely on incident radiation impinging on some form of detector sensitive to particular wavelengths of energy. Most remote sensing instruments currently in orbit (and a large portion of those planned for the near future) are passive sensors. They are utilised for multispectral Earth observation, Earth radiation budget measurement and for profiling atmospheric constituents (Table 1).

| Source of radiation | Sun |
|---|---|
| Target | Atmosphere, clouds, Earth's surface |
| Sensor detects | Reflected, scattered, emitted radiation from Earth's surface, clouds, atmosphere |
| Earth-Atmosphere budget | Radiation from the sun is predominantly shortwave ($< 0.4\mu$m ) ultra-violet and visible) |
| | Radiation returning from Earth's surface or atmosphere is mainly longwave ($> 0.4\mu$m ) energy and heat fluxes: visible near and thermal infra-red and microwave) |

**Table 1.** *Passive sensor systems*

**Active sensor systems**

These systems are sensitive to surface roughness, moisture content, surface electrical properties and target motion. They operate independently of solar illumination and are (largely) unaffected by clouds and precipitation. Active systems can be used in an imaging mode (*e.g.* Synthetic Aperture Radar SAR) or non-imaging mode for range detection (radar altimetry) or surface roughness related phenomena (*e.g.* wind scatterometry). Due to the nature of radar signals, the imaging mode of these systems is subject to geometric distortions and a degree of image noise ('speckle') which need to be rectified for subsequent processing (Table 2)

| Source of radiation | The active sensing instrument sends out its own pulses of energy |
| --- | --- |
| Target | Earth's surface (and atmospheric particles, such as rain, if shorter microwave wavelengths are selected) |
| Sensor detects | Reflected, re-emitted returning radiation from Earth's surface. Variation in (phase and polarisation) characteristics of returned energy from known outgoing pulse allows identification and measurement of different objects and phenomena. |

**Table 2.** *Active sensor systems*

## 3.3   Spectral and atmospheric effects

Solar energy impinging on the Earth is not constant nor at the same energy level at all wavelengths. Furthermore, different molecules in the atmosphere selectively absorb or transmit certain wavelengths which further inhibits spectral zones where measurements can be made successfully. One further limitation is the fact that surface materials also reflect, absorb and emit radiation depending on wavelength so that the selection of appropriate spectral regions becomes critical to being able to make measurements of distinguishable geophysical parameters.

The presence of clouds and haze is often a restrictive influence on many remote sensing measurements. This has stimulated a drive to develop passive and active microwave systems which operate at wavelengths capable of penetrating clouds to different degrees and also of operating without the presence of incident sunlight.

The key then to successful remote sensing information extraction is the selection and monitoring of specific wavelengths of the electromagnetic spectrum for the purposes of identifying objects and phenomena with specific reflectance or emittance characteristics. Figure 1 (from Lillesand and Kieffer, 1994) summarises the opportunities and restrictions for this approach.

Not unsurprisingly, therefore, a great number of instruments have been developed to enable appropriate measurements of geophysical parameters to be made. These have been application driven for the large part and have focused on meteorology and Earth, or natural resources, observation and monitoring. Predominant amongst these are Meteosat/GOES satellites, the long established NOAA series and the Landsat and SPOT series of satellite systems.

As both technology and knowledge of these systems have improved over the last twenty years or so, both spectral and spatial resolutions have been refined to improve the precision and breadth of possible observations. This trend looks set to continue into the next century the development of many more instruments with increasingly specific objectives. The reader is directed to the recent, excellent text by Kramer (1994) for details of the capabilities of this wealth of instruments already in operational or experimental use as well as those planned for the next decade.

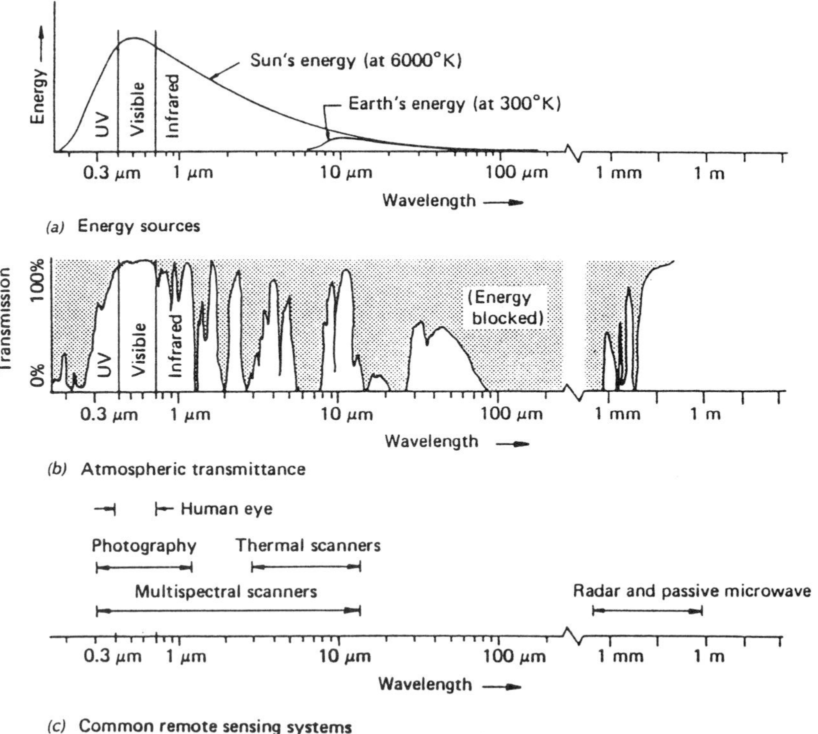


**Figure 1.** *Spectral characteristics of energy sources, atmospheric effects and sensing systems (Lillesand and Kieffer 1994)*

# 4 Data availability

## 4.1 Current data sources

The number of operational and pre-operational Earth observation satellites continues to grow, and with it the amount of data collected. Increasingly common are the number of non-imaging remote sensing instruments that are also carried onboard these satellites. Undoubtedly more and more data are becoming available and an increased sophistication in the types of data products has been seen in recent years. This has been accompanied by significant developments in information availability to assist the user to access these data and products. We list below the major, currently operational Earth observation systems.

### NOAA meteorological satellite

The AVHRR sensor has been flown on the NOAA TIROS-N series of meteorological satellites since June 1979. There are two operational platforms orbiting at any one time.

Also carries TOVS (atmospheric sounder profiler) instrument for weather, environmental and geophysical observations. ARGOS capabilities for accessing ground information. Recent failures of NOAA-10 and NOAA-13. NOAA-14 launched successfully in 1994. NOAA programme to merge with ESA METOP in the late 1990's.

### Geostationary meteorological satellite

Meteosat, GOES, GIMS series providing frequent weather information and acting as relay satellites for global weather observations and forecast broadcasts. Programmes ongoing into the next century. Meteosat to become MSG (Meteosat second generation).

### Landsat series

Platforms 4 and 5 nearing the end of design life but still a mainstay of high resolution Earth observation activities in many application areas. Landsat 6 failed in 1993 and Landsat 7 is expected to launch in 1997. Besides continuing the TM data series, Landsat 7 will also carry a 15m spatial resolution channel operating in a panchromatic mode across the visible wavelengths.

### SPOT series

A new generation of (European) Earth observation satellites providing improved cartographic capabilities as well as thematic EO applications. Three systems launched to date. SPOT 4 will carry the new 'VEGETATION' instrument (similar to AVHRR) allowing simultaneous multi-scale studies of surface phenomena for the first time.

### GEOSAT and TOPEX/POSEIDON

Operational 'active' systems dedicated to altimetry. GEOSTAT mission driven by military requirements and ended in 1990. CNES TOPEX/POSEIDON mission launched in 1992 continues the collection of altimetry data worldwide. These instruments can be used to assist scientific applications on geodosy, sea-level change, ocean circulation etc.

### Russian systems (RESURS-O, MK, KFA etc)

RESURS-O satellite series with similar capabilities and objectives to Landsat. Programme planned to run to the mid 1990's and believed to be still operating. High resolution (ex-military) data from RESOURCE-F series that carry KFA, MK, DD5 camera systems is now available in analogue and digital form with resolution of 2–20 metres.

### IRS series

Indian systems with capabilities similar to TM now being marketed by EOSAT (who also operate Landsat) since the failure of Landsat 6. New IRS instruments (including high resolution panchromatic imaging expected in orbit in the 1990's.

**ERS series**

ERS mission is the first European Earth Observation remote sensing satellite (Meteosat for weather). Mission launched in 1991 and viewed as a great technical success providing a new source of observations from improved active instruments. New generation of thermal imaging radiometer (ATSR) providing better information on atmospheric effects and surface temperature accuracies. ERS-2 launched in 1995.

## 4.2   New data sets and products

Exciting new developments in the availability of EO data concern (a) the flow of data from the former Soviet Union military satellites and (b) the generation of large time series of thematic data products by international organisations such as the International Geosphere Biosphere Programme (IGBP). High resolution Russian (ex-military) satellite data is now available commercially in analogue and digital form through an organisation called RESOURCE-F WORLDMAP. There is a growing number of WORLDMAP distributors. WORLDMAP is a joint venture between the Russian State Centre PRIRODA and JEBCO Information Services (USA) for the exclusive marketing and distribution of digital versions of PRIRODA's RESOURCE-F series satellites that carry camera systems with resolutions of 2–20m.

Two Russian agencies are licensed to deal in the satellite photography. The State Centre PRIRODA provide data from the RESOURCE F-1, F-2 and F-3 satellites. These include data from the KFA-2000, KFA-1000 and KATE-200 cameras. All data from these systems are sold through PRIRODA. Sovinformsputnik (SIS) provide data from converted military satellites including the KVR-1000, TK-350 and DD-5 camera systems. All other companies that distribute Russian data must go through PRIRODA or SIS eventually. Most other companies that advertise sale of Russian data are agents of Soyuzcarta, the Russian marketing arm of PRIRODA in the past. Soyuzcarta still has the rights to sell analogue data (prints). In the future, Soyuzcarta will be an agent for sale of digital data with requests handled by an American agency called WORLDMAP. These new data sets offer remarkable spatial resolution further expanding the capabilities of remote sensing for mapping and monitoring more local scale objects and phenomena. The following table indicates the typical spatial resolution of these systems.

| Camera System | Mode | Spatial Resolution (m) |
|:---:|:---:|:---:|
| KFA-200 | Multispectral | 25–30 |
| MK-4 | Multispectral | 12–14 |
| TK-350 | Panchromatic | 10 |
| KFA-1000 | Multispectral Panchromatic | 8–10 |
| KFA-3000 | Panchromatic | 2–3 |
| KVR-1000 | Panchromatic | 2–3 |
| DD-5 | Panchromatic | 1 |

**Table 3.** *Modes and Spatial Resolutions of Russian ex-military systems*

## IGBP World Data Sets

The International Geosphere Biosphere Programme (IGBP) is an initiative established in 1986 under the aegis of the International Council of Scientific Unions. The objective of IGBP is to contribute to observations and understanding of global change phenomena. The research was established in 1990 and this has led to the creation of a number of core projects:

- International Global Atmospheric Chemistry (IGAC)

- Global Change and Terrestrial Ecosystems (GCTE)

- Biospheric aspects of the hydrological cycle (BAHC)

- Land ocean interactions in the coastal zone (LOICZ)

- Joint Global Ocean Flux Study (JGOFS)

- Past Global Changes (PAGES)

There are a further three IGBP integrating activities that service these core projects:

- GAIM - Task Force on Global Analysis, Interpretation and Modelling to advance the predictive understanding of the Earth System

- IGBP-DIS - Data and Information Service to provide guidance on the gathering, management and exchange of data

- START - System for Analysis, Research and Training to provide for a world wide network of research centres to strengthen regional research capabilities and inter-disciplinary scientific expertise, particularly in developing countries.

Although the respective activities work in concert to improve and integrate data availability from all scientific observation systems, in terms of EO data, the IGBP-DIS is perhaps the most important of these activities to consider. The overall aim of the IGBP-DIS is to improve the supply and management of data and information to attain IGBP's scientific goals and support the numerous working groups. There are three main activities of the IGBP-DIS:

- Data set development
    Assessment of needs
    Determination of availability
    Ensure processing and preparation of data sets

- Data dissemination
    Advise core projects on data distribution and transfer
    Assistance in establishing networks

- International data coordination
    Collaboration to ensure that IGBP data needs are met
    Assessment of existing data bases

A particular example of the type of work being done by the IGBP-DIS to increase the availability of EO data is the global 1km project of the Land Cover Working Group (Townshend 1992). IGBP members agreed a programme to collect (NOAA AVHRR) data for 18 months from April 1992 as a near global data acquisition activity. They undertook the coordination of data acquisition by NOAA, USGS, (EROS Data Centre ) and ESA utilising a global network of ground receiving stations. Data archiving is carried out at EROS Data Centre as part of it 'Distributed Active Archive Centre' functions (DAAC) and also at ESA. Global time series measurements are now becoming available including 10 day NDVI composites. This represents one of the first truly global time series data sets available. Processed using a consistent suite of algorithms, users can examine past trends and build them into studies incorporating new data from similar instruments.

All the initiatives described above demonstrate not only the increased availability of data and a drive to facilitate improved access to those data but also, more importantly, the ability to handle the large data volumes involved in a realistic and meaningful way. This indicates the potential for establishing data bases and networks of users who can access consistent time series of data from which to extract a host of geophysical observations. These developments represent, therefore, the desire and ability to launch, manage and access the wealth of EO data that can be expected in the next decades and which are described below.

## 4.3  Improving data access

Data from most of these EO systems are commonly referred to now by the term 'product level'. This indicates the degree of post processing that has been applied to the original data to derive some form of enhanced or thematic product. Raw data is usually denoted as Level 0 although there is a degree of variation between systems on the various levels available. For example, SPOT provide data at five distinct product levels (Table 4).

The Committee on Earth Observation Systems (CEOS) is making efforts to encourage standardisation of data formats (and thus product levels). CEOS has been set up to organise the world-wide coordination of Earth observation satellite programmes. Comprising three distinct types of member organisation (full member, observer and associate) CEOS membership represents 20 space agencies (with their own EO systems), 5 with planned EO systems and 10 international user organisations respectively. The main aims of CEOS are to optimise EO satellite activities via international cooperation, support its members and the international community (the focal point for international coordination of such activities) and promote exchanges of political and technical information with a view to harmonising EO satellite systems

Consistent, standard product levels are important because they are often a basic piece of information for locating, selecting and obtaining the appropriate data for a given application. Quality of measurement can also be assessed so that users have confidence in the observations that have been made and the instruments that were used to obtain them. Consistency is expected to become increasingly important as the number and duration of operational satellite programmes increase. This is highlighted in particular by the development of new (global) IGBP products as described above.

Many satellite operators and data suppliers now offer access to 'browse systems'

| Level | Processing | Product Examples |
|---|---|---|
| 1A | Raw image ingested, detector normalisation, radiometric correction | Basic data set<br>Panchromatic and XS data |
| 1B | Detector normalisation, radiometric correction, systematic geometric correction (Earth rotation) | Basic Data set<br>Panchromatic and XS data<br>Panchromatic and XS data merged<br>Merging consecutive along-track scenes |
| 2A | Corrections based on satellite attitude data, image data rectified to given cartographic projection but without GCPs | Variable scene size<br>XS data resampled at 20 metres<br>Panchromatic data resampled at 10 metres |
| 2B | Image data rectified to given cartographic projection using GCPs | Variable scene size, directly registered with a map (GCPs required from map scales of 1:25,000 - 1:100,00)<br>XS data resampled at 20 metres<br>Panchromatic data resampled at 10 metres<br>Pixel to pixel joining of two level 2 scenes |
| S | Image data rectified to given cartographic projection using GCPs or reference image | Variable scene size, directly registered with a map or reference image<br>Pixel to pixel joining of two level 2 scenes<br>Multi-temporal data sets |

**Table 4.** *Typical SPOT product levels*

(increasingly on-line via Internet) which can search data bases of acquired data, present listings of available products and even display sampled imagery for visual inspection. Catalogues of the various products can be scanned to locate the observations of interest to a user. All of these developments are geared to improving access to EO data and the meta-data which describes it. In the future it is expected that many of these catalogue systems will be fully inter-operable offering context search facilities for sophisticated data and products.

Particular initiatives include ESA's Guide and Directory Service (GDS) and the EC Centre for Earth Observation (CEO). Based at the ESA facilities at ESRIN, Italy, GDS already provides on-line information about ESA's programmes and missions as well as reference material regarding mission products. Using a technology known as 'hyper-media' full text documents, images, graphics, audio and animated image sequences are gathered together in the system. The GDS service also provides a network entry point for a number of other information systems (e.g. ESA-IRS, LEDA, etc.) which can be located and accessed. CEO is a programme still under development but with more wide reaching implications.

The EC and ESA have agreed to combine their expertise (with that of member states) to increase the use and value of EO data by establishing a coordinated, decentralised European EO network called EEOS, the European Earth Observation System. CEO is the EC's contribution to EEOS and it is intended to build on established institutions and existing or planned networks and projects. Launched by the EC Joint Research Centre (JRC) in 1992, the CEO intends to improve user access to (and use of) EO related data and services by:

- encouraging better communication between users and user communities,

- stimulating the creation of high-level products (where and when necessary),

- improving data standardisation and quality assurance,

- coordinating design and operation of future decentralised archives and databases.

The goal is to design and implement CEO by the end of the Fourth Framework Programme in 1998. During that year it is also expected that there will be a dramatic rise in the number of operational EO satellites making more data available. Europe, USA and Japan intend to embark on major programmes of Earth observation missions designed to take EO into the next century and provide long term time series data from a variety of instruments that have been shown to offer reliable and much needed, geophysical observations.

## 4.4 Future missions

The number of EO missions and programmes is set to increase markedly in the next twenty years. Numerous reference materials (*e.g.* Kramer 1994, Jane's 1994, CEOS 1994, NASA 1993) provide detailed information on many of the initiatives to be expected. Several are particularly worthy of mention although the reader is directed to cited texts for more comprehensive information.

### Seastar/SeaWiFS

The Sea-Viewing Wide Field-of-view Sensor (SeaWiFS), originally intended to fly on Landsat 6, has been designed to be incorporated as a separate (commercially operated) mission (known as SeaStar) in the American EOS Polar Platform missions. The operation of the SeaStar satellite (including data processing and distribution) will be coordinated by Orbital Science Corporation. It is the first sensor specifically designed for observing ocean and marine phenomena since the demise of CZCS. This new sensor will provide data from 8 bands of data in the visible and near infrared suitable for measurements of ocean colour, ocean biology and ecology as well as phytoplankton concentrations.

### RADARSAT

RADARSAT was planned to be the first Canadian remote sensing satellite and the world's first operational SAR platform (as opposed to the ERS experimental missions). Canada is at the forefront of SAR technology and applications and has used these techniques operationally for many years, particularly for ice monitoring. Collaboration between the private sector, provincial governments and the Canadian Space Agency has led to a new initiative in data capture and dissemination with both public and private sector sharing the development and risks of the mission.

The mission was scheduled for launched during 1995 and has a nominal five year lifespan. The design of the RADARSAT platform will also allow a new initiative in SAR data collection as the SAR instrument will operate at a variety of beam selection modes which allow different swath widths, resolutions and incidence angles. These modes are the result of the unique beam forming capabilities of the SAR instrument. Multi-look data, wide swaths, variable incidence angles, fine resolution 'spotlight' beams, scanning beams are all possible with the RADARSAT antenna. The flexibility to switch rapidly between modes during overpass has been developed to meet the optimum requirements of both research and operational users. These include the demand for near real time product delivery.

Based on the experience of ERS-1 to date, it is likely that users can expect a great deal of new capabilities and observations from RADARSAT. The complementarity of the RADARSAT and ERS-2 missions is also to be anticipated optimistically, as a source of new observations and a continuation of the data archive already established through the outstanding performance of ERS-1 to date.

## MTPE-EOS Mission to Planet Earth / Earth Observing System

'Mission to Planet Earth' Earth Observing System (MTPE-EOS) is a NASA initiated concept to use both space and ground based measurement systems to provide the scientific basis for understanding global change. The goal is to launch a series of dedicated missions (the EOS) in polar, low inclination orbits to provide continuous observations over at least a 15 year period from the late 1990s onwards. MTPE intends to utilise the capabilities offered by complementary system developed by ESA, NASDA and NOAA to provide an integrated suite of instruments and a comprehensive data and information system (EOSDIS). Since inception the programme has been subject to budgetary constraints which have led to changes in the mission profiles. However EOS is still intended to become a reality by the turn of the century in tandem with ESA and EUMATSAT developments such as ENVISAT, METOP and so on.

The proposed launch date of EOS is 1998 with a prospective mission length of 15 years. NASA began conceptual studies in 1982. Coordination with ESA, Japan and Canada began in 1986. A series of six polar orbiting satellites are planned using 3 identical satellites each with a five year design lifetime. There are far too many proposed missions to document here and the reader is directed to Kramer (1994) and NASA (1993) for detailed information. However, it is suffice to point out that they encompass all the major spaceborne remote sensing sensor types including multispectral imaging radiometers/spectrometers, passive and active microwave instruments (SAR, altimeter, etc.) and passive microwave sounders as well as new developments such as sensors for atmospheric chemistry and laser based sensors (LIDAR, ranging, etc.). These sensors are combined together in the EOS concept as groups focused on application themes. These are:

| Mission Series | Planned observations/applications |
|---|---|
| EOS-AM (morning pass) | Clouds, radiation, and aerosols<br>Land and ocean surface temperature<br>Vegetation and ocean phytoplankton<br>Global biological productivity<br>Tropospheric chemistry |
| EOS-COLOR (afternoon pass) | Ocean phytoplankton and biological production |
| EOS-AEROSOL | Aerosols and atmospheric chemistry |
| EOS-PM (afternoon pass) | Clouds, radiation and aerosols<br>Precipitation and humidity<br>Snow cover and sea ice<br>Land and ocean surface temperature<br>Vegetation and ocean phytoplankton<br>Global biological productivity |
| EOS-ALTIMETRY | Ocean circulation and sea level<br>Ice sheet elevation<br>Land surface elevation<br>Cloud layering |
| EOS-CHEMISTRY | Tropospheric and stratospheric chemistry<br>Aerosols<br>Solar radiation |

## ESA European Polar-Orbiting Platform Programme (EPOP/Columbus)

This is the European initiative to provide for the data needs to address environmental research, process measurement and monitoring needs into the 21st century. Like the American EOSDIS programme this will comprise a Polar Platform supporting a variety of environmental observation and measurement sensors. This programme also involves separately flown satellite platforms and sensors. The Earth Observation Programme has four main elements:

- An ERS-1 follow-on mission (ERS-2)

- A Solid Earth gravity mission (ARISTOTELES)

- A Meteosat Second Generation (MSG)

- A series of Polar Orbit Earth Observation Missions (POEM)s

Polar Orbit Earth Observation Missions (POEM) are the first of the ESA Earth observation programme missions based on the Columbus Programme Polar Platforms and will be launched in the late 1990s. The POEMs are designed to provide data for specific disciplines of environmental monitoring, to this end two series of polar platforms will be distinguished: the M-series (meteorology, ocean and climate), and the N-series (Earth resources and atmosphere).

Building on experience with the ERS and NOAA series the POEM-1 mission will comprise two dedicated mission series, ENVISAT and METOP to cover the requirements of both environmental and meteorological mission objectives. ENVISAT is designed to monitor the Earth's environment on a range of scales using instruments developed on the ERS series and from a suite of other pre-operational instruments largely related to atmospheric chemistry and radiation budgets. ENVISAT will also carry the MERIS instrument, a programmable high resolution imaging spectrometer capable of making observations suitable for ocean, atmosphere and land applications.

METOP is the European contribution to the continuation of the current dual (morning and afternoon pass) NOAA platform configuration beyond the year 2000. NOAA intend to focus their attention on the afternoon pass in the future and ESA/EUMETSAT will fill the gap by sharing the service provision with NOAA. This will ensure the continuity of the mission profile and extend the time series of low resolution (morning pass) data for the operational requirements of the meteorological community and also those of global cooperative programmes such as IGBP METOP is expected to complement ENVISAT (providing enhanced monitoring capabilities) and continue the trend towards long term data sets for consistent observations into the future.

## 4.5   Data rates and ground segment issues

The dramatic rise in the number of operational EO satellites is only surpassed by the associated (enormous) data rates that can be expected. Multi-dimensional and multi-sensor data, from numerous platforms and at different times of day will yield a huge amount of observations. The EOS missions, for example, are expected to generate level-0 data at a rate of the order of 200–300 Gigabytes per day. A similar amount can be expected from the combination of all other non-EOS missions which include POEM, RADARSAT, SPOT, Landsat, *etc.* (NASA 1993). It is as well that so many programmes currently exist, or are being promoted and developed, to demonstrate and establish techniques for handling and disseminating such large data volumes.

However, even these programmes may prove insufficient. With the anticipated data rates from the EOS mission complement given above there has been a pressing need for the development of sophisticated and complex distributed ground segment capabilities. Further development will be required in light of the logistical difficulties which have had to be overcome in managing and processing the (comparatively) low data rates involved in the IGBP-DIS operations.

A further ground segment is that of the cost of access to these data. Commercially operated systems (Landsat, SPOT, etc.) charge a (relatively high) premium for their data sets. Licences to receive these data at a ground station are also expensive capital costs. In the future, the US government policy on data costs looks set to be revised to favour reduced costs to users. Data from Landsat 7 seem to be the most likely candidate for a more relaxed data pricing policy in the near future. However, in Europe, there appears to be a contrary trend. Meteosat data, until now freely available to users with their own (relatively low cost) ground stations are likely to be broadcast in an encrypted form in the near future. Decryption licences and equipment will need to be obtained and this may prove a disincentive for some users.

Undoubtedly, we are on the verge of a dramatic shift in data availability which will test man's ability to process and act on information to a high degree. However, a cautionary note is that, despite the many remarkable efforts to coordinate and manage large amounts of data to date, much of the EO archive already established remains largely untapped. Also, the issue of what degree of freedom of access to future data sets is not resolved nor a consistent policy established as yet. If we are to fully realise the potential of the projects envisaged in the future, at least as much effort as has gone into developing the instruments and tools will need to be concerted into processing, analysing, and disseminating the massive volumes of information that are about to cascade down from the heavens.

# References


Asra G and Dokken D J [eds], 1993, EOS Reference Handbook, NASA

Barrett E C and Curtis L F, 1982, Introduction to environmental remote sensing, (2nd edition), Chapman and Hall

Cracknell A P, and Hayes L W B, 1993, Introduction to remote sensing, Taylor and Francis

Cracknell A P, 1994, Remote sensing and the determination of Geophysical parameters for input to global models [in] Foody G and Curran P [eds], Environmental remote sensing from regional to global scales, Wiley, pp.167-180

Kramer H, 1994, Observing the Earth and its environment, Springer-Verlag

Lillesand T M and Kieffer R W, 1994. Remote sensing and image interpretation, Wiley

Townshend J R G, 1992, Improved Global Data for Land Applications, IGBP Global Change Report No. 20, IGBP

# Contemporary European Initiatives for the Utilisation of Space-based Data in Environmental Monitoring

D R Sloggett

Earth Observation Sciences Ltd.,
Farnham, UK

## Abstract

Heightened public and political awareness of the impact of man's activities on the environment have led to a number of activities aimed at increasing our understanding of the physical, chemical and biological processes that exist at the heart of the Earth's ecosystem. In broad terms these activities have two major objectives. The first is to increase our ability to understand the current state of the environment in order that short-term responses, designed to improve the environment, are implemented as efficiently as possible. The second is to develop improved predictive techniques that will help us understand the socio-economic consequences of any changes that might occur, such as a rise in sea levels or alterations to climate patterns. This improved understanding will assist governments, working at local, regional and global levels, to formulate policies that will encourage the sustainable development of natural resources in the future.

In order to increase our knowledge of the processes that control our ecosystem, we need to increase our observations of the Earth at a variety of scales, from the local to the global, using *in-situ*, aircraft and satellite instruments. Analysis of these observations will allow the development of improved models of the processes, at a range of scales which will provide the foundations on which future predictions and policies can be made. Increasingly it is being accepted that such observations have to be multi-disciplinary in nature, in order to understand the coupling between the physical, chemical and biological processes involved. This is also reflected in one of the key trends in scientific research to emerge in the last decade, *i.e.* the move towards inter-disciplinary investigations into, *inter alia*, the coupling between the ocean and the atmosphere, the land and the ocean and the exchanges that occur between the atmosphere and the land.

Space has long been recognised as an important place from which to view the Earth using satellite-based instrumentation. Satellites, such as the NOAA polar orbiting spacecraft (with AVHRR, the ubiquitous Advanced Very High Resolution Radiometer instrument), Landsat, SPOT, J-ERS1 and the European ERS-1, have increased our understanding of the interactions between some elements of our ecosystem. Notable achievements have included the preparation of global vegetation maps, observations of activities in the tropical rain forests, monitoring of the famous El Niño or southern oscillation, studying the by-products of several volcanic eruptions in the atmosphere, and the assessing of the extent of the ozone thinning that is occurring in northern Europe and in the southern hemisphere. Many of these activities are providing important insights into the way our ecosystem operates. However, access to these datasets is often complex. There is an urgent need to increase the accessibility of the observations to the widest possible scientific community to encourage their widest possible use in interdisciplinary research programmes.

These research activities, and the planned launch of a large number of satellites over the next decade, such as the European Envisat spacecraft, require the development and commissioning of an infrastructure that provides scientists with means of accessing the data and models arising from these activities. The infrastructure must provide facilities, on a global, regional and local basis: to advertise the location of datasets, information and models; to search for specific instances of data, information and models that are relevant to a particular research activity; and to access those items identified in the search process.

This paper outlines some of the elements of a global data network and highlights important European initiatives aimed at contributing to its development. It also uses an example of a Coastal Information Service (CIS) to illustrate key characteristics of the network which provides access to the research and operational user community. The paper also offers suggestions as to key areas where developments should be undertaken to help improve the accessibility of data, such as the introduction of techniques designed to screen datasets and create high level (meta-data) descriptors of the contents of the data and the development of multi-disciplinary thesauri designed to create applications inter-operability when searching datasets stored in different geographical locations.

# 1   Introduction

## 1.1   Background

The thinning of the ozone layer over parts of northern Europe, the increased levels of carbon dioxide reaching the atmosphere in the last century, the effects of acid rain on the forests of the Baltic countries and southern Germany, and the changes in the quality of water in some of Europe's maritime basins, (*e.g.* the Baltic, Adriatic and Mediterranean), have highlighted the ability of mankind to alter the balance of the Earth's complex ecosystem.

This heightened awareness of the effects mankind is able to have on the environment is illustrated by a number of developments. These include: increased public awareness of the issues; the growing number of international agreements and protocols related to the climate, such as those agreed in Montreal and at the Rio Earth Summit; increased levels of scientific research and cooperation between scientists across the world, such as the activities carried out by the Intergovernmental Panel on Climate Change (IPCC) and the International Geosphere-Biosphere Programme (IGBP); and the increased political emphasis on the concept of sustainable development of economies and the socio-economic consequences of these changes. Agreements related to the environment are not a new political response to public pressure. The 1978 Barcelona Convention covering the monitoring of discharges into the Mediterranean Sea was one of the first agreements to follow from the 1972 Helsinki accords in which the foundations for environmental management were laid through the linkage of the security of states with the need to consider the local environment. In the future, concern also exists about the socio-economic implications of climatic change, such as increased sea levels, extreme climate events, modifications to patterns of agriculture and fisheries, and increased desertification.

For some years research scientists have been aware of trends in climate patterns, such as the periodic advance and retreat of ice, from examination of Antarctic ice cores. Some of these are believed to be induced by subtle variations in the orbit of the Earth around the Sun. This alters the level of energy received by the Earth's ecosystem from its main source. In trying to identify if mankind's activities are having an effect upon our environment there is a need to establish the background levels of change, which are part of natural processes, in order to distinguish any fingerprint or signature that is indicative of anthropogenic activity.

In trying to detect small changes in the physical, biological and chemical processes that control our planet research scientists need access to a time series of observations on a variety of scales from local, through regional, to global. Contemporary observations of many key geophysical parameters are limited in time and spatial coverage. Over many parts of the world, such as the Southern Ocean, observations are restricted to point measurements. These limitations restrict scientists who are trying to use real world observations, in tandem with models of the processes that control our ecosystem, to forecast future climatic changes.

Information on the implications of future climate change is of interest to a wide range of decision makers concerned with establishing policy on, *inter alia*, emission levels of harmful gases, the levels and fate of pollutants reaching the coastline, future coastal defence projects, and sustainable development of resources such as the tropical rain forests. What is needed is an infrastructure of instruments and data management facilities that will enable policy makers and planners to have access to high quality datasets and information about the current perspective on the environment and predicted changes and their socio-economic implications.

This fusion of real world observations—derived from a number of sources, predictive models, and socio-economic forecasts—will provide scientists and decision makers with the information they require to derive policies that are based upon the principle of the most effective use of resources. This paper considers such an infrastructure. It takes account of the proposed development of the Global Environmental Data Network, the Global Ocean Observing System and the Global Climate Observing System and, through an application in the coastal zone, illustrates the type of infrastructure that could be established over the next decade to provide information services to research groups and decision makers around Europe.

## 1.2 Contemporary environmental pressures

Today there are many pressures that exist in our environment. Mankind, through economic and social activities, has the potential to become a major forcing function within the global ecosystem. Studies initiated through the auspices of the IPCC have highlighted the potential difficulties that might arise for mankind if the present rates of increase of emission of carbon dioxide remain unchecked. These studies have considered a number of different scenarios for future emissions. The first of these is the IPCC 'business-as-usual scenario' in which man-made carbon dioxide emissions rise from 7–8Gtons per year at the turn of the century to over 20Gtons per year by the year 2100. Other scenarios consider varying degrees of controls on emissions of gases such as carbon dioxide ($CO_2$), methane ($CH_4$), chlorofluorocarbons (CFCs), carbon monoxide (CO), and nitrogen oxides ($NO_x$); one such scenario envisages emission levels at 50% of the 1985 levels by the middle of the next century.

In addition to the levels of emission of potentially harmful gases into the atmosphere, mankind has also been disturbing the balance of nature in other ways. The rapid depletion of the tropical rain forests is steadily reducing the biomass that is available for photosynthesis. In areas which have been cut there is evidence of significantly increased levels of soil erosion. Evidence is also growing of the by-products of intensive farming in the coastal zone; witness the higher levels of nitrates being recorded in many European maritime basins. Population growth is another factor which is causing localised pressures, as is economic activity in coastal regions, where the spillage of oil from tankers is seen to be an on-going problem.

Results from studies (see Houghton *et al.* , 1990 and 1992) have revealed a range of potential changes that might occur in the environment. These include: alterations in agricultural practices; more frequent occurrences of extreme climatic events; increases in sea level, as a result of melting ice; higher recorded levels of skin cancer, as a result of the decreasing concentrations of ozone in the atmosphere; and relocation of fishing grounds, due to modifications in ocean and coastal circulation patterns.

Politicians, through the Montreal accord concerning the emission of CFCs and Agenda 21 of the Rio Summit, have recognised the need to address these potential changes and to quantify their potential socio-economic implications for mankind. They have decided to take action to develop sustainable policies concerning the exploitation of renewable and non-renewable resources and to adopt measures to reduce emission levels of harmful gases, such as carbon dioxide.

In order to refine and monitor the effect of such policies, environmental managers require access to data on a variety of scales. These range from the global scale (where access to concentrations of phytoplankton enable scientists to postulate the level of carbon dioxide draw-down into the ocean on a seasonal and annual basis) to the local scale, where riverine discharges need to be monitored for sources of phosphates and nitrates that can have a major impact in a limited area. Applications vary from monitoring the dispersion of sewage outfalls, to observing the area affected by the emissions from a power station, through land-use studies (including urbanisation), to agriculture and studies of the coastal zone.

## 1.3  Natural phenomena

The Earth is a complex planetary ecosystem where physical, chemical and biological processes sustain a diverse range of life forms and habitat. Many of these processes arise as a direct result of the large quantities of energy that we receive from the Sun and, to a lesser extent, forces (such as tides) that arise due to the gravitational interaction of the Earth with the Moon and the planets. Evidence derived from For example, Antarctic ice cores, going back over 160,000 years, show that the Earth's temperature closely tracked levels of methane and carbon dioxide in the atmosphere. Whilst the exact nature of the mechanisms that create this high degree of correlation is poorly understood they do reveal links to subtle alterations in the orbit of the Earth around the Sun. The theory behind these observations is known as the Milankovitch hypothesis. This suggests that the level of radiation received by the Earth is time-dependent with sub-periods of 96,000, 40,000 and 21,000 years due respectively to (a) the periodic variation of thee eccentricity of the Earth's orbit, (b) the periodic variation of the angle of inclination of the Earth's axis to the ecliptic (referred to as the obliquity of the eccentric) and (c) the periodic precession of the equinoxes.

These differences in the levels of energy entering the planetary system effect the processes that control our ecosystem, such as ocean circulation patterns, climate and sea levels. They also create a background of continuous climate change that occurs as a result of natural forces. In addition to these periodic and relatively slow changes to our climate system, events such as large scale volcanic eruptions, earthquakes and severe floods can have devastating short term implications on the environment and a local economy. As an example of the potential economic consequences of climate change we may consider the El Niño oscillation in the southern hemisphere: here clear evidence exists of the local impact on fisheries. Such examples illustrate the need to understand these processes in more detail, to forecast their occurrence and likely effects (risk assessment), and devise suitable approaches to minimise their economic and environmental consequences.

## 1.4  Anthropogenically induced changes

In recent years anthropogenic changes have been recognised as having the potential to significantly increase the rate of change of warming of the Earth—over and above that already occurring due to natural phenomena. Reports published by the IPCC (see Houghton *et al.* , 1990 and 1992) suggest that global sea levels have risen by 10–20cm and the mean surface air temperature has increased by 0.3–0.6°C over the last 100 years. These changes are broadly consistent with those that would be expected from climate models using known levels of carbon dioxide emissions since the time of the industrial revolution. However, these changes are also within the bounds of natural climate variability. Our challenge is to isolate the changes that are caused by natural causes and those which are linked to anthropogenic forcing. To do this we need access to a long time series of observations made on a global basis.

Concerns over the potential long term impact of anthropogenic-induced changes to our planet at local, regional and continental scales have resulted in governments all over the world approving budgets for long term space missions that are capable of monitoring any changes that might occur. These space missions will deploy a variety of sensor systems. Some will be linked to previous missions. Others will be steps into new domains where repetitive global measurements have previously not been available. These sensors, operating globally at spectral and spatial scales hitherto not available, will monitor changes in atmospheric chemistry, biodiversity, sea levels, sea surface temperature, ice sheets, tropical rain forests, and the extent of deserts and coastal regions. The scale, frequency and resolution of these observations poses significant challenges for technologists. Their task is to design the ground-based infrastructure that allows operational, research and commercial users to access the data in ways that will enable them to use it effectively.

# 2   A global environmental monitoring network

## 2.1   Global networks

Internationally, governments are showing signs of recognising that the formulation of effective and sustainable environmental policies requires access to long term datasets, predictive models and socio-economic forecasts, that describe and assess the interaction of the physical, chemical and biological processes that control our planetary ecosystem.

Whilst important, long-term monitoring is not the only use of environmental data that can be made by decision makers. Often access to data is required in near-real-time in order to respond to disasters, such as flooding events, volcanic eruptions, tidal surges that are coincident with low pressure areas, and large scale pollution incidents. To support effective decision making, data and information derived from in-situ, satellite and airborne sensors, needs to be integrated with models to provide nowcasts or forecasts of the development of the event. In this way decision makers can be provided with information that enables them to take effective decisions in a timely way.

Initiatives, such as the Global Environmental Data Network (GEDN), the Global Ocean Observing System (GOOS) and the Global Climate Observing System (GCOS), are designed to be part of a global monitoring network that would be able to provide secure archives for historical data and access to near-real-time observations of the environment in those applications where it would be useful, such as disaster monitoring.

Whilst many initiatives exist to help increase access to data on a global scale the agencies charged with their development face many problems. Budgetary limitations pose the most serious constraints on the planners. The lack of any significant new sources of finance for these projects requires that the maximum use be made of existing and planned infrastructure developments on a national and regional scale; a case in point being the development of proposals for the EuroGOOS project; (see Flemming *et al.* 1994). Global networks will evolve from the interoperability of regional networks that are being established in, for example, Europe, Japan and the United States.

## 2.2   The European earth observing system (EEOS)

The joint ESA and European Commission initiative on the EEOS is hoped to be one element of the GEDN. At a recent Workshop in Cambridge, organised by the British Association of Remote Sensing Companies (BARSC), Sloggett (1994a) suggested a potential mission statement for the EEOS. This was:

> *'The European Earth Observing System is the infrastructure that enables the operational, research or commercial use of information derived from Earth observation data by organisations responsible for the development of policies or activities concerning the sustainable exploitation of renewable and non-renewable resources in Europe and its international economic and cultural partners.'*

This proposed mission statement highlights a number of key points that are important for the development of the EEOS element of the GEDN. The first of these recognises that the EEOS should provide an infrastructure that can supply a wide range of different user communities. The second point highlights the utilisation of information derived from the data. The third point emphasises the role of the EEOS in formulating effective approaches to resource management and the last point draws attention to the need to work within an international context. The four elements of the proposed mission statement establish some necessary drivers from the design and development of the EEOS.

In common with our international partner's needs to build upon existing projects and infrastructure, the EEOS will have its roots in a number of established activities. These include the current ESA ground segment for missions such as ERS-1, NOAA AVHRR, Landsat and Coastal Zone Colour Scanner (CZCS). In the case of the ERS-1 mission important facilities include the Processing and Archiving Facilities (PAFs), the user services at ESRIN and the network of ground station facilities; see Sloggett (1989). In the case of the other missions the archives of CZCS, Landsat and AVHRR data at ESRIN are also important.

ESA is also planning to build upon these facilities as it creates the ground segment for Envisat. Once developed this ground segment will also become part of the EEOS. It is therefore possible to envisage that ESA will be responsible for establishing the data provider side of the EEOS, concerned with the access to data derived from satellites. They may also provide the facilities required to support the near-real-time users of the EEOS where data will be turned into information once it reaches the ground.

With ESA providing access to the satellite data, and in some cases the near-real-time applications infrastructure, the European Commission has to address the facilities required to support the users who wish to gain access to non-satellite-based data. This is one of the roles of the proposed Centre for Earth Observation (CEO) project planned within the Fourth Framework Programme of the European Commission.

## 2.3   The centre for earth observation project

The European Commission and ESA have agreed that the element of the EEOS that will be provided by the Commission is the CEO project. This project is currently being coordinated by a team drawn from all over Europe that is located at the Joint Research Centre at Ispra, Italy. The project is currently in a pathfinder phase where a number of studies are being undertaken to build upon a feasibility study contract that was undertaken by an industrial team in 1993. These more focused study programmes are based upon a two-pronged attack on the user's requirements for access to data and information that is available from a CEO system. The first of these has taken four application-based areas covering the coastal zone, marine radar altimetry, forestry and agriculture and the atmosphere and is studying the service provisions that should be made available by the CEO to each of these communities. The second line of attack is to undertake a series of horizontal studies, reflecting economic and regional groupings of countries, to analyse more general requirements for access to Earth observation (EO) data.

In undertaking all of these studies the CEO team hopes to build up a more complete understanding of the characteristics of the infrastructure required to encourage and support the widest possible use of EO data, and the information derived from it, across Europe. This desire reflects concerns within the European Commission on the level of exploitation of EO data that has been achieved to date. It also recognises the increased volumes of data that are going to be available over the next decade as missions such as Envisat are launched.

In seeking to establish an infrastructure designed to encourage greater use of EO data the CEO team has to balance the needs of operational uses of information derived from EO data with those of the research community. This poses some interesting problems for the CEO team. At one end of the spectrum there will be a need to exploit data from a variety of sources to generate information in both near-real-time and off-line. At the other end of the spectrum there is a need to support research groups who wish to access data sources, such as meteorological data over a test site. One typical example would be to have access to atmospheric pressure, temperature, wind speed and direction information. These are important geophysical parameters for a wide range of applications subjects.

Operational users, such as agencies of the European Commission, will require a range of services from the CEO. The near-real-time services include monitoring oil spills and Harmful Alga Bloom (HAB) events, flooding, active volcanoes and forest fires. The off-line monitoring networks would still cover some of these applications areas. For example it is possible to envisage that the frequency of monitoring of an area, such as a volcano or coastal region, would increase when indications, derived from other sources, showed increasing probability of an event occurring. Interferometric SAR observations of the swelling of a volcano prior to its eruption, indications of the existence of the right conditions for a HAB to occur, and knowledge of high levels of rainfall could all be used as external indications of the need to activate a near-real-time service to the appropriate authorities.

One of the founding concepts for the CEO project outlined by Simpson and Elkington (1993) has been the concept of a *federated* architecture. This envisages an infrastructure where any research group, operational or commercial user can join the CEO as an equal partner both in terms of data provision and data consumption. In publishing their own datasets a user adds value to the network by allowing other users to access measurements of the environment that have been made, often in a restricted area. Research scientists gain value from comparing and contrasting results they obtain with those observed by other groups working in similar fields. Often they are looking for validation, or confirmation, of their own findings and conclusions. In those cases where the findings may differ they seek to understand the physical, chemical and biological processes that might explain contrasting results. It is envisaged that the CEO can be an important framework that will support research groups working together in this way. The aim must also be to ensure that the maximum benefit is obtained around Europe in this regard as budgetary pressures require that limited resources are spent wisely.

Another important development in contemporary science is the need for researchers that have traditionally been associated with a specific field of science to understand external forcing functions and factors that effect their own area. In the coastal zone, interfaces exist with the atmosphere, the oceans and the land. Research groups working in the coastal zone have to understand the effects each of these interfaces can have as they seek to understand the processes involved in their application. As we consider the Earth as a global ecosystem it is important to recognise that this is linked with land-air, ocean-air and land-ocean interfaces that control the movement of natural and anthropogenic sources of materials through our environment. This is one area where the CEO project can help provide important new interfaces that cross the boundaries of traditional disciplines.

## 2.4   Some fundamental principles for the CEO project

In analysing the likely final operational and research goals for the CEO it is clear that there are a number of fundamental principles that must be taken into consideration. The first of these is the recognition that operational services will require the integration (or fusion) of data from several sources in order to provide access to reliable information sources. Often the approach outlined in this paper of using models and observations of the environment in tandem requires that the models use data from a number of sources, notably meteorological data. In areas such as forecasting crop growth, the movement of oil on the surface of the sea, predicting the area affected by a release of chemicals from an industrial plant, and deciding upon the scale of floods that might occur in a given river basin, there is a common problem, namely access to up-to-date meteorological data. Our analysis suggests that this is one of the fundamental data sources that will be required to support operational and research activities within Europe.

A second principle is that of adopting standards to the way in which data is documented and catalogued within the CEO environment. The creation of a federated or even semi-federated architecture of data suppliers and consumers and information consumers requires that a high degree of interoperability be established within the network that supports the CEO. In attempting to implement such an infrastructure one is bound to be constrained by a large number of factors, such as the existing investments in database systems and technologies made in a number of the data suppliers that might join the network.

A third principle is the need to recognise the requirement for locally-based services that can take account of local phenomena. Variations in bathymetry, meteorological conditions, upwelling, the position of the thermocline, concentrations of nutrients and other sources of food, levels of salinity and oxygen in the water and the local nature of aerosols in the atmosphere all represent examples of very localised effects that must be taken into consideration when generating information from data. If, for example, we take the case of measuring accurately the levels of chlorophyll in the water using an ocean colour instrument, such as the CZCS, we need to take account of localised atmospheric conditions (ozone concentrations, aerosols, air pressure *etc.* ) in order to remove the 90% of the signal that originates from the atmosphere and access the 10% that is derived from the ocean surface.

# 3   Environmental research & operational services

## 3.1   The role of research groups

Research in several of the applications areas discussed so far has highlighted the potential use of EO techniques in monitoring the environment. The European Association of Remote Sensing Laboratories (EARSeL) is one of the leading pan- European bodies coordinating research into the environment across a broad range of applications areas. The research undertaken by member laboratories often involves studies carried out using a combination of *in-situ*, aircraft and satellite data over a key test area. Results are analysed to gain insights into the physical, chemical and biological processes involved in the region concerned. The outcome of the research is then often expressed in the form of models of the basin and regional scale effects. These models can then form the basis of an operational infrastructure which uses satellite, *in-situ* and airborne observations of specific instances in time to initialise the model and enable nowcasts to be made of the future development of a process being studied. This linkage of intensive monitoring of the processes involved in a local area, through model development and exploitation, is a classic evolution of scientific understanding into operational use of observations from a number of sources. It also highlights key ingredients of the CEO infrastructure.

## 3.2   Priorities for the European Commission

The European Commission (1992), through its paper COM (92) 360 Final concerning future space policy and actions has established its priorities for the use of EO data. These priorities have been agreed with the European Parliament. They aim, through a number of initiatives contained within the Fourth Framework Programme, to demonstrate the role of EO data in the daily lives of the European tax payers, who have funded space research to date. The emphasis is on the operational uses of EO data that assist in the development of policies and procedures for the effective deployment and use of resources in the environment.

Contemporary public opinion on the state of the environment is often linked to the pictures that we see on television each night and to the public's perception of localised 'changes' in our environment, such as water rationing due to a lack of rain, 'warmer' summers, and extreme events—such as the famous storm that occurred in the United Kingdom in the late 1980s.

In order to respond effectively to increasing public concern on the state of our environment, governments, their agencies and pressure groups require access to quantified and validated observations of the environment on a routine basis. Research groups need to know of the results of other groups work in order to decide what new vistas to explore in their own local activities. Government agencies require long term operational access to information that provides quantifiable insights into changes that are occurring in the environment. This is where EO has a real role to play.

The effective use of data from a variety of sources, such as satellites, *in-situ* observations and aircraft is fundamentally dependent upon the existence of an infrastructure that provides users with an ability to easily access data stored in a number of geographically disparate locations, see Black *et al.* (1992). The Centre for Earth Observation (CEO) programme is one element of such an infrastructure. Elkington and Simpson (1992) have recognised this need and proposed the development of what they referred to as a *federated* architecture.

This argument has been developed further by Sloggett *et al.* (1996) in the course of the OCTOPUS study contract undertaken for the Space Policy Unit of the Commission's DG XII. Some of the ideas from the study were presented at a BARSC workshop on 'Exploiting the data explosion' in Cambridge (September 1994) and developed into a full scale proposal of a Coastal Information Service (CIS) at an EARSeL workshop on the coastal zone in Delft (October 1994).

## 3.3   The information requirements of decision makers

Decision makers require access to quantified information about the nature of the changes in our environment. Agencies, such as the recently formed European Environmental Agency (EEA) are mandated to collect information on the state of the environment and report this to the European Parliament in order for them to enact policies that will encourage sustainable development of key resources (see EEA 1994a).

Recently particular concerns have been voiced over the increasing level of development that is occurring within the coastal regions of Europe - with over 50% of the population living within 50 km of the coast and serious socio-economic pressures likely to occur along the southern rim of the Mediterranean Sea.

The European coastline is over 143,000 km long and is very diverse. Ranging from the northern Arctic shores of Norway to the the Mediterranean Sea and from the eastern Atlantic to the Black Sea, the European coastlines are subject to a variety of pressures, both from the climate and from mankind. The Dobris Assessment (see EEA 1994b) and the European Land-Ocean Interaction Studies (ELOISE) Science Plan, (see Cadee *et al.* 1994) have highlighted some of the specific details of the pressures on the coastal zone.

The EEA Dobris assessment highlights the lack of an effective coastal management strategy and points to increasing levels of coastal pollution, eutrophication, the conflict of uses in the coastal zone, over-exploitation of resources, and lack of control of off-shore activities. The ELOISE Science Plan highlights similar areas with an emphasis on the development of sustainable approaches to the management of the coastal zone.

However, in order to derive an effective approach policy makers require access to accurate and quantified assessments as to the extent of the problem and the socio-economic consequences resulting, in order that they can make informed decisions that encourage a sustainable approach. One way to do this would be to develop within Europe a Coastal Information Service that provides access to the information required by policy makers.

## 3.4   A coastal information service

Given the emphasis upon the coastal environment it is appropriate to consider the role satellites might play in providing operational services to users with responsibility in coastal management. Their interests can be broadly classified into one of two areas. The first group is concerned with near-real-time use, for example sediment transport. The second group are interested in long term changes such as in levels of eutrophication, pollution, sea level and water quality.

Satellite-based instruments have been shown (see ESA, 1994), to offer an important source of data on coastal regions. Sensors such as the Along Track Scanning Radiometer (ATSR) offer important insights into the thermal structures that exist in the sea. This is known to be of importance for applications in fisheries. Ocean colour sensors, such as those proposed for the SeaWiFS and Envisat missions, will provide access to information on water quality and sediment transportation processes in coastal regions. Radar altimetry has been shown to provide important insights into tidal processes and currents. Synthetic Aperture Radar (SAR) data has also been shown to provide important insights into local bathymetric conditions, pollution and wave structures.

It is possible to consider establishing a European-wide Coastal Information Service that combines satellite data, *in-situ* measurements, models of the physical, chemical and biological processes in the coastal region, and airborne sensors. This combination or fusion of data would provide a range of information streams to end user agencies. Some would provide near-real-time services, providing for example information on the location and extent of a pollution event, whilst others would provide long term statistical summaries that would be of use to agencies, such as the EEA, whose task is to take a slightly longer term look at trends, as part of a process of quantifying the state of the environment. The Coastal Information Service envisaged would be one example of the types of services that could emerge from the EEOS and CEO projects.

## 4   A coastal pollution monitoring network

In formulating the approach to the CEO, the European Commission project team has considered very carefully the methodology that should be adopted for the current pathfinder phase of the project. Instead of adopting a traditional system engineering approach the CEO project team has decided to embark upon a number of experiments or demonstration activities that connect together data providers, such as satellite ground stations and their associated archives, processing chains (that are capable of extracting accurate geophysical parameters from a combination of data providers), and end users of the information derived through this process.

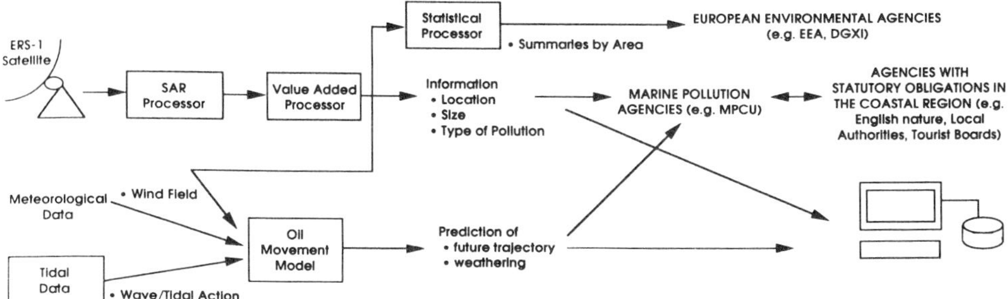


**Figure 1.** *A schematic of a European Pollution Monitoring Network*

An example of a potential demonstration activity will illustrate this philosophy. Sloggett (1994c and 1996) has outlined the concept of a European Oil Spill Monitoring Network as one element of an overall CEO infrastructure; see Figure 1. It is envisaged that this is based upon a number of existing facilities that are able to interoperate through the infrastructure provided by the CEO. A single ground station would be used to receive raw ERS-1 SAR data. The raw data would be reduced to an image using a SAR processor at the ground station. This would then be passed, within the satellite ground station, to an automated processing capability where oil spills would be reliably identified. Information on the location and size of a spill would be passed directly to end-user pollution monitoring agencies. Figure 2 illustrates the transmission over a network to national modelling centres with access to meteorological and coastal information (tides, currents, bathymetry, *etc.* ). Such centres would be capable of forecasting the trajectory of the slick and assessing its potential threat to beaches, wildlife reserves, fisheries and aquaculture facilities and areas of outstanding natural beauty, *etc.* (see Figure 3). End user agencies, such as those charged with coastal protection, would receive this information and integrate it, within a workstation environment, with data describing local habitats. In the UK, the Royal Society for the Protection of Birds (RSPB) is known to have created such a database of the main bird wintering areas and breeding grounds in our coastal waters. This database, in addition to those created by organisations such as the Institute of Terrestrial Ecology (ITE) and within the European CORINE programme (see Wyatt, 1994), would provide an excellent basis for such an analysis of the sensitivity of each coastline.

Research reported by Sloggett (1994c and 1996) is providing evidence that such an oil spill monitoring infrastructure may be feasible. Algorithms have been devised that locate areas in the sea with reduced radar cross sections. These are then screened using contextual data within an overall framework based upon evidential reasoning. This screening process is the major component of the algorithm, as it removes areas of reduced radar cross section that are not oil slicks. These arise due to, *inter alia*, wind shadowing effects, surface films (such as fish oils and algae), and wakes from ships in the area. These areas could, if not eliminated in the process, create false alarms that would be reported to the agencies concerned. This would not be acceptable to them as they require reliable information on the location and size of spills.

A key parameter that is also of great interest to the agencies concerned, and one that effects directly their response strategy, is the type of oil. Experiments reported to date by Sloggett (1994c and 1996) have shown that it is difficult to envisage how the

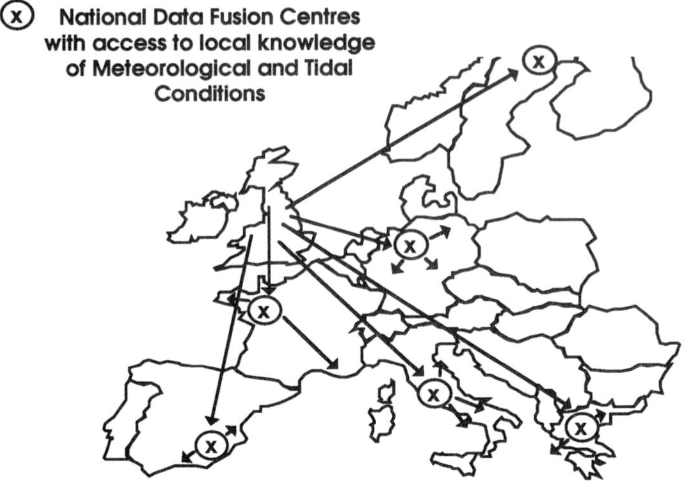


**Figure 2.** *An illustration of information through National Value Added Centres*

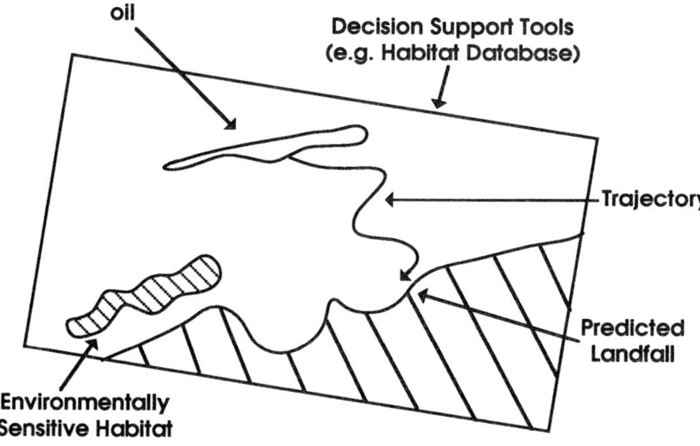


**Figure 3.** *An illustration of the local information required in a Pollution Monitoring Network*

type of oil could be directly derived from the ERS-1 SAR observations. With its single frequency of operation and single polarisation it is not a sensor with a wide spectral coverage. However, Sloggett (1996) has suggested that it might be possible to infer the type of oil through a combination of models and observations. This research is just starting. Its aim is to discover features within a slick that might indicate, albeit on a coarse scale, the type of oil; for example in the range light, medium or heavy crude. It is possible that by combining the way the oil is dispersing with a knowledge of the local wind/wave forcing we can assess what type of oil would respond to such an environment.

This combination of satellite data, an oil spill prediction model and access to local tidal and meteorological data contains the classic ingredients of a demonstration activity that would form one element of an overall CEO capability in the future. This integration of data to produce a reliable information stream has been referred to as a Value Added Centre (VAC) which, in many ways similar to the CEO, is not a centre *per se* but an amalgam of data, services and projects working together in concert to supply an end-user with a very specific stream of information.

# Acknowledgements

The author would like to acknowledge many helpful conversations with Dr Martin Sharman of DG XII, Dr Luigi Fusco of the European Space Agency and Dr Mark Elkington and Dr George Simpson of Earth Observation Sciences in the formulation of the ideas expressed within this paper.

Many of the ideas discussed herein resulted from the author's work on the OCTOPUS Study which was conducted for the European Commission. The contributions to this study of Dr Vittorio Barale, Dr Roland Doerffer, Dr Simon Boxall, Dr Jim Aiken, Dr Meric Srokosz, Dr Martin Bohle-Carbonell and Mr Alan Cross are warmly acknowledged. The author would also like to record his appreciation for contributions made by Dr Rœland Allewijn of the Rijkswaterstaat in The Netherlands and Dr Robin Vaughan of the University of Dundee for their work on, and promotion of, the EARSeL Special Interest Group on water management and water quality.

# References

Black T J *et al.* , 1991, An assessment of the European technical capability for the treatment and interpretation of satellite Earth observation data: *Working paper two, Ref: TN-91/319/2.0.*

Cadee N, Dronkers J, Heip C, Martin J-M and Nolan C, 1994, European Land- Ocean Interaction Studies Science Plan, *European Commission Publication - EUR 15608 EN.*

European Commission 1992, The European Community and Space: Challenges, Opportunities and New Actions, *COM (92) 360 Final, Brussels, 23rd September 1992.*

European Environmental Agency 1994a, Putting Information to Work, EEA Brochure and Mission Statement, Copenhagen.

European Environmental Agency 1994b, Europe's Environment - The Dobris Assessment.

European Space Agency 1994, Proceedings of the First ERS-1 Pilot Programmes Workshop in Toledo, Spain, June 22-24 1994.

Flemming, *et al.* 1994, The Case for EuroGOOS, Third Draft, 15th March 1994.

Houghton J T, Jenkins G J and Ephraums J J, 1990, Climate Change - The IPCC Scientific Assessment, World Meteorological Organisation, United Nations Environment Programme, Cambridge Press.

Houghton J T, Jenkins G J and Ephraums J J, 1992, Climate Change - The IPCC Scientific Assessment, World Meteorological Organisation, United Nations Environment Programme, Cambridge Press.

Simpson G and Elkington M, 1993, CEO Discussion Paper: A Federated Network Approach to the Management of Earth Observation and Related Data, Earth Observation Sciences Internal Paper.

Sloggett D R 1989, Satellite Data: Processing, Archiving and Dissemination, Ellis Horwood Ltd, Chichester, Sussex.

Sloggett D R 1994a, CEO - A European Framework for Applications and Service Development, BARSC Workshop - Exploiting the Data Explosion, Cambridge.

Sloggett D R *et al.* , 1994b, The OCTOPUS Study Final Report, A Report for the European Commission (DGXII).

Sloggett D R 1994c, Automatic satellite-based oil slicks detection and monitoring system. Proc. First ERS-1 Pilot Projects Workshop, Toledo, Spain, 22-24 June, pp395-402.

Sloggett D R 1996, An Automated Approach to the Detection of Oil Spills in satellite-based SAR Imagery, The Determination of Geophysical Parameters from Space eds. N.E.Fancey I.D. Gardiner and R.A. Vaughan, Proceedings of NATO ASI held in Dundee August 1994—this volume

Wyatt B 1994, Coastal Biomes, Private Communication.

# Remote Sensing Activities at the Natural Resources Institute for Measurement of Geophysical Parameters

Ian D Downey

Environmental Science Group
Natural Resources Institute, UK

## 1 Introduction

The Natural Resources Institute (NRI) is an executive agency of Britain's Overseas Development Organisation (ODA). Based in Chatham, Kent the NRI employs over 400 staff in the UK and overseas and has expertise in three main areas: Resource Assessment and Farming Systems, Integrated Pest Management and Food Science and Crop Utilisation.

As an integral part of the British government's overseas aid programme, the principal aim of the institutes work is to alleviate poverty and hardship in developing countries by increasing the productivity of their renewable natural resources. Remote sensing offers one means of providing information to address these problems as identified at the UNICED 1992 Earth summit and subsequently as part of Agenda 21 "Appropriate technologies must be made available to developing countries or integrating environmental considerations into development policy and practice".

The Environmental Science Group at NRI has developed (in collaboration with Bradford University, Reading University TAMSAT Group and Silsoe College) an approach known as Local Application of Remote Sensing Technology (LARST). One of the principal outputs of this approach has been the development of robust low cost satellite data receivers for NOAA and Meteosat.

These receiver systems are installed in the host country and operated by national or regional institutes to provide rapid, direct access to EO data in-country thereby allowing the direct application of geophysical observations (or more usually their indirect proxies) in improving natural resource monitoring and management. The LARST Systems are based on simple principles that ensure their sustainability and longevity of operation

| System Characteristics | Operational Drivers |
| --- | --- |
| Robust hardware | Applications driven |
| Easy-to-use software | Needs of local managers |
| Low cost PC based systems | Not technology led |
| Low cost printing | Encourage dissemination |
| Free real time data | Locally sustainable operation, daily outputs, local validity |
| Locally sustainable technology | Useful products for decision makers |

The LARST approach utilises satellite observations of geophysical parameters directly, such as cloud top temperature and sea surface temperature as well as information derived from the satellite orbital position. In addition, these observations can be used (as a proxy) to provide indications of other phenomena which cannot be measured directly from satellite observations but which manifest themselves in a phenomenon which can be detected, such as pest habitats from NDVI, potential fish stocks in upwelling regions and agricultural encroachment from fire (hot spot) detection.

In the remainder of this paper we present three case studies which illustrate the activities at the NRI.

# 2 Weather forecasts and rainfall estimation: Ethiopia

Following a request from the Ethiopian National Meteorological Services Agency (NMSA) for improved rainfall monitoring, an ODA programme was originated to develop national rainfall and vegetation monitoring. The need for the project arose from a realisation that information sent to Ethiopia from the regional centre in Nairobi was insufficient and often untimely; development of a national capability was urgently required.

The programme provided NMSA with a Meteosat satellite primary data user system (PDUS) to acquire imagery and MDD forecast messages from Rome and Bracknell, a high resolution receiver for NOAA orbiting satellites, and (under separate ODA funding) a Media Presentation System to provide TV forecasts, together with training and technical support. The programme was developed within the context of the Inter-Governmental Authority on Drought and Development (IGADD, a body largely concerned with early warning and food security) with the intention that resulting data should be made available to the regional early warning unit and other national and international agencies. Wider objectives were to provide the Government of Ethiopia with effective rainfall and crop status estimates and monitoring to assist in early warning of abnormal conditions.

Satellite data are received and processed locally to generate a number of products to meet these objectives. Images of cloud top temperature are captured using the LARST systems. There is a physical relationship between cloud top temperature, the duration of cold clouds and the amount of rainfall. Rainfall estimation models based on local conditions are used to predict precipitation from the location of cold clouds over ten day

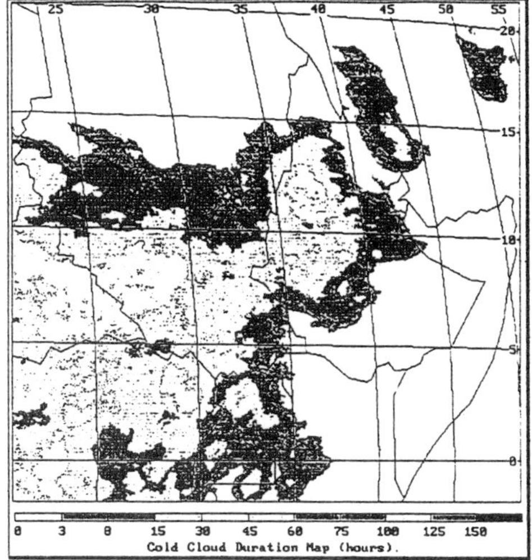



**Ten-day Cold Cloud Duration Map, Horn of Africa.**

Second Dekad, August 1990

Images of cloud temperature are captured using the LARST dish antenna and Meteosat receiver system and are used to produce rainfall estimates for the same ten-day period (see below).

**Ten-day Rainfall Estimation Map, Horn of Africa.**

Second Dekad, August 1990.

Rainfall estimates are produced from Meteosat Cold Cloud Duration (CCD) images.

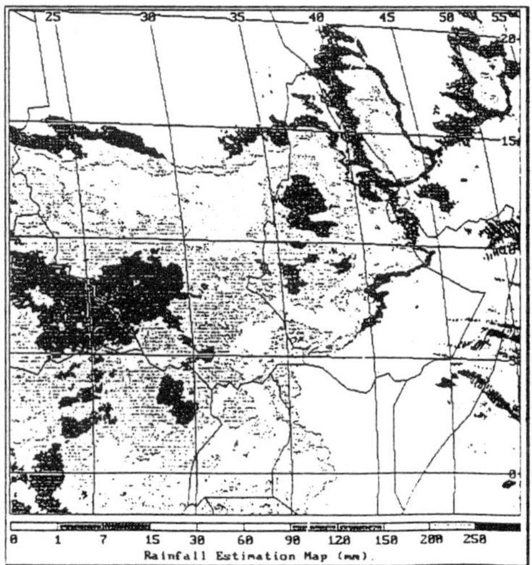



**Figure 1.** *Meteosat* CCD *map and rainfall estimates for Horn of Africa region, 1990*

periods. These dekadal rainfall estimates (calibrated with local rain gauge networks) are used to monitor seasonal variation and provide forecasts of crop growing conditions in important food growing areas of Ethiopia (Figure 1).

Similarly, dekadal time series images are acquired and processed to assess variation in NDVI. The ability to monitor moisture and vegetation vigour using these techniques is important for assessing seasonal (*e.g.* drought) conditions and providing early warning of anomalies which may indicate threats to cropping areas such as drought or pest invasion (*e.g.* vegetation conditions suitable for locust infestation).

The programme (based on measuring relatively few direct geophysical parameters from satellites) is now completely self sustaining and the NMSA data products pro-

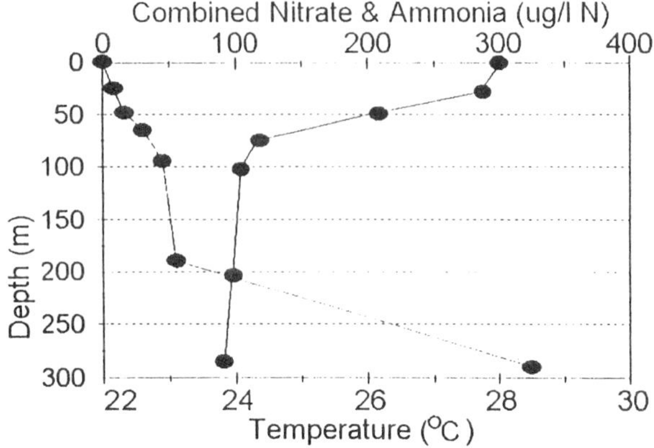


**Figure 2.** *Lake Malawi temperature and nutrient profiles.*

vide vital information to assist national and regional planning for drought and famine preparedness in East Africa.

# 3    Monitoring lake surface temperature — Malawi

Lake Malawi is the third largest lake in Africa and provides a valuable resource for Malawi, Tanzania and Mozambique. The lake represents 25% of Malawi's land area and stretches the distance equivalent to that of Edinburgh to Southampton. The lake is used as a fresh water source, a method of transportation and as a vital source of protein-rich fish stocks for local populations. It is also the world's first inland marine national park established expressly for the specific protection of fresh water fishes.

The UK/SADC Pelagic Fish Resource Assessment Project was established and aimed to assess the abundance of offshore fish stocks and their potential for exploitation. The project also aimed to measure limnological conditions to enable management decisions to be supported if a fishery was developed. Besides collecting *in situ* data on fish and lake conditions, regular satellite data were collected to provide direct access to measurements of the lake surface unavailable by any other means. A ship-borne sampling survey takes two weeks to traverse the full length of the lake.

The satellite data collection system installed by NRI provided repetitive and real time data which were used to make quantitative estimates of lake surface temperature and to extract qualitative information on sediment loads and chlorophyll presence. Measuring such geophysical phenomena provides important indications of the physical, biological and chemical condition of the lake. The fisheries potential of Lake Malawi and other similar water bodies is vitally dependent on the surface and near surface conditions where the bulk of primary production occurs. Although primary production occurs near the surface, nutrients are held in bottom waters (Figure 2).

These can be brought to the surface in two ways, by breakdown in thermal stratification of the water column allowing nutrients to diffuse upwards, or by upwelling events which rapidly bring nutrients to the surface waters. Both phenomena are characterised

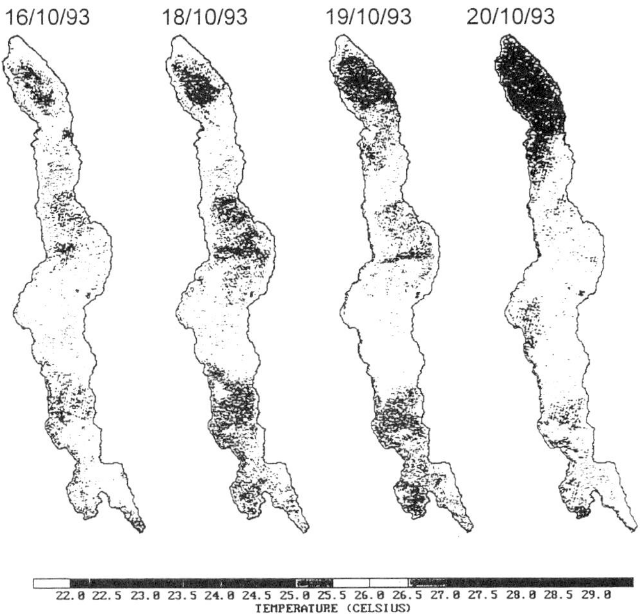


**Figure 3.** *Images of a thermal upwelling event detected in Lake Malawi*

by lake surface cooling. By monitoring the lake surface temperature, therefore, studies of nutrient cycling (and thus phytoplankton production) can improve fish stock location and management.

To achieve operational, reliable and accurate lake surface monitoring, the satellite data were calibrated with surface observations made from research vessels. This involved a number of stages:

- Calibrate raw channel data into radiance units.

- Convert radiance values to brightness temperatures.

- Correct for brightness temperature deficit by using spit-window atmospheric correction algorithms.

- Compare results with actual lake surface temperatures.

Methods and algorithms were developed and tested in-country to establish year round estimates of lake surface temperature of the order of $\pm$ 0.5 K.

The project has contributed to improved understanding of remote sensing applications for fisheries research in the SADC region. The automatic data capture and processing system developed provides regular printed maps of surface temperature with a temporal and spatial frequency unavailable by any other means. Using the automated system, UK/SADC project scientists are now able to examine the evolution of important lake behaviour events and study the seasonality in lake surface structure. A large scale upwelling was located and monitored in October 1993 (Figure 3), an occurrence that would have brought vital nutrients up to the surface layers. This event, which would have been missed without the use of the NRI LARST system, led to a subsequent increase in the survival of certain pelagic fish larvae in the northern lake region. The implications of this kind of information for local populations are profound (Wooster *et al.* 1994).

# 4   Detecting and monitoring fire events

Fire is a key indicator of anthropogenic activity and associated biomass burning. Clearance of vegetation for agriculture, logging access, settlement, hunting, poaching, and even some tree species regeneration all involve fire to some extent. The presence of too many fires over time can cause biodiversity loss and, furthermore, uncontrolled and natural wild fires consume vast areas of vegetation causing significant biomass loss. NRI has significant experience of operational and experimental fire detection in Africa and southeast Asia.

Besides the physical loss of biomass to these various forms of consumption, burning releases large amounts of gases (*e.g.* $CH_4$, $CO$, $CO_2$, $N_2O$, $NO$ ), volatile compounds and aerosols into the atmosphere which can have effects on climate, weather and alteration of the energy balance at the Earth surface (Justice 1994).

The most practical and feasible means, by far, to quantify and monitor these biomass burning events at the regional or global scale is to utilise Earth observation remote sensing and GIS technologies. NOAA satellite data offer one of the most practical tools for detecting and monitoring biomass burning activity (Robinson 1991). The data from NOAA satellites have been available for over 10 years without interruption. The 12 hour repeat coverage capability is a major advantage of NOAA platforms and the only practical way to attempt to detect short lived fire events operationally and to provide enough observations over time to extract longer term trends of fire activity. The wide area coverage and local reception capabilities also make NOAA data ideal for early warning operations.

This fire information can be directly useful. Regular satellite overviews of daily fire activity during the burning seasons provides a unique measure of the scale, location, timing and (likely) intensity of these events. Areas where burning is considered a problem can be identified easily and such information provides important inputs to developing sound natural resources and fire management policies. Preliminary verification of a joint project between the Brazilian Space Agency (INPE) and the Brazilian Institute of Natural Resources (IBAMA ) showed that over 90% of the fires identified by AVHRR were matched by field located burn scars or active fires within 500 metres of given AVHRR fire locations (Justice *et al.* 1993). This corresponds with NRI experience of operational fire detection in the Ivory Coast, Central African Republic and Indonesia.

Various aspects of fire can be sensed, including the energy released by active fires, smoke, char and fire scars. The mid-IR window ($3$–$5\mu$m) is particularly interesting because it is near the spectral maximum for radiative emissions for objects radiating at temperatures found in fires, and in a region of reduced solar and low terrestrial radiation. Channel 3 is centred at the spectral emission maximum for objects radiating at ground

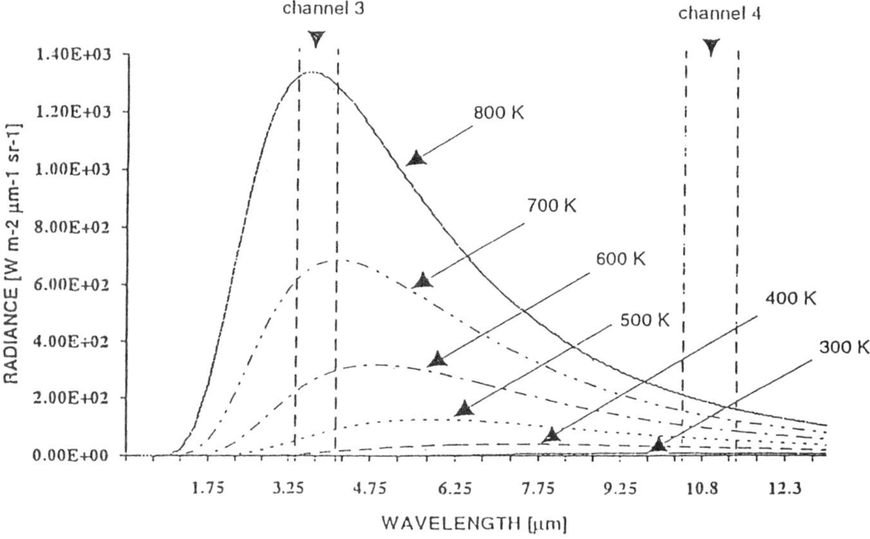


**Figure 4.** *Planck radiances for blackbody temperatures from 300K to 800K. For a given increase in temperature the increase in area under the channel 3 portion of the curve is much greater than under the channel 4 portion (Kennedy et al. 1994).*

temperatures of the order of 800K. This channel is ideally sensitive to intense sources of heat in ambient conditions and is used to detect burning fires which tend to saturate the signal in the 3.55–3.93$\mu$m region observed (Kennedy *et al.* 1994). Channel 3 is also very much less affected by haze and water vapour. Figure 4 shows the location of channels 3 and 4 in relation to black body temperatures and demonstrates the suitability of channel 3 for detecting hot spots.

Most fire detection algorithms do not require atmospheric correction for successful implementation. However, some approaches have a degree of implicit cloud screening tests as part of the fire detection process (Kaufmann 1990, Kennedy *et al.* 1994). In the case of fire detection itself, the simple use of DNs or the calculation of brightness temperatures is usually sufficient to achieve the objective.

Saturation thresholds for AVHRR instruments have ranged from 322 to 331 K (Robinson 1991). In practice, NRI has found that applying appropriate thresholds can be difficult because of diurnal variations. Different thresholds may need to be applied at night and during the day (*i.e.* different overpass times) within the same ecosystem. They may also need to be altered seasonally and inter-annually. Thus different thresholding algorithms based on absolute thresholds will exclude detectable events or include spurious noise (Robinson 1991). This means that local conditions influence fire detection and that a universal threshold is unrealistic and inappropriate. Fire detection is therefore achieved by a number of methods, all of which utilise the dominating effect of hot fires in channel 3 signals at some stage.

The simplest algorithm is to retain all pixels which are saturated (or near saturated) in channel 3. The detection of fires would be triggered by setting a brightness temperature threshold. In most cases, this saturation would indicate unusually hot features and could be assumed to represent a fire. However, this method is susceptible to confusion with surface hot spots (such as extensive bare soils) and bright clouds and may not be suitable for regional scale fire studies.

Multi-channel threshold algorithms have gained a great deal of support in recent years as they have been shown to be regionally robust and simple to implement. A wide range of multi channel threshold criteria exist, developed from work by Kaufman et al (1989, 1990), Langaas (1989, 1992, 1993), Belward *et al.* (1994), Kennedy *et al.* (1994). A series of tests are employed in pixel-by-pixel $T^B$ thresholding, all of which must be passed for a pixel to be classified as a fire pixel. As in the case of simple channel 3 thresholding, all the published multi-channel thresholds have been developed for specific regions and are based on empirical results. The algorithms usually take the form:

1. $T^B(3) > k_1$ $\longrightarrow$ (candidate fire pixels must be hot)
2. $T^B(3) - T^B(4) > k_2$ $\longrightarrow$ (channel 3 must be hotter than channel 4)
3. $T^B(4) > k_3$ $\longrightarrow$ (candidate fire pixels should not contain strongly reflective clouds which can saturate channel 3)

where $k_1$ is a brightness temperature (or DN) close to channel 3 saturation, $k_2$ is the difference between brightness temperatures in channels 3 and 4 and $k_3$ is a brightness temperature that delineates cold clouds. The selection of values for $k_1$, $k_2$ and $k_3$ vary between regions. The (region specific) constants are assumed representative of the entire fire season, and of all image pixels remaining after the application of cloud masking algorithms.

The approach of Kaufman *et al.* (1990) utilises AVHRR channels 3 and 4 and was developed for cool forest environments in Brazil. Modifications of Kaufmann's (1990) approach (Kennedy *et al.* 1994) seek to assure that high radiant temperatures in channel 3 exceed those in channel 4 in order to distinguish between fires and warm surfaces. The relationship also includes top of atmosphere (TOA) reflectance data from channel 2 where fire and burnt areas will tend to have low reflectance due to the presence of ash cover. Fire detection is triggered by satisfying these criteria:

- $T^B(3) > 320K$;     $T^B(4) > 295K$;     $T^B(3) - T^B(4) > 15K$
- channel 2 top of atmosphere reflectance $< 16\%$

This method has been proposed as one of the most appropriate automated approaches to study fires throughout the ecosystems of West Africa (Kennedy *et al.* 1994).

Sub-pixel temperature estimation algorithms are applied to pixels where it is possible to measure the brightness temperatures of the channels. This technique allows the calculation of two subpixel areas (target zone and background) within an unsaturated pixel. By comparing a specified target threshold temperature with the calculated target temperature it is possible to identify a fire in the pixel. It is necessary to specify the background temperature in these algorithms and this can be done by using a look-up table of background temperatures related to the zone of interest or by calculating background temperatures using the responses of channels 3, 4 and 5 of surrounding pixels. Given the specification of the AVHRR and high variability in fire regimes, it is considered difficult to investigate the contribution of subpixel fires on an operational, regional basis (Justice *et al.* 1993).

Other techniques which rely on spatial analysis or contextual methods can be utilised. They analyse the spatial variability of thermal signals in a pixel neighbourhood. The contextual approach is based on the same idea of a hierarchy of $T^B$(channel N) tests as described above. However, using this algorithm, the brightness temperatures of potential fire pixels are checked against the brightness temperatures of cloud-free (assumed non- burning) neighbouring pixels. This takes into account the spatial-temporal thermal variability of the background which is not easily dealt with in either single channel thresholding or pixel-by-pixel multi-channel thresholding (Justice and Dowty 1993).

NRI experience of fire detection covers several operational and experimental projects in a number of (largely tropical) regions of the world. These are all in areas where fire and biomass burning has a direct impact on the environment and thus on the local populations. We give several examples.

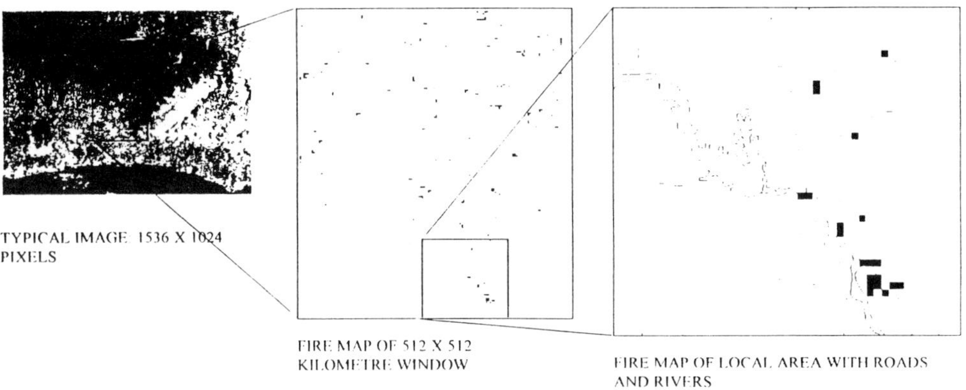


**Figure 5.** *Map of fire activity in Côte d'Ivoire processed on the* LARST *system*

## Côte d'Ivoire

Collaborative work between NRI and the Joint Research Centre (JRC), of the European Commission, has been conducted at the Lamto experimental station, in the savannah zone of Côte d'Ivoire (Figure 5). This has shown that a PC based HRPT receiving system could be used operationally, close to the burn zone, within several hours of arrival and provide images from the NOAA 9, 10, 11 and 12 satellites useful for fire management. By linking satellite observations to ground observations in real time some of the key problems of characterising fire events, their impact and management can be addressed (Belward *et al.* 1993). The availability of *in situ* AVHRR imagery, in real time, offers important new possibilities for fire management.

Satellite overpasses were accurately predicted using the HRPT receiver and analysis system and suitably preprocessed imagery made available within 5 minutes of data capture. The results showed that grassland savannah fires as small as 50 m in length can be detected, that only flaming fires (not smouldering fires ) are seen, and that the afternoon overpass corresponds well to the period of peak fire activity in the region. Progress to date indicates that local reception of NOAA satellite data is of immense practical value and is a cost effective way of obtaining information on vegetation fires at national or local scale which enables better fire awareness and management.


## Central African Republic

Collaborative work has been conducted between NRI, JRC and ORSTOM in Bangui, CAR to assess the extent and detectability of fire events. Successful deployment of in-country AVHRR receiving equipment provided data on the progression of burning across CAR during the burn season. This work was supported by calibration from controlled fires set in the field and airborne video coverage during the burn to gain information on fire size and burn pattern in this ecosystem.


## Indonesia

NRI have a NOAA data receiver installed in Palangkarya, Indonesia as part of an ODA technical project looking at tropical forestry management and how to improve its effectiveness. Part of the programme includes fire detection and mapping during the burn season and NRI personnel are regularly collecting and archiving data.

NRI has developed and successfully deployed low-cost, robust receiver stations capable of interpreting the real-time High Resolution Picture Transmission (HRPT) data which is continually transmitted from the orbiting satellites. Field experiments in West Africa have successfully demonstrated the utility of these systems to provide up to the minute information on forest clearance by fire. The system can be used for large area overview and for local response. The data are received and processed on PC (micro-computers) to provide maps of vegetation fires *while they are still burning.*


# 5   Conclusions

NRI has an ongoing research and development programme to develop systems and methodologies which provide the means for developing countries to make appropriate, sustainable use of satellite determined geophysical parameters. Developments such as the LARST approach have enabled a large installed user base in many countries to operate satellite receiving stations for direct environmental monitoring.

These capabilities (illustrated above) provide resources to assist in:

- Better resource monitoring

- Improved disaster preparedness, early warning and response

- Greater degrees of self reliance

- Ongoing technical support and training

For the future, many of the issues discussed in the session on data availability are of direct relevance. Improved data, collected consistently over time will provide the foundation for many environmental applications to be realised in developing countries. The role of high resolution systems needs to be further developed in line with changes in data access and pricing. Furthermore, the capabilities of active systems have great relevance for many parts of the tropics where cloud cover is a key limitation to optical remote sensing. The question of data and equipment cost is always a key consideration in developing countries as well as enabling the uptake of the technology in an effective manner.

The focus at NRI will continue to be on appropriate use of these systems to address achievable goals to realise direct, tangible benefits. There are many immediated applications that could be assisted by access to remotely sensed geophysical data provided the cost and sustainability of the methods are carefully considered.

# References


Belward A S, Grégoire J-M, D'Souza G, Trigg S, Hawkes M, Brustet J-M, Serca D, Tireford J-L, Charlot J-M and Vuattoux R (1993), In situ, real-time fire detection using NOAA/AVHRR data, Proceedings 6th AVHRR Data Users Meeting, Belgirate, Italy, June 1993, 333-339

Belward A S, Kennedy P J, Grégoire J-M, (1994), The limitations and potential of AVHRR GAC data for continental scale fire studies International Journal of Remote Sensing, 15(11):2215-2234)

Franca J-R A, Brustet J-M, Fontan J, Grégoire J-M, Malingreau J P, (1993b), A multi-spectral remote sensing of biomass burning in West Africa during 90/91 dry season, Presented at the XVM-EGS General Assembly, May 1993, Wiesbaden, Germany

Justice C O [ed.] (1994) African savannas and the global atmosphere research agenda, IGBP Global Change Report No. 31, IGBP

Justice C O and Dowty P (1993), IGBP-DIS Satellite fire detection algorithm workshop technical report, IGBP-DIS Working Paper No 9 Presented at NASA/GSFC, Greenbelt, Maryland, USA Feb 25-26

Justice C.O., Malingreau J.P., and Setzer A. (1993), Satellite remote sensing of fires: potential and limitations [in] Fire in the environment: the ecological, atmospheric, and climatic importance of vegetation fires [ed] P.J. Crutzen and J.G. Goldammer, Wiley and Sons, New York, 77-88

Kaufman Y., Tucker C.J. and Fung I. (1989), Remote sensing of biomass burning in the tropics, Advances in Space Research, 9:265-268

Kaufman Y.J., Setzer A., Justice C., Tucker C.J., Pereira M.C. and Fung I. (1990a), Remote sensing of biomass burning in the tropics, [in] Fires in tropical biota, [ed.] Goldammer J.G., pp. 371-399, Springer-Verlag, Berlin

Kaufman Y., Tucker C.J. and Fung I. (1990b), Remote sensing of biomass burning in the tropics, Journal of Geophysical Research 95(D7): 9927-9939

Kaufman, Y. J. and B. N. Holben, (1993) 'Calibration of the AVHRR visible and near-IR bands by atmospheric scattering, ocean glint and desert reflection' International Journal of Remote Sensing, 14, (1)21-52.

Kennedy P.J (1992) Biomass burning studies: the use of remote sensing, Ecological Bulletins 42:133-148

Kennedy P.J., Belward A.S. and Gregoire J-M. (1994), An improved approach to fire monitoring in West Africa using AVHRR data, International Journal of Remote Sensing, 15(11):2235-2255.

Langaas S. (1989), A study of spectral characteristics of burnt areas in Senegal, with recommendations for an AVHRR based bushfire monitoring methodology, Dept. of Surveying, Agric. Univ. of Norway, Aas

Langaas S. (1992), Temporal and spatial distribution of savanna fires in Senegal and The Gambia, West Africa, 1989-1990, derived from the multi-temporal AVHRR nigh images, Int JOURNAL Wildland Fire, 2:21-36

Langaas S. (1993), A parameterised bispectral model for savanna fire detection using AVHRR night images, International Journal of Remote Sensing, 14(12):2245-2262

Robinson J.M. (1991), Problems in global fire evaluation: is remote sensing the solution? in *Global Biomass Burning* edited by Levine J.S., MIT Press, pp.67-73

Robinson, J. M. (1991) Fire from Space: Global fire evaluation using infrared remote sensing. International Journal of remote Sensing, 1991, Vol 12 No. 1, pp 3-24.

Wooster M.J., Sear C.B., Copley V.R. and Patterson G. (1994) Monitoring Lake Malawi using real time remote sensing, Final Report (Remote Sensing Component) of the UK/SADC Pelagic Fish Resource Assessment Project, NRI, 125pp.

# Synthetic Aperture Radar:
# Principles and Applications

J Askne

Chalmers University of Technology
Göteborg

## 1  General Introduction

The principle of the synthetic aperture radar (SAR) goes back to the 1950s and had its break through with the US satellite SEASAT in 1978. In spite of its limited lifetime of 3.5 months a large number of images were produced and gave the necessary time for analysis over the years until the European ERS-1 satellite was launched in 1991. This satellite, which is still working and which was joined by ERS-2 during 1995, has been producing an immense set of SAR data and given a breakthrough for the interferometric SAR technique.

This chapter consists of four sections describing some general basics about microwave remote sensing and SAR in particular, followed by technical principles. Basic aspect on how to use the SAR sensor in a qualitative and quantitative manner is included in the third section followed by an overview of some of the applications.

This has largely been based on course material and research in the Remote Sensing Group at Chalmers University of Technology, Göteborg, and the contributions from L Ulander, A Carlström, J Hagberg, H Israelsson, and Y Sun are gratefully acknowledged. The experience of the group is based on investigations associated with ESA AO-projects within the PIPOR project for sea ice, participation in the ESA Fringe group and the ESA Calibration group as well as other projects all financially supported by the Swedish Space Agency. The reliance on our own research activity has of course limited the perspective but has been necessary in order to produce the lecture material. The notes may serve as an introduction to a fascinating and fast evolving field, and for further information the readers are referred to articles in journals like the International Journal of Remote Sensing, IEEE Transactions on Geoscience and Remote Sensing, to conference proceeding such as the two ESA ERS-1 meetings in Cannes 1992 and Hamburg

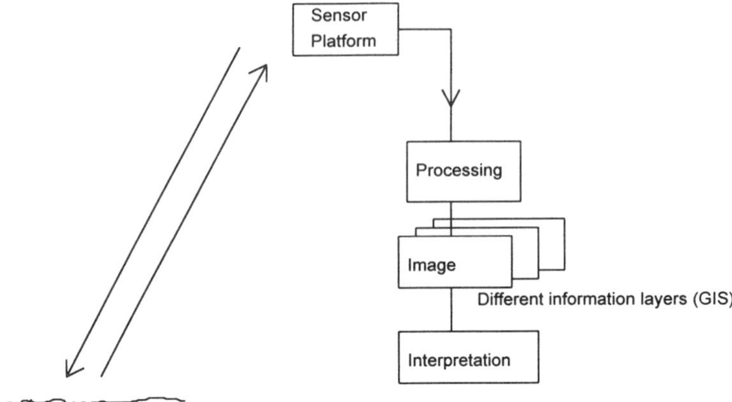


**Figure 1.** *Illustrating the basic concept of a sensor system used for remote sensing of a ground surface.*

1993 and to proceeding from IGARSS conferences. General material is also produced by ESA, such as a CD-ROM guide to ERS-1 including image examples on various applications and such as the ESA earth observation quarterly as well as ERS-1 specific information.

## 1.1   Some basics of microwave remote sensing satellites

### Remote sensing systems and observations

The originally motivation for the SAR sensor was ocean and sea ice applications, but the interest for land applications has increased considerably during recent years. The independence of sunlight and the ability to look through clouds and produce images with a resolution of the order of 30m is of large interest for many applications. This section will include some general comments on remote sensing using microwaves followed by a survey of satellite SAR systems and finally returning to what can be measured by means of SAR sensors.

In Figure 1 we have included an illustration of a sensor system including a ground surface/target observed by an active or passive sensor system. The detected signal is processed and possibly an image is formed. In order to interpret the information we may need auxiliary (extra) information from other sensors or from registrations of the properties of the surface. Such extra information may be a digital elevation map (DEM) over the area, it may be information where we have built up areas, forest areas, agricultural fields and various information of the areas such as size, normal crop etc. The information may be combined using a Geographical Information System (GIS).

The sensor system or a measurement system in general should be able to detect, differentiate and recognise properties on the basis of spatial properties such as form and geometry and temporal properties. For that reason the sensor should have a high resolution as well as radiometric and geometric accuracy. A signature is also recognised by its properties of reflectivity, spectral properties and polarisation properties. This is

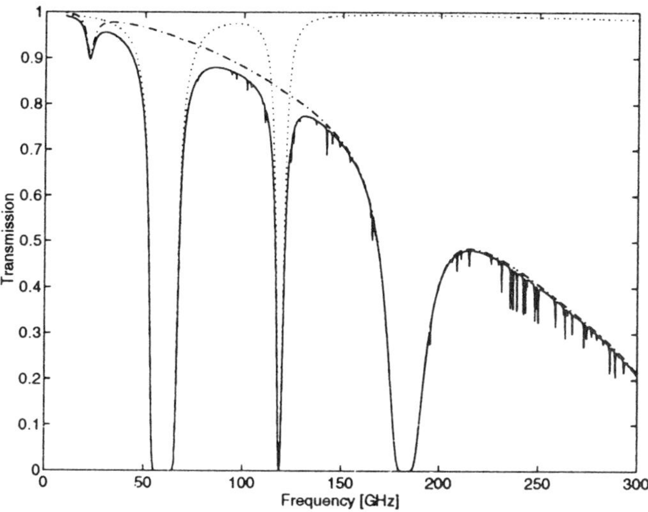

**Figure 2.** *Illustrates atmospheric transmission in the microwave part of spectrum. Dotted line: oxygen contribution, dash-dotted line: water contribution, solid line: contribution from all resonance lines in atmosphere.*

possible in the microwave domain where the wavelength is reasonably large compared to variations of many properties of interest to measure. Therefore we need to develop our understanding of the interaction between the wave and the medium and we may also meet the objectives by developing more advanced multisensor, multifrequency, multipolarisation systems with higher resolution and accuracy.

## 1.2 Microwave instruments

Microwave observations from space have taken place at least since 1968 when the Soviet satellite Cosmos 343 was launched with nadir-looking radiometers. The interest in microwave sensors is partly due to their ability to look through the atmosphere without being influenced very much. This is illustrated by the atmospheric transmission over wavelengths of interest for remote sensing, see Figure 2.

The development of radar technology goes back to World War 2 and by tradition from this time the different wavelength regions have been denoted as given below in Table 1.

The interaction between the electromagnetic wave and the medium varies with the wavelength. Normally the penetration into the earth is larger at longer wavelength and resonances may occur between natural objects and an electromagnetic wave with a wavelength of the same order as the dimensions of the medium.

| Band | Frequency GHz | Wavelength cm |
|---|---|---|
| P-band | 0.225–0.39 | 133–77 |
| L-band | 0.39–1.55 | 77–19 |
| S-band | 1.55–3.90 | 19–7.7 |
| C-band | 3.90–6.20 | 7.7–4.8 |
| X-band | 5.75–10.9 | 5.2–2.8 |
| Ku-band | 10.9–18.0 | 2.8–1.7 |
| Ka-band | 18.0–36.0 | 1.7–0.8 |

**Table 1.** *Microwave frequency bands*

## 1.3  SAR space programme

### US SEASAT

The synthetic aperture radar, SAR, is an instrument that can produce microwave images of the earth from space with resolution comparable to or better than optical systems. The method was proposed in the early 1950s by Carl Wiley. In late 1971 the use of synthetic aperture radar for sensing of the ocean surface was discussed, and plans were formulated by the NASA program office for a SAR-mission. The SEASAT-program was conducted for a total cost to NASA of $77M, excluding the launch vehicle. The development was accomplished in about 30 months after the go-ahead including satellite development, sensor integration and tests - an achievement probably not possible today. The satellite was launched on 26 June 1978 but a power failure three months after launch (10 October) brought the mission to an end. Still so much data was produced that until today probably not all has been analysed. SEASAT was in many ways a remarkable satellite. It was the first satellite with an imaging SAR system used as a scientific sensor, and the first satellite with a scatterometer (after tests on Skylab). SEASAT was also equipped with a radar altimeter and the Scanning Multichannel Microwave Radiometer, SMMR, also launched on Nimbus-7, c.f. below. SEASAT was a satellite designed for oceanographic applications. It was a satellite ahead of its time and it seems like we will have to wait until the polar platforms in order to obtain a sensor package as complete as that on SEASAT.

It took a long time after SEASAT for further satellites equipped with synthetic aperture radar to be launched. The first US classified satellite with a SAR, Lacrosse, was launched in 2 December 1988, and the resolution is thought to be about one metre. Not many details are known about this and further military satellites. Instead the US programme has been based on shuttle missions for limited times.

### US SIR-A and SIR-B

An L-band SAR was launched for a one week shuttle mission, SIR-A (Shuttle Imaging Radar), in 1981 and another, SIR-B, for one week in 1984 with somewhat improved capability

## US SIR-C/X-SAR

The SIR-C/X-SAR Shuttle flight in April 1994 (a second flight in August 1994) was the first spaceborne SAR system to simultaneously acquire images in multiple frequencies and polarisations at L and X-band. It is a joint US-German-Italian project flown onboard the space shuttle Endeavour. SIR-C can electronically steer its radar beam to vary incidence angle between 15 and 55 degrees and acquire polarimetric images. X-SAR which operates in X-band with a fixed VV polarisation is pointed mechanically. The ground swath is 15–90km depending on antenna beam orientation, and the resolution is 10–200m.

## USSR ALMAZ

COSMOS-1870 is regarded as the first USSR SAR mission and as a prototype for ALMAZ. The satellite was launched in 1987 and the mission ended in 1989. Typically a low orbit around 300km is used. The SAR operated in S-band (9.6cm) with a spatial resolution of 20–25m and a swath width of 20km. ALMAZ-1 (March 31,1991 - Oct. 17 1992) was the first Soviet SAR satellite with images commercially available. The radar wavelength was again 9.6cm and HH polarisation was used. The resolution was 15 to 30m depending upon range and azimuth. The swath was 40km within a range of 350km obtained by roll-tilt and the angle of incidence varied between 38° and 58° . The orbital altitude was 300km and it covered latitudes up to 78° at intervals of 1–3 days. A passive microwave and IR system was also included.

## ESA ERS satellites

The ESA preparations for ERS-1 goes back to at least 1981 and the satellite was launched in July 1991. The core instruments consist of a C-band active microwave instrumentation acting as a SAR or a scatterometer and a radar altimeter. These instruments are complemented by an along-track scanning radiometer and a microwave radiometer as well as the passive laser corner retro-reflector. The precise range and range rate equipment, PRARE, failed directly after the launch of ERS-1. GOME is the Global Ozone Monitoring Experiment sensor developed for ERS-2 and the only difference in instrumentation between ERS-1 and ERS-2.

## AMI

AMI stands for Active Microwave Instrument (AMI) at 5.3GHz which combines the functions of a Synthetic Aperture Radar, SAR, a wave mode and a wind mode scatterometer. The aim is to measure wind fields and wave spectrum over the oceans and taking all-weather high resolution images over polar caps, coastal zones and land areas. The SAR and the wind scatterometer modes cannot be used simultaneously and normally the wind scatterometer is used over ocean and, the SAR over ice and land. The SAR can be switched on for 10–15 minutes per orbit and since the system is currently using the reserve travelling wave tube, the switch on time is limited to 10 minutes. As at 1994 some 450,000 SAR $100 \times 100 \text{km}^2$ images have been acquired since launch.

**Radar Altimeter**

A Radar Altimeter, RA, operating at 13.7GHz with the aim to determine altitude, significant wave height, ocean surface wind speed, and various ice parameters.

**Along-track Scanning Radiometer and Microwave Radiometer**

The core instruments are complemented by an Along-Track Scanning Radiometer and Microwave Sounder, ATSR/M. The ATSR is a scanning three channel infrared imaging sensor operating at 1.6, 11 and 12m; (another channel at 3.7m is now defunct). This instrument allows the same spot on the Earth's surface to be viewed with a spatial resolution of 1×1km at angles of 0 and 55 degrees. This permits a more accurate correction for atmospheric temperature measurements than previously as long as the atmospheric along track variations can be considered negligible. The RMS. accuracy is better than 0.5K over a 50×50km square with 80% cloud cover. The Microwave Sounder is a nadir looking two-channel radiometer operating at 23.8 and 36.5GHz. It is designed to measure the vertical column water vapour content and cloud liquid of the atmosphere within the 22km footprint of the radar altimeter. In conjunction with the ATSR measurements the tropospheric range corrections for the altimeter can be accurately determined with a predicted precision of 2cm.

## NASDA JERS-1

The Japanese Earth Resources Satellite JERS-1, launched in 1992, is equipped with an optical sensor and a synthetic aperture radar. The L-band system (1,275GHz/20kHz) has a 12×2.2 m large horizontally polarised antenna and a 75km swath. The off-nadir nominal angle is 35 degrees, and the resolution is 18m in range and 18m (3 looks) in azimuth. SAR images have been commercially available for some time.

The optical sensor is a multi-spectral radiometer with a very high geometric resolution (18×24m) with the capability of taking stereoscopic imagery.

An important aspect is the possibility for onboard recording of 20 minutes of sensor data or over 72Gbits, in order to cover areas outside the coverage of ground stations. The nominal lifetime for the mission is two years. Due to some antenna unfolding problems at launch time the SAR is operated with limited output power.

The expected lifetime is two years and the satellite has a 44 days (westward) repeat cycle and an altitude of 568km.

## 1.4 Future SAR satellites

### RADARSAT

RADARSAT is a Canadian satellite expected to be launched in late 1995. It has an expected lifetime of 5 years. RADARSAT has a rather advanced and complete SAR as payload although only at one frequency, C-band, and one polarisation, HH. The objectives are to produce better ice and weather forecasts, to reduce navigational risks and make resource exploration safer and more effective. The satellite will also be quasi-operational by providing surveillance of Canada's high Arctic, scanning the region once

| Mode | Swath Width (km) | Resolution range×az (m) | Looks | Incidence angles |
|---|---|---|---|---|
| Standard | 100 | 25×28 | 4 | 20–49 |
| Wide swath(1) | 165 | (48–30)×28 | 4 | 20–31 |
| Wide swath(2) | 150 | (32–25)×28 | 4 | 31–39 |
| Fine resolution | 45 | 11-9×9 | 1 | 37–48 |
| SCANSCAR(narrow) | 305 | 50×50 | 2–4 | 20–40 |
| SCANSCAR(wide) | 510 | 100×100 | 4–8 | 20–49 |
| Extended(high) | 75 | 20-19×28 | 4 | 50–60 |
| Extended(low) | 170 | 63-28×28 | 4 | 10–23 |

**Table 2.** *SAR Modes of RADARSAT*

per day, to permit merging the data with other spatial data for the same area and providing the information to users (such as ships) within a few hours of the satellite's passage. Ocean applications include improved weather and sea state forecasts, detection of ships and oil spill. Land applications include identification of geological features important in mineral and petroleum exploration, soil moisture and forest changes.

Orbit: RADARSAT is to be launched into a circular, sun-synchronous orbit with an altitude of 792km and an inclination of 98.6 . The satellite will circle the earth 147/24 times in 24 hours and the orbit typically will cross the equator around 10:00 and 14.00. The repeat cycle will be 24 days with sub-cycles of 3 and 7 days.

Instrument properties: flexibility in the design of the SAR is provided by two features. The beam forming is obtained by an adjustable phase of the elements across the antenna width. The ground range resolution is varied by using two pulse widths of about 17.3 and 11.6MHz with resolutions in the range from about 20 to 30m, and by an additional bandwidth of 30MHz to provide the finer resolution required for certain applications. The SAR can be switched on for 28 minutes per orbit and there are two tape recorders on board which will be capable of 10 minutes of recorded data. RADARSAT provides 5 different SAR modes, illustrated in Table 2 and Figure 3.

## ERS-2

ERS-2 is almost identical to ERS-1; the only exception being a new instrument for the Global Ozone Monitoring Experiment, )GOME). ERS-2 will be launched in 1995.

## ALMAZ-1B

The next ALMAZ satellite is planned for 1996;it will include SAR systems working at X-, S-, and P-band. There will also be an X-band real aperture radar. Roll of the satellite will be used to cover a larger swath. The resolution will be 5–7m for X-band, 5–7, 15, or 15–40m (depending upon mode) for the S-band and 22–40m for the P-band equipment. The real aperture radar will have a 300×2000m resolution. Other sensors are planned on the spacecraft. For further details see Kramer (1994).

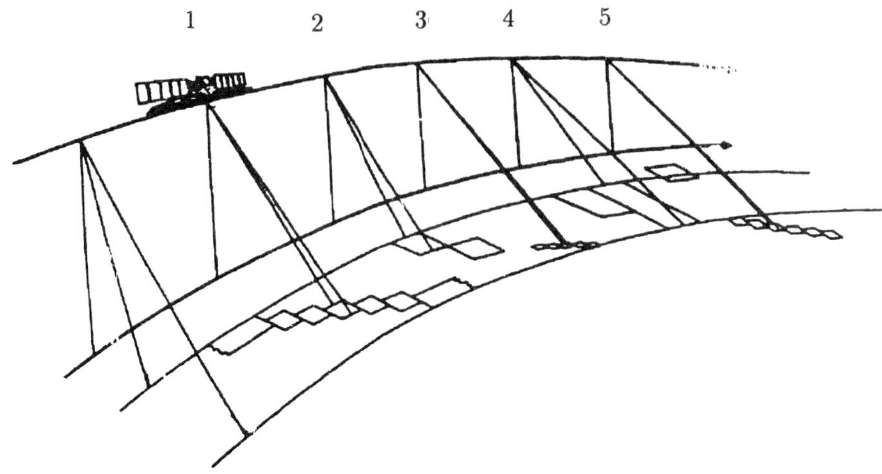

**Figure 3.** *Illustrating different* SAR-*modes of* RADARSAT. *The five scanning modes are (from left to right): (1) standard beams, (2) wide swath beams, (3) fine resolution beams, (4) a scan bar and (5) extended coverage at high and low incidence.*

## ENVISAT ASAR

ENVISAT is the next generation European SAR-equipped satellite, planned to be launched in 1998. The advanced SAR, ASAR, has a large number of different swath and polarisation states. The image mode consists of seven overlapping swath options with swath widths of 100 km and incidence angles covering the range 15° to 45° . Either VV or HH polarisation can be used. The spatial resolution is 30m and the effective number of looks is approximately 5.

The wide swath and global monitoring mode uses electronic beam-steering for a very fast multiplexing between subswaths so that an overall swath width of 440km can be achieved in order to monitor large areas. The wide swath mode has a resolution of 100m using two looks. By using a 1km resolution mode the data can be stored onboard and the monitoring being independent of ground station coverage.

ASAR also offers a possibility for alternating the polarisation between VV and HH by interleaved looks in one single pass. The swath is the same as for the image mode, but the radiometric resolution is reduced. Note that the phase between VV and HH cannot be determined in this mode.

## 1.5 RADARSAT II

The RADARSAT program is planned to continue with a RADARSAT II to be ready to be launched in the 1999/2000 time frame and to be essentially a copy of RADARSAT I.

### Polar platforms

The SAR is a very demanding instrument concerning power, weight, and data rate and there may also be problems of incorporating it together with other instruments due to electrical disturbances. The US future satellite platform is presently under debate. The SAR which had been discussed for a long time was a very complex one including L, C, and X-band of which two may have polarimetric capability. Presently no plans for a SAR system is included in EOS (Earth Observation System). A 'free-flyer' mission (launched by the shuttle) based on SIR-C/X-SAR is under discussion for 1996 in order to obtain a lifetime of about one year.

The European contribution to the polar platforms consists of ENVISAT-I (launch in 1998) and METOP-I (launch in 2000) of which ENVISAT is equipped with SAR (see above). Several different missions are included in the Polar Earth Observations Program including meteorology, ocean/ice, climate, atmospheric chemistry, land.

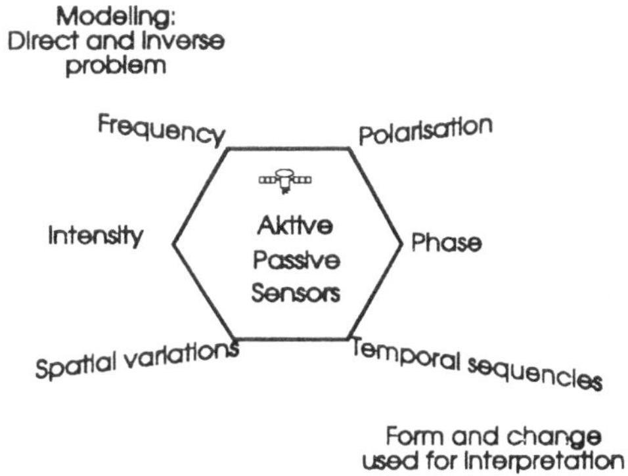


**Figure 4.** *Illustrating the different cornerstones of observables with active and passive microwave sensors used on satellites. Intensity, frequency and polarisations response is used with modeling to analyse the direct and inverse problem while temporal and spatial variations including the phase information from interferometric measurements are used in the image interpretation*

## 1.6 Observations with active microwave sensor systems

From remote sensing images information can be obtained based on different properties as illustrated in Figure 4.

A certain frequency and polarisation is transmitted in the case of an active sensor and the intensity at the same polarisation (or possibly the orthogonal polarisation) is received and analysed. In the case of active sensors two images can be recorded over the same area and the phase difference between the images studied (SAR interferometry to be described later). The intensity and polarisation of the various targets as well as the phase information can be used in modelling of what happens in the interaction between the microwave and the medium and help us in understanding the properties of the target.

When an image is formed by point observations over the swath of the sensor the spatial variations on various scales are giving specific image information which is very important for understanding what kind of areas/objects that have been imaged. Some areas have very specific temporal changes, *e.g.* an agricultural area during the growing season, and by comparing a number of images over the same area different areas can be classified. We use form and change for interpretation.

Normally we wish to observe a number of observables on a global basis for a long period of time. These will be used to derive maps of the mean fields, as well as to measure temporal and spatial fluctuations. It is fundamental to make this type of measurements during a long time under as similar conditions as possible. On the other hand there are measurements that will focus on specific problems and which will require detailed *in situ* measurements together with the satellite measurements. Results of these measurements will be used in model developments of the processes. The goal must be to develop from simple models describing a particular phenomenon to predictive models that allow us to infer the responses of the earth to changes in the environment.

Measurements with other sensors and experience from earlier measurements often have to be combined with the actual measurements to solve the measurement problem.

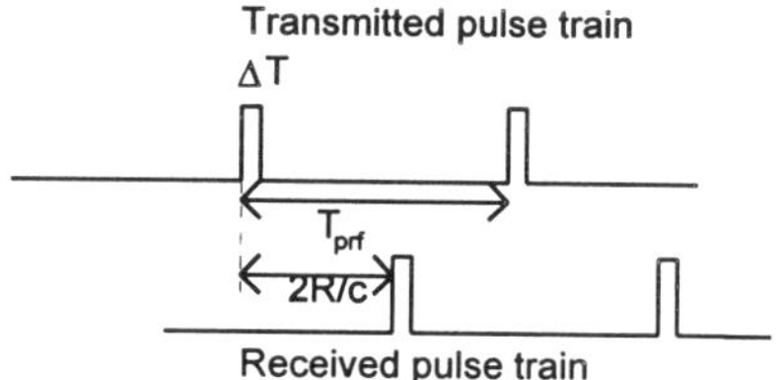


**Figure 5.** *Illustrating transmitted and received signal of a pulse-modulated radar*

# 2 Technical principles

## 2.1 General radar properties

**Basic radar principles**

The range measurement is based on the fact that the signal echo is received after a delay of $T = 2R/c$, where R is the distance to the scattering object and $c$ is the speed of the electromagnetic pulse. The range resolution is determined by the pulse width $\Delta T$ of the pulse and we obtain $\Delta R = c\Delta T/2$. Note the factor 2 caused by the radar pulse going back and forth. So far we have considered a single pulse, but in practise we use a pulse train with a pulse repetition frequency, $PRF = (T_{prf})^{-1}$. This means that we cannot be absolutely sure if the radar echo is caused by one pulse or another (Figure 5).

If we use a side looking airborne radar, see Figure 6, there is a projection factor in the range resolution expression depending upon the incidence angle $\theta$ of the beam $\Delta r = c\Delta T/(2\sin\theta)$. The field of view of the system is determined by the antenna

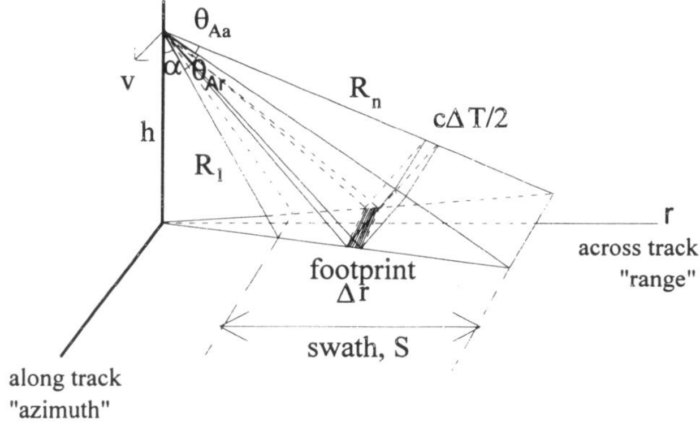


**Figure 6.** *Side Looking Airborne Radar, SLAR - basic principle*

beam and we should use a long antenna, $D_a$, in the flight (azimuth) direction in order to produce a good angular resolution, $\theta_{Aa}$, as long as we do not use a synthetic aperture radar. In order to cover a wide swath we need a small antenna, $D_r$, across the flight direction, with an aperture angle of $\theta_{Ar}$. The swath, S, is determined by $A_r$. The resolution is given by $\Delta_r = c\Delta T(2\sin\theta_i)$ where $\theta_i$ is the incidence angle, the angle between the incident wave and the vertical of the surface. We also have a condition on $T_{\mathrm{prf}}$ as two pulses following after each other should not give return pulses at the same time from the swath. That means that the difference between the far range and the near range, $\Delta R = R_f - R_n$, must satisfy

$$cT_{\mathrm{prf}} > 2\Delta R \tag{1}$$

In order to obtain a good range resolution we need a short pulse width. However a short pulse width means a system with a large bandwidth as well as high amplitude as the pulse energy i.e. (pulse power times $\Delta T$) determines the detection possibilities. To avoid these difficulties one can use a chirp radar (Figure 7).

In the case of a chirp radar the frequency of the transmitted frequency is varying linearly with time. The pulses illuminate a relatively wide area on the ground but each

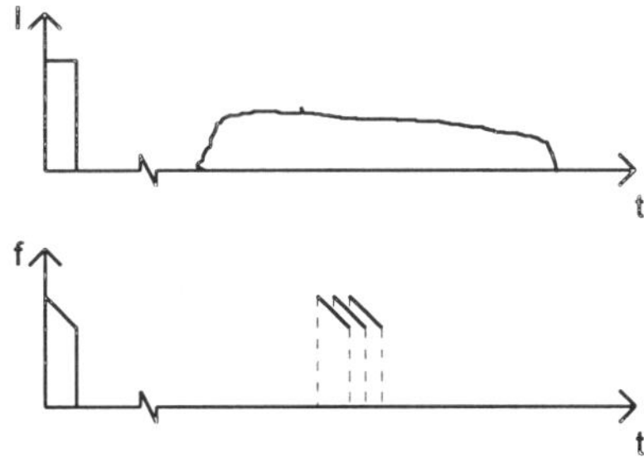


**Figure 7.** *Illustration of a normal pulse radar and a chirp radar*

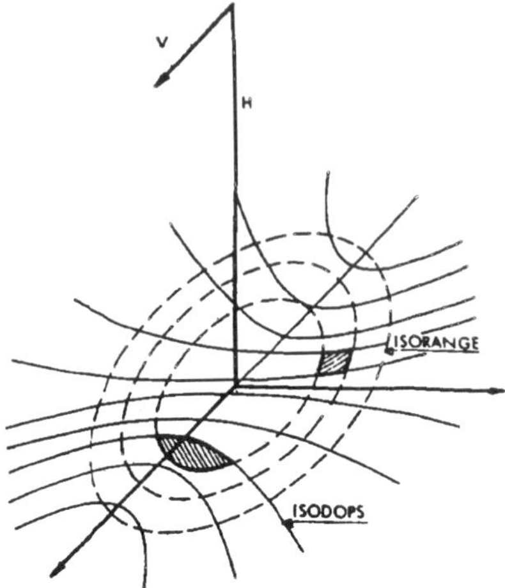


**Figure 8.** *Illustration of isodop and isorange curves*

point within that area returns a signal with a frequency varying linearly with time. By correlating the return signal with a frequency modulated signal a sharp peak is obtained for a distance related to the time offset. The resolution depends on the ability to sample sufficiently often the returned frequency varying signals not to be aliased by the sampling rate. In the case of the ERS-1 SAR system a $37\mu$s chirp pulse is transmitted and after filtering is reduced to an effective pulse length of 64ns.

In order to measure the velocity of the scattering object we use the Doppler effect. The phase of the return radar wave is given by $\Phi = \omega_0 t - 2kR$ and $\delta\Phi/\delta t$ determines the instantaneous angular frequency. When $R = Ro + vt$ we obtain for the instantaneous angular frequency $\omega_{\text{inst}} = \omega_o + \omega_D$ where $\omega_D$ is the Doppler shift $\omega_D = -2kv$. We finally obtain that $f_D = -2v_r/\lambda$ with $v_r$, the radial velocity, and $\lambda = c/f_0$. With a radar on board a satellite we obtain echoes from ground with Doppler shifts according to what apparent radial velocity a certain scatterer at ground has. Points with the same Doppler shift are connected by isodop-curves, see Figure 8, while points with the same distance are connected by circular isorange curves. The isodops are given by:

$$x^2 + y^2 = R^2 - h^2$$

and since

$$f_D = -2\frac{v_r}{\lambda} = -2\frac{v}{\lambda}\frac{x}{R}$$

we obtain

$$x^2 \left( \frac{4v^2}{\dfrac{v}{f_D\lambda}x} - 1 \right) - y^2 = h^2 \,. \tag{2}$$

The isodop and isorange curves are used to locate scattering objects on the ground. In the forward direction the resolution can be obtained by frequency filtering. With

$f_D = -2v_r/\lambda = -2v\sin\theta_i/\lambda$ we obtain different Doppler shift for different $\theta_i$ or $f_D = (2v/\lambda)\cos\theta_i\Delta\theta_i$. Consequently the ground resolution,

$$\Delta x = \Delta\theta_i R = (\lambda R/2v)\Delta f_D/\cos\theta_i,$$

is determined by the frequency resolution $\Delta f_D$ ($v$ is the velocity of the airplane at the height $h = R\cos\theta_1$ and $\theta_i$ is the incidence angle of the wave, a plane earth approximation is used).

## Measurements of radar cross section

We will now analyze the scattering of radiation from the ground using a radar from a radar transmitting a power denoted by $P_t$ and with an antenna gain G and an effective receiving area $A_e$. At the ground the power density is reflected by an effective area $A_g$ and the received power is

$$P_g = \frac{P_t}{4\pi R^2}GA_g \tag{3}$$

If the ground surface reacts as an antenna concentrating the backscattered radiation $P'_g$ with an antenna gain $G_g$ the received power at the satellite is

$$P_r = \frac{P'_g}{4\pi R^2}G_g A_e \tag{4}$$

The relation between the transmitted and received radiation then is

$$\frac{P_r}{P_t} = \left(\frac{P'_g}{P_g}G_g A_g\right)\frac{GA_g}{(4\pi r^2)^2} = \sigma\frac{\lambda^2 G^2}{(4\pi)^3 R^4} \tag{5}$$

where we have introduced the scattering cross section $\sigma$ of the ground. The scattering cross section is dependent upon how large the physical surface giving rise to the scattering is,(see Figure 9) and for an extended area we can introduce a normalised scattering cross section per unit surface $\sigma^0 = \sigma/A_g$, where $A_g$ represents the geometric surface giving rise to the scattering.

Let us now determine the radar cross section, $\sigma$, for a square metallic surface (side length a). In the monostatic case, i.e. the transmitter and receiver are placed at the

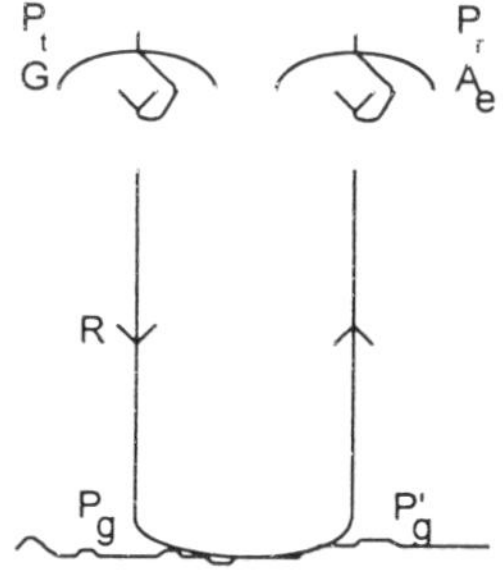


**Figure 9.** *Definition of radar cross-section*

same place, $\theta = 0$, and

$$\sigma = G_g A_g = \frac{4\pi A_g{}^2}{\lambda^2} = \frac{4\pi a^4}{\lambda^2} \tag{6}$$

while in bistatic case $(\theta \neq 0)$ we have

$$\sigma = \frac{(1 + \cos\theta)^2 \pi a^4}{\lambda^2} F^2(\theta) \tag{7}$$

with

$$F(\theta) = \frac{\sin\left(\dfrac{ka}{2}\sin\theta\right)}{\dfrac{ka}{2}\sin\theta} \tag{8}$$

For small $\theta$ we may for convenience approximate $F(\theta)$ with a Gaussian, $F^2(\theta) \propto \exp\left[-4\ln 2(\theta/\theta_A)\right]$, with $\theta$ small and $\theta_A = 0.9\lambda/a$.

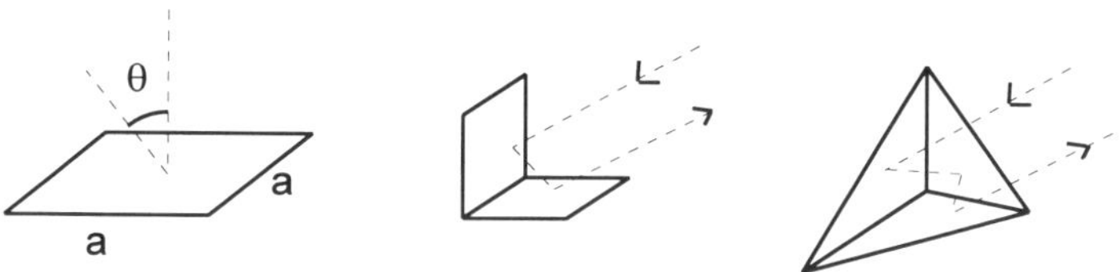


**Figure 10.** *Plane, dihedral and trihedral radar reflectors*

The trihedral corner reflector (Figure 10) has the ability to reflect the incident radar wave back to the radar. It has a broad radiation diagram, and trihedral reflectors are extensivly used to calibrate radars. A dihedral corner reflector effect is often found in nature. The relection between ground and the stem of a tree or a building or other construction often can result in strong backscatter.

In Equation 5 we defined the radar cross section $\sigma$. Solving for $\sigma$ we obtain:

$$\sigma = \frac{(4\pi)^3 R^4}{\lambda^2 G^2} \frac{P_r}{P_t} \tag{9}$$

Consequently $\sigma$ can be found by measuring the ratio $P_r/P_t$ and the distance R, supposing the system parameters $\lambda$ and G are known. The radar cross section of a point target is a hypothetical area intercepting that amount of power which, when scattered isotropically, produces an echo equal to $P_r$ as received from the object. For extended areas the normalised radar cross section $\sigma^0$ is used as will be discussed later.

In remote sensing we are interested in the back scatter from extended targets, $\sigma^0$, but in order to know the correct surface to use for normalising the scatter we have to know the incidence angle and the local terrain slope. If these are not known we should instead speak about the radar brightness, $\beta^0$, (Raney *et al.* (1994)). In a SAR image the quantity $\beta^0 \Delta A_z \Delta r$ is the observed radar reflectivity per pixel if $\Delta A_z$ is the pixel distance in azimuth and $\Delta r$ is the pixel distance in range. The parameter $\beta^0$ may be described as the (mean) radar brightness or (mean) radar reflectivity per unit pixel area. The radar reflectivity $\sigma^0$ of a given surface observed under some specific conditions such as soil moisture, roughness, incidence angle, slope angle etc, may be considered as an intrinsic property, and in principle we can only provide an estimate of the reflectivity. This is generally implied when we use $\sigma^0$.

## Speckle

An important effect in connection with the radiometric resolution of a radar is *speckle*. Radars transmit phase coherent waves. Echoes from different part of a reflecting body within a resolution cell will have different phases and we obtain constructive or destructive interference between the different components, producing the speckle effect. A short mathematical description is given below.

The reflected field characterised by an arbitrary phase can be written:

$$E = x \cos \omega t + y \sin \omega t = A \cos(\omega t - \Phi) \tag{10}$$

where $x = A \cos \Phi$ and $y = A \sin \Phi$ are the field amplitudes in phase and out of phase with the transmitted wave. A large number of random scatterers produce according to the central limit theorem a normal distribution of $x$ and $y$ with the variance $s^2$:

$$p(x) = \frac{1}{\sqrt{2\pi}s}e^{-x^2/2s^2}, \qquad p(y) = \frac{1}{\sqrt{2\pi}s}e^{-y^2/2s^2}. \tag{11}$$

$x$ and $y$ are independent and thus $p(x,y) = p(x)p(y)$. We also have that the signal power $A^2$ is given by $A^2 = x^2 + y^2$, and want to determine the statistical distribution for the field amplitude $A$ and phase $\Phi$. We have assuming randomly oriented phases that $p_A(A)dA = p(x,y)2\pi A dA$ and

$$p_A(A) = \frac{A}{s^2}e^{-A^2/2s^2} \tag{12}$$

This is the Rayleigh probability distribution which we obtain for the amplitude of the scattered wave. A good approximation to a Rayleigh distribution tends to occur for as few as six to ten scatterers within a resolution cell. If we determine the mean amplitude for the Rayleigh distribution we obtain for the mean field amplitude and its variance:

$$\mu = \overline{A}^2 = \int A\, p_A(A)dA = s\sqrt{\pi/2}$$

$$\sigma^2_A = \overline{A^2} - \overline{A}^2 = \left(2 - \frac{\pi}{2}\right)s^2 \tag{13}$$

or $\sigma_A/\mu = 0.523$. This is why we speak about multiplicative noise—if the amplitude of the scattered wave increases the speckle will also increase. Figure 11 shows an illustration of the probability functions illustrating the histogram we expect to obtain in an image of a smooth surface where all the variation is caused by the speckle. For SAR imaging we use multiple look imaging to reduce the speckle and the distribution for the N-look image (Elachi C, 1987) is also given in Figure 11.

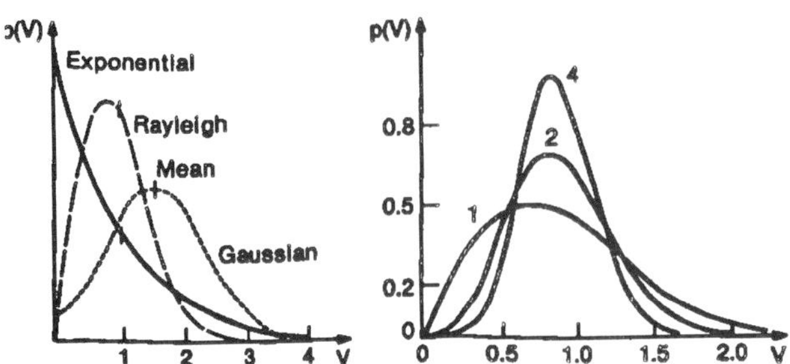

**Figure 11.** *Illustrating in (a) the Gaussian and Rayleigh probability functions and in (b) the Gamma distribution for a N-look image (from Elachi)*

## 2.2   Synthetic aperture radar, SAR

A special type of active microwave sensor, an imaging radar, is called the Synthetic Aperture Radar, SAR. The unique feature of this radar is that it uses the forward motion of the spacecraft to synthesize a much longer antenna, which in turn, provides a high ground resolution. In order to make use of the forward motion, both the amplitude and phase of the return signal have to be recorded. The timing measurement is used to discriminate individual cells across the satellite track while the Doppler-induced variations in the frequency of the return signal are employed to provide the along track resolution.

Sensors like the SAR (as well as high resolution optical sensors) produce data rates in excess of 100 megabits per second. This data stream puts large demands on the data processing facilities.

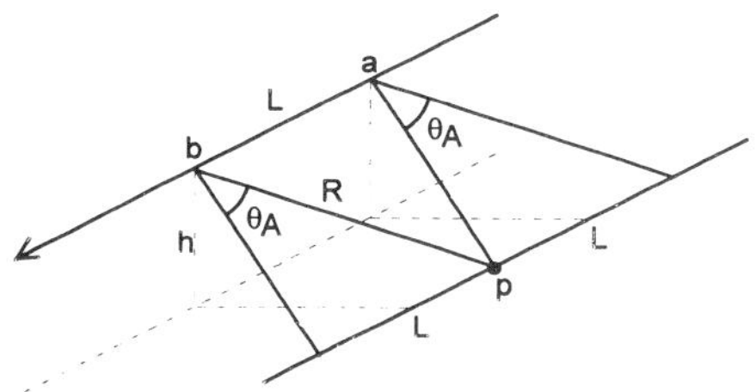

**Figure 12.** *Basics of SAR principle*

**Basic SAR concepts**

A very simple, and not completely correct description of the resolution of a synthetic aperture radar is the following (see Figure 12). Assume a target, $p$, is illuminated by the antenna on board the satellite, which will be the case over a distance of $\theta_A R$, where $\theta_A A$ is the beamwidth of the antenna is approximately $\lambda/D_a$, ($D_a$ is antenna size in the along track direction). Now assume that we can properly process the scattered radiation taking into account the change in the satellite position by means of the accompanying Doppler shift. If so we would synthesize an effective antenna resolution of SAR with a length of $L = \theta_A R$ and the effective beamwidth would be $\gamma \approx \lambda/L$ and the resolution would be $\gamma R = D_a$. (the more careful analysis below yields approximately $D_a/2$, where the factor 2 is caused by the radar wave propagating back and forth, *cf* the mirror effect). This means that we obtain a resolution of the same order as the physical antenna size *independent* of satellite height.

A more correct analysis can be based on the two way Doppler shift obtained earlier, $f_D = -2v_r/\lambda$. According to Figure 12 $v_r = vx/R$, where $x$ is a coordinate along the track from the sub-satellite point and $x_{\max} = \theta_A R/2$ for points illuminated by the antenna. Consequently the limiting values of $f_D$ are $\pm v/D_a$. The available Doppler bandwidth therefore is $2v/D_a$ and the temporal resolution due to the Doppler bandwidth is the inverse, $\Delta\tau = D_a/2v$. The spatial resolution then is $v\Delta\tau v = D_a/2$. The difference of a factor 2 between the two presentations is due to the fact that we could not properly take into account that in the radar case the pulse is propagating back and forth.

Yet another possibility would be to analyse the radiation from the antenna and the Doppler shift in the scattering by a simple mathematical analysis. The radar transmits and receives with the same antenna, characterised by the antenna beam width, $\theta_A$. The radar antenna total gain is then proportional to $F_n^2$. Let us now approximate the antenna gain in the along track direction by a Gaussian lobe $G^2 = G_0^2 \exp\left[-8\ln 2(\theta/\theta_A)^2\right]$. The return signal has a phase determined by $\exp i(\omega_0 t - 2k_0 R)$, where $R$ is the distance to the ground, $R = \sqrt{R_0^2 + v^2 t^2} \approx R_0 + v^2 t^2/2R_0$ , where $R_0$ is the closest distance to the ground. Due to the motion effect we obtain a backscattered signal described by a chirp in signal with $\alpha = 2k_0 v^2/R_0$ in the following expression (neglecting the phase term associated with $R_0$)

$$E_r \propto G^2 \exp\left[i(\omega_0 t - 2k_0 R)\right] = \exp\left[8\ln 2\left(\frac{\theta}{\theta_A}\right)^2\right]\exp\left[i\left(\omega_0 t \frac{\alpha}{2}t^2 - 2k_0 R_0\right)\right] \qquad (14)$$

We now change from the angular dependence $\theta$ to time by using $\theta = vt/R_0$, *cf* Figure 12, and introduce the quantity $\Delta t = R_0\theta_A/(v\sqrt{2})$. We then obtain a complex chirp signal

$$E_r \propto \exp\left[-4\ln 2\left(\frac{t}{\Delta t}\right)^2\right]\exp\left[i\left(\omega_0 t - \frac{\alpha}{2}t^2 - 2k_0 R_0\right)\right] \qquad (15)$$

If this signal is going through a pulse compression filter where the pulse is mixed with a time varying local oscillator to compensate the chirp, the time variation after the filter is given by

$$\exp\left[-4\ln 2\frac{\alpha}{\Delta\omega^2}t^2\right] \qquad (16)$$

where $\Delta\omega = 8\ln 2v/(R_0\theta_A)$. If, on the other hand, we had a synthesized antenna with a halfwidth of $\gamma$ it would be represented by an expression of the same sort as Equation 14 above, i.e.

$$\exp\left[-8\ln 2\left(\frac{\theta}{\gamma}\right)^2\right] \qquad (17)$$

and we then obtain for $\gamma$, again using $\theta = vt/R_0$

$$\gamma = \frac{2\ln 2}{\pi} \frac{\lambda}{\theta_A R_0} \tag{18}$$

If we now compare with the beam width for a physical antenna with constant illumination, $\theta_A = 0.9\lambda/D$, we see that the effective length of the synthesized antenna is of the order twice the distance the satellite has travelled while it at the same time illuminates the target. Similarly we obtain that the synthetic resolution is given by $R_0\gamma = 2\ln 2 D/(0.9\pi) \approx 0.5D$.

The result gives the impression that by reducing the size of the antenna the resolution is increased! However, there is a lower limit to the antenna size. and there is a trade off between the antenna size in azimuth, $D_a$, and azimuth resolution, on one hand and the antenna size "in range", $D_r$, which determines the swath on the other hand. We will demonstrate that with the following derivation.

An actual SAR uses a series of pulses instead of a continuous signal. The pulse repetition frequency, PRF, should be high enough to sample the Doppler spectrum. With $f_D{}^{\max} = v/D_a$ we obtain, according to the Nyquist sampling theorem, that $PRF \geq 2v/D_a$ (for a satellite with $v = 6.4$km/s and a 10m antenna this means 1280 Hz). However, we also have a maximum value of the $PRF$ as the pulses reflected from the near edge and the far edge of the swath will be received simultaneously which limits the swath and sets the maximum value of the $PRF$. According to Equation 1 we have $cT_{\mathrm{prf}}/2 > \Delta R = (h/\cos\theta_i)\theta_A \tan\theta_i$ assuming a plane earth and neglecting the variation of the depression angle $\alpha$ over the swath. With $\theta_A \approx \lambda/D_a$ we obtain

$$PRF = \frac{1}{T_{\mathrm{prf}}} \leq \frac{cD_r}{2\lambda h}\frac{\cos^2\theta_i}{\sin\theta_i} \tag{19}$$

The two constraints on the PRF imposes a limit on the antenna area (for a satellite the earth curvature has to be taken into account)

$$D_r D_a > \frac{4\lambda h v \sin\theta_i}{c\cos^2\theta_i} \tag{20}$$

## SAR processing

The price to pay for the increased resolution is the high data rate from the satellite sensor to the ground station where the data has to be processed to produce an image. The spectrum of the received signal is determined by the pulse width used for the range resolution, see Figure 13a. Due to the pulse repetition procedure, the spectrum consists of a series of lines separated by the PRF. Each line is Doppler broadened by $B_d = 2v/D_a$, see Figure 13b. The SAR swath is divided in $N_r$ range cells corresponding to the range resolution, and the echo signal is moved into a memory of Nr elements. Similarly the azimuth footprint $\theta_A R$ is divided into $N_a$ azimuth cells corresponding to the azimuth resolution. After reception of $2N_a$ number of echoes it is possible to generate a synthetic image of an azimuth length equal to the synthetic aperture length. The azimuth lines are mixed with the signal from a processing filter (an oscillator which

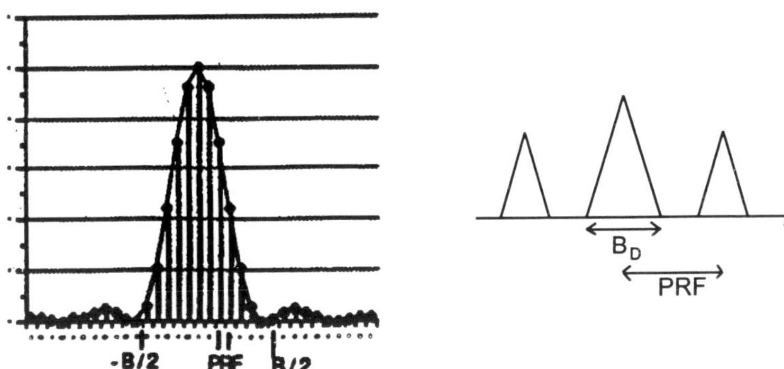



**Figure 13.** *(a) Illustrating the full pulse spectrum and (b) Doppler broadened individual lines*

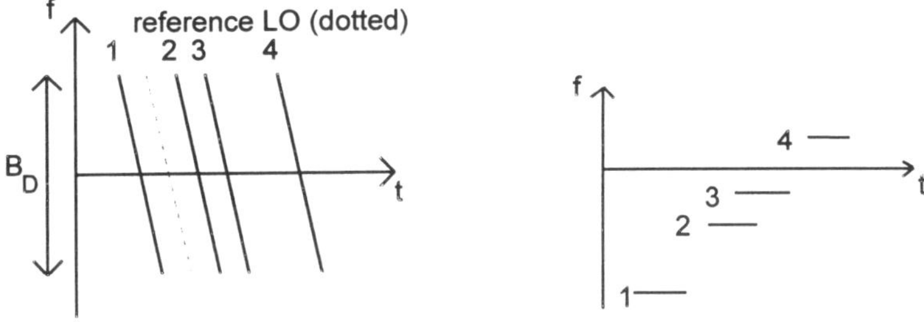



**Figure 14.** *(a) Doppler history of individual targets at the same range and (b) Frequeny response after mixing with LO*

produces a signal varying in time and frequency like $v(t) = \exp i\alpha t^2)$. We then obtain a constant frequency difference between echoes from different targets at the same range, see Figure 14.

The targets can now be separated by frequency filtering. From section 2.2.1 we have that $f_D = -2v_r/\lambda = -2x/\lambda R$ or $x = \lambda R f_D/2v$. This means that $\Delta x = \lambda R \Delta f/2v$ and the resolution is determined by how narrow the bandpass filter is. The minimum value is determined by $\Delta f = v/L$ with $L = \theta_A R$ as $L/v$ is the time the target is seen. We will obtain a separation between targets along the azimuth line if their distance on the ground is greater than $x_a = Da/2$. This is the maximal azimuth resolution.

SAR processing places a very high demand on data transfer rates and computation speed. As an example we can consider ERS-1 moving with a speed of 6.7km/s (as measured on the ground) and with an azimuth sampling distance of 3.94m and with 5616 range samples per second. The radar echo from each element is to be represented by an inphase and out-of-phase component (I & Q), each one represented by 5 bits. This means a data flow of $6700\times5616\times(5+5)/3.94 = 96$Mbits/s, which together with some further information gives a data flow of 105Mbits/s. To calculate one range line we have to use data from the entire synthetic aperture length of approx 4km. Real time processors for satellite SAR are presently not available. A modern design by Norsk Data is producing one SAR scene covering $100\times100$km in 8 minutes with 25m resolution by using four vector units each with 80Mflops. A real time processor has to produce the image in 15 seconds and if a maximum resolution image is needed we need something which is more than 300 times faster than the present design. A more realistic alternative would be a low resolution image, say $1\times1$km which means about two times faster than the present design but much smaller. The design of a processing algorithm and processor architecture for a synthetic aperture radar is a challenge, in particular if it is to be a satellite born one.

## Speckle reduction by multi-look processing

The problem of speckle in radar imaging was introduced earlier and one way to reduce speckle is to use multi look processing. Instead of using the full Doppler band width we divide the Doppler band width using N bandpass filters (N-look image). This means that echoes from a larger number of targets are incoherently added and the noise reduced. An alternative (but not practical due to the increased computation load) is to average adjacent pixel values in a single look image. (Note in a radar image the radar return from a certain range and along track distance is represented by the value of the pixel. Either the intensity of the radar signal, proportional to the backscattering coefficient, $\sigma^0$, or the amplitude ($\sqrt{\sigma^0}$) is used in a so called intensity or amplitude image. The pixel distance can be chosen arbitrarily but should be less than the radar resolution.)

## SAR power considerations

The received signal to noise ratio (SNR) is the ratio between the received signal power to the thermal noise power. Generally, for good image quality it is desirable to achieve an SNR of at least 3:1. The SNR for the SAR is obtained from (1.5) with $\sigma = \sigma_0 x_a x_r$, with $P_n = kT_R B$ and $x_a = \lambda RB/(2v)$

$$SNR = \frac{P_t G^2 \lambda^3 \sigma_0 x_r}{2(4\pi)^3 R^3 v k T_R} \tag{21}$$

$T_R$ is the receiver noise temperature and $P_t$ is the averaged transmitted power (as we have to transmit a number of pulses to build up the aperture, $N = PRFL/v$ determined from the peak transmitted power by ($\tau$ is the output pulse length)

$$P_P = P_t/(\tau PRF) \tag{22}$$

From Equation 21 we conclude that the $SNR$ increases with frequency with constant antenna size as long as $\sigma_0$ is independent of frequency, which is normally not the case. From Equation 21 we also obtain the so called noise equivalent backscatter cross section,$\sigma_n^0$, which is the value on $\sigma_0$ when $SNR = 1$.

| SAR antenna size | 10m long, 1 mwide |
|---|---|
| frequency | $5300 \pm 0.2$ MHz (C-band) |
| bandwidth | $15.5 \pm 0.1$ MHz |
| long pulse duration | 37.1 $\mu$s |
| compressed pulse length | 64ns |
| PRF | 1640–1720 |
| peak power | 4.8 kW |
| polarization | VV |
| signal sampling window | 299 s (99km telemetered swath) |
| A/D complex sampling | 18.96 Msamples/s |
| I & Q quantisation | 5 bits for on-board range compr. |
| | 6 bits for on-ground range compr. |
| radiometric resolution | 2.0 dB at $\sigma^0 = -18$ dB |
| swath width | 100km |
| incidence angle | 23° at mid-swath |
| resolution (rad. res. 2.5 dB) | 25°$\times$ 22m (range azimuth) |
| data rate | $<105$ Mbps |

**Table 3.** *SAR parameters for ERS-1*

## Special SAR concepts

ERS-1 is the first remote sensing satellite of the European Space Agency, ESA. The satellite is equipped with a number of sensors including a SAR with properties as given by Table 3. The next generation of ESA satellites with a SAR is ENVISAT, which will have a more advanced SAR onboard with a larger possible swath but then decreased resolution. Some applications however, need high resolution. One way to increase the resolution is to use the technique of a SPOT-SAR or Spotlight SAR. If the antenna can be continuously pointed at the target P it is possible to view the target over a much larger arc. This allows a longer synthetic aperture and therefore a finer azimuth resolution according to the first and simple description of the SAR-concept. In theory the resolution will approach the ultimate resolving power of the wave, i.e. $\lambda$, but we loose at the same time area coverage.

Polarimetric SAR technique has been used in the US space shuttle. In the polarimetric case the SARs transmit both vertically and horizontally polarized field components and also receive both the components. Not only amplitude but also phase is determined. This means that a complete description of the scattering object can be done and we can characterise the polarised, coherent scattered component completely and also determine the inchoherent, unpolarised component. ENVISAT A SAR will use VV and HH measurement but will not measure the phase between the components.

Interferometric SAR, INSAR, has attracted much interest recently. Since the SAR system is coherent, it is possible to use it in an interferometric mode. Two images of the same area gathered from two slightly different across track positions interfere and produce phase fringes that can be used to accurately determine the off-nadir angle, $\theta$, to each resolution cell. The two images can be obtained from two different SAR-systems

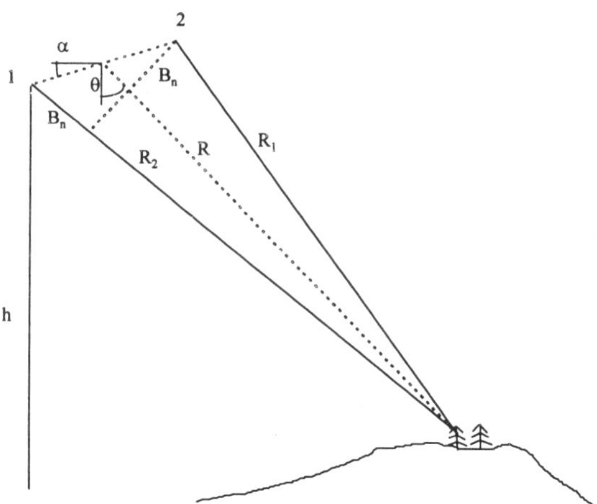

**Figure 15.** INSAR *geometry, perpendicular to the line of motion*

at the same time or from the same system at two different parallel passages over the area. The system geometry is shown in Figure 15. The satellite track separation, called the baseline, is described by $B_n$ and $B_p$, the baseline normal and perpendicular to the line of sight. All the scatterers in one resolution cell contribute to the return signal and we obtain speckle by the randomness of the phases between the different scatterers in two neighbouring resolution cells.

The pixel values, $s_i$, in an interferogram are produced from two complex SAR images by multiplication of every pixel, $g_{1,i}$, in one image with the complex conjugate of every corresponding pixel, $g_{2,i}$, in the other. Thus

$$
\begin{aligned}
s_i &= g_{1,i} g_{2,i}{}^{*} \\
&= a_{1,i} \exp\left[-i\frac{4\pi}{\lambda} R_{1,i}\right] a_{2,i}{}^{*} \exp\left[i\frac{4\pi}{\lambda} R_{2,i}\right] \\
&= |a_{1,i} a_{2,i}{}^{*}| \exp\left[i\frac{4\pi}{\lambda}(R_{2,i} - R_{1,i} - \Phi_{\text{noise},i}\right] \\
&= |a_{1,i} a_{2,i}{}^{*}| \exp\left[-i\frac{4\pi}{\lambda} B_{p,i} - i\Phi_{\text{noise},i}\right] \\
&= |s_i| \exp\left[-i\Phi_{\text{noise},i}\right]
\end{aligned}
\tag{23}
$$

where $a_{1,i}$ and $a_{2,i}$ are the complex pixel amplitudes, $R_{1,i}$ and $R_{2,i}$ are the two slant range coordinates, $B_{p,i}$ is the baseline parallel to the line of sight as given in Figure 15 and $\Phi_{\text{noise},i}$ is the phase noise that is due to speckle decorrelation and thermal noise. Starting from the phase in Equation 23 and by assuming that the scene is stable (the scatterers not moving), it is possible to derive a linearised expression for the variations of the interferogram phase

$$
\Delta\Phi = \frac{4\pi}{\lambda} B_n \Delta\theta = \frac{4\pi B_n}{\lambda R \tan\theta}\Delta R + \frac{4\pi B_n}{\lambda R \sin\theta}\Delta z + \Phi_{\text{noise}} + 2\pi n
\tag{24}
$$

where $B_n$ and $R$ are defined above, $\Delta\theta$ is the difference in elevation angle, $\Delta R$ is the slant range difference and $\Delta z$ is the altitude difference between pixels in the interferogram.

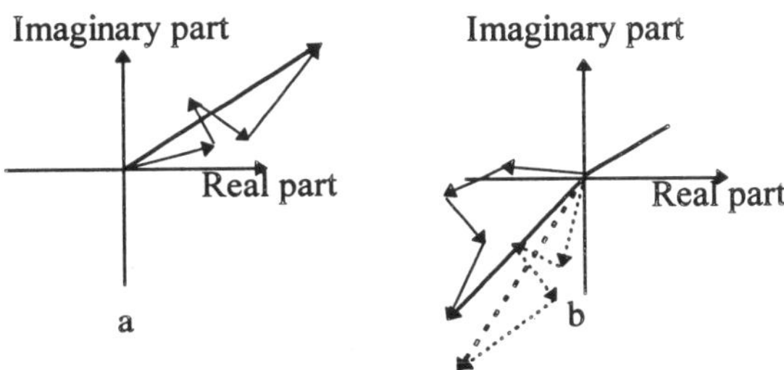


**Figure 16.** *Contribution to return signal from four different scatterers within one resolution cell viewed from two slightly different across track positions. The dotted line indicates the effect of different slant range and the full line the effect including speckle.*

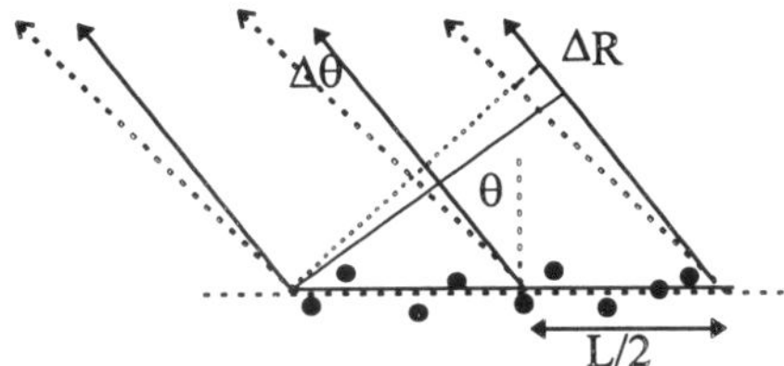


**Figure 17.** *Scatterers within one range resolution cell from two slightly different angles.*

The first term in Equation 24 is purely a systematic effect that easily can be removed in the processing by applying a correction factor, the 'flat earth' compensation. In the second term there is a direct relation between the phase and the altitude in the image. The last term represents the phase ambiguity induced by the modulo $2\pi$ phase registration. The ambiguity has to be removed in the processing by adding the correct integral number of $2\pi$ to each measured value. This is called phase unwrapping. If the $2\pi$ ambiguities are removed this phase difference can be used to calculate the off-nadir angel $\theta$.

It is possible to increase the phase resolution by increasing the baseline up to a certain limit where the coherence is lost (baseline decorrelation of speckle). This limit can be calculated from the Rayleigh criterion that says that the phase difference between the speckle from the same resolution cell in the two images must be less than $2\pi$, otherwise the coherence is completely lost (*cf.* the properties of an antenna beam generated by an antenna with size L). If L is the size of a resolution cell, we obtain ($\theta$ is the incidence angle)

$$R_2 - R_1 = \Delta R = (L/2)\cos\theta\,\Delta\theta \tag{25}$$

and the demand according to the Rayleigh criterion in order to obtain coherence between the two registrations is that $\Phi_{\text{speckle}} < \pi/2$ or $\Delta R < \lambda/4$, or in other words

$$\Delta\theta \leq \frac{\lambda}{2L\cos\theta} \tag{26}$$

This means that the maximum distance between the satellites across the sight line, the baseline B, *cf.* Figure 15 is

$$B \leq \frac{R\lambda}{2l \cos \theta} \tag{27}$$

For ERS-1 ($\lambda = 5.66$cm, $R = 850$km, $L = 25$ m, $\theta = 23°$) the maximum baseline over flat terrain is about 1.3km. This value will be influenced by surface topography.

The INSAR technique is also very sensitive to small shifts (of the order of $\lambda$) of the scattering surface. In this case a term

$$\Delta\Phi_{\mathrm{mov}} = \frac{4\pi}{\lambda}\Delta\eta$$

where $\Delta\eta$ corresponds to a coherent change (movement) of all scatterers within the resolution cell along the line of sight. This makes it possible to study minute changes e.g. in association with earthquakes.

# 3    Qualitative and quantitative aspects of SAR

## 3.1    Introduction

In order to determine properties of the earth's surface in a qualitative manner using SAR we have to emphasise some specific aspects concerning the sensor properties such as viewing aspects and speckle noise together with definitions of measured properties and calibration of the sensor. In order to obtain quantitative results we must use image analysis to combine the point measurements with the context of the surrounding points. We must also use modelling of expected properties to support the interpretation as we presently have only one channel SAR sensors available in space.

## 3.2    Basic definitions

In connection with the measurement of $\sigma$, the basic concept describing the strength of scattered radar signals, it is necessary to introduce some terminology (see also Section 5).

**Spatial properties**

*Spatial resolution*: the discrimination between the echoes from neighbouring targets, how close two targets can be and still be detected as two separate targets.

*Spatial accuracy*: the accuracy with which each pixel can be given geographic latitude and longitude values.

**Radiometric properties**

*Precision*: the degree to which it is possible to discriminate between different radar cross-section values from the different intensity levels in a SAR image (relative radiometric accuracy).

*Absolute accuracy*: the accuracy by which radar cross-section values can be associated with intensity levels in a SAR image.

*System dynamic range*: the range of cross section values for which the stated accuracy and precision is achieved.

*Noise-equivalent $\sigma^0$*: the normalised radar cross section (in dB) which generates a signal equal to the system noise level.

**Phasimetric properties**

*Precision*: the degree to which it is possible to discriminate between different phase values between different channels in a polarimetric or interferometric radar system.

## 3.3   Calibration

For an SAR we have three kind of calibration: geometric, radiometric and phasimetric. The calibration may be an absolute or relative one providing the relation between the instrument output and a physical quantity or a reference quantity (*e.g.* a trihedral) respectively.

**Geometric calibration**

The aim of the geometric calibration is to determine the absolute location of the pixels in an image. For this purpose we need information of the position and velocity of the sensor platform (index $s$) as well as the target (index $t$). The SAR measurements are based on two equations:-

$$\text{range equation:} \qquad R = \tfrac{c}{2}T$$

$$\text{Doppler equation:} \qquad f_D = \frac{2}{\lambda}\frac{(R_s - R_t)}{R}(v_s - v_t).$$

For localisation of the pixels on the earth we need an earth model consisting of

$$\text{a reference ellipsoid:} \qquad \frac{x_t^{\,2} + y_t^{\,2}}{R_e^{\,2}} + \frac{z_t^{\,2}}{R_P^{\,2}} = 1$$

$$\text{a Digital Terrain Model:} \qquad z_t = F(x_t, y_t).$$

There are some error effects due to the special side looking radar geometry, and related to the conversion of the measured slant range to the ground range. When the surface is not flat, but we have topographic features, we obtain foreshortening when the front side of a hill seems to be shorter than the back side, see Figure 18(a).

In the extreme case we obtain layover, when the top of the hill is closer to the radar than the foot of the hill, and radar shadow, when part of the back side of the hill is not illuminated, see Figure 18(b).

If we assume a plane earth we have a displacement in the ground plane of $\Delta x = h/tg\theta_i$, where $h$ is the height and $\theta_i$ the incidence angle. For $h = 1000$m and $\theta_i = 23°$ we have $\Delta x = 2356$m . Note that determination of the normalised radar cross section is dependent on the slope of the surface by a factor relating the area in the slant range

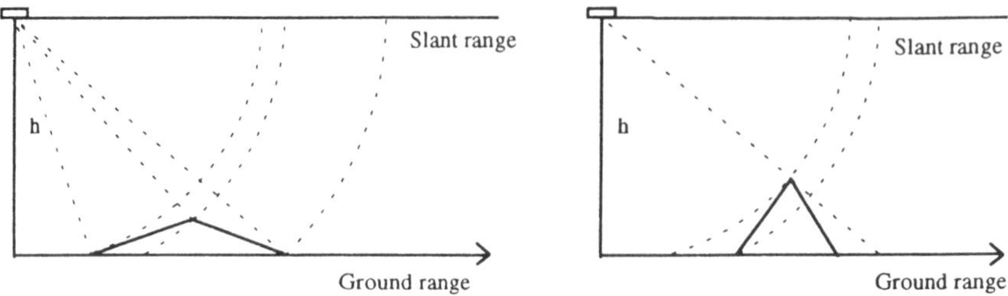



**Figure 18.** *(a) Foreshortening in side looking radar image; (b) layover and radar shadow in a radar image*

image to the corresponding area on the ground surface. Beside the geometric factor we have an incidence angle dependence of $\sigma^0$ which is harder to correct for. (Slope angles can be determined by INSAR.).

Geometric distortions can be caused by sensor properties and localisation, atmospheric effects on radar pulse propagation, terrain effects and processor errors as well as resampling of the image to a map projection.

## Radiometric calibration

Radiometric calibration of the SAR image is performed by the following equation (Laur 1992, Ulander 1992) assuming no local slope to correct for (*cf* radar brightness $\beta^0$)

$$\langle \sigma^0 \rangle = \frac{f}{k} \left\{ \langle DN^2 \rangle - \langle DN_n{}^2 \rangle \right\} \tag{28}$$

where $\langle\ \rangle$ denotes spatial averaging (over 50–500 pixels), $K$ is the calibration constant, $f$ is a range-dependent correction, and $DN$ and $DN_n$ are the pixel values in the area of interest and noise, respectively. The definition of the range correction $f$ for all $FD$ images is according to

$$f = \left( \frac{R}{R_{\text{ref}}} \right) \frac{\sin \theta}{\sin \theta_{\text{ref}}} \frac{G^2(\vartheta_{\text{ref}})}{G^2(\vartheta)} \tag{29}$$

where $R$ is the slant range ($R_{\text{ref}} = 847$km), $\theta$ is the local incidence angle ($\theta_{\text{ref}} = 23°$ ) and $G$ is the antenna gain pattern as a function of $\vartheta$, the elevation angle from nadir: ($\vartheta_{\text{ref}} = 20.35°$ is the antenna boresight angle). The ERS-1 fast delivery SAR images are not provided with sufficient information to calculate $\theta$ and $\vartheta$. Instead, the nominal values have to be used in this case. Full information is available with the PRI product. The elevation gain pattern of the ERS-1 SAR antenna has been measured by ESA in flight and the calibration constant $K$ has been determined. $K$ depends on which PAF has produced the image.

Note also that $K$ and $G$ like $\sigma^0$ are normally given in dB and have to be converted to natural units by the following equation

$$P \quad (\text{in natural units}) = 10^{P(\text{db})/10} \tag{30}$$

where $P$ is either $K$ or $G$.

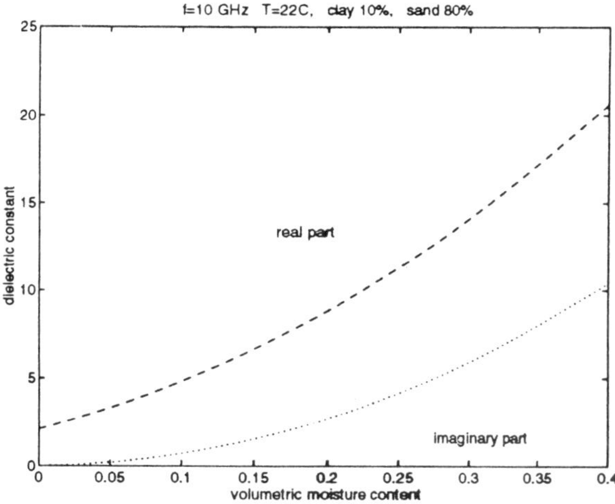


**Figure 19.** *The $\varepsilon_r$ dependence on soil moisture for a sand surface*

## Phasimetric calibration

The phase is important for polarimetric and interferometric SAR measurements. In the latter case phase delays through clouds and due to atmospheric humidity can be a serious problem requiring more investigation. By using several images over the same area the problem can be identified.. Interferometric calibration also involves establishing the absolute coherence which requires compensation for system effects and baseline decorrelation.

## 3.4  Modelling of radar cross sections

### Dielectric contrast

The scattering of the radar wave by a target such as a rough surface is determined by the dielectric properties of the surface and its roughness parameters in a rather complicated fashion. The dielectric properties are given by the dielectric constant $\varepsilon_r$, which is closely related to the moisture content. The relation between moisture content and $\varepsilon_r$ has been examined by a number of authors. Hallikainen *et al.* (1985) have developed a semi-empirical relation for vegetation and soil surfaces, where the real and imaginary parts of the dielectric constant have been modelled as functions of the volumetric moisture content, Figure 19.

Also the temperature and the soil composition (percentage of sand, silt and clay) influence the strength of the dielectric constant and is accounted for in the model. The dielectric constant for complex media is important to be able to determine as function of interesting properties, *e.g.* soil moisture, in order to interpret radar measurements.

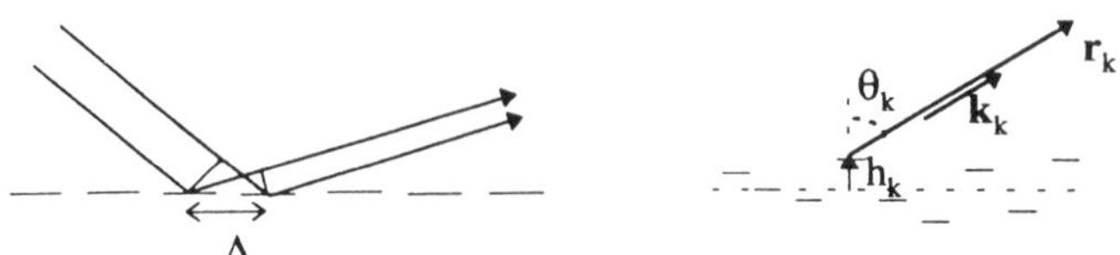

**Figure 20.** *Scattering from a periodic surface; (b) scattering from a rough surface.*

## Scattering from a periodic surface

The plane surface scatters radiation only in the forward direction, while the periodic surface scatters in a number of directions given by the demand on constructive interference for reflections from two points separated by the periodicity L (other kind of reflections are not repeated constructively and are then destroyed by destructive interference). This is illustrated in Figure 20(a).

We obtain constructive interference in all those directions (*cf* optical phase gratings) where the difference in path length is a multiple of a wavelength. This leads to a generalised form of Snell's law ($k_x$ is the $x$-component, i.e. along the surface, of the propagation constant $k = w/c$)

$$k_x^{\text{refl}} = k_x^{\text{inc}} + n\frac{2\pi}{L} \tag{31}$$

For the special case of back-scattering we have $k_x^{\text{refl}} = -k_x^{\text{inc}}$ and with Equation 31 we obtain

$$\Lambda = \frac{n\lambda}{2\sin\theta_i} \tag{32}$$

which is the important relation ($n = 1$) for radar reflections from the ocean surface. This relation is related to Bragg scattering in crystals, and is called the Bragg-resonance. For ERS-1 we obtain ($\theta_i = 23°$, $\lambda = 5.3$cm) that $\Lambda = 6.8$cm, which is the wavelength of the capillary/short gravity waves responsible for the scattering from oceans.

## Scattering from a rough surface

Let us now consider what happens if the surface is rough. To describe some of the principal effects on an incident wave due to the height distribution of the surface, we will consider N scatterers with individual heights, $h_k$, relative to a mean surface (neglecting any correlation between the scatterers) Figure 20(b). The total electric scattered field is the sum of the individual field components, all with their individual phases (the factor 2 is due to that the radar wave is propagating back and forth):

$$E_s = \sum_1^N u_k e^{-i2\mathbf{k}_k \cdot \mathbf{r}_k} \tag{33}$$

The radar cross section is determined by the scattered power:

$$E_s E_s^* = \sum_1^N u_k e^{-i2\mathbf{k}_k \cdot \mathbf{r}_k} \sum_1^N u_k e^{+i2\mathbf{k}_k \cdot \mathbf{r}_k} \tag{34}$$

Sufficiently far away from the surface we may neglect the differences between the individual scatterers and normalise their amplitudes, $u_i = 1$, $\mathbf{k}_i = \mathbf{k}$, $i = 1.....N$. With $\mathbf{r}_k = \mathbf{R}_k + \mathbf{h}_k$ we obtain for the mean scattered field

$$\langle E_s E^*{}_s \rangle = \sum_1^N \sum_1^N e^{-i2\mathbf{k}\cdot(\mathbf{R}_t - \mathbf{R}_p)} \left\langle e^{+i2\mathbf{k}\cdot(\mathbf{h}_k - \mathbf{h}_p)} \right\rangle \tag{35}$$

We introduce $\delta_{kp} = 2\mathbf{k}\cdot(\mathbf{h}_k - \mathbf{h}_p)$ and can assume that $\delta_{kp}$ is small and independent of $k$ and $p$ and therefore can be denoted only $\delta$. With $\delta$ small we can use a series expansion and obtain

$$\left\langle e^{-i\delta} \right\rangle = \left\langle 1 - i\delta - \delta^2/2 + i\delta^3/3! + ... \right\rangle = 1 - \left\langle \delta^2 \right\rangle/2 + \approx \exp\left\{ -\langle \delta^2 \rangle/2 \right\} .$$

In this derivation we have used that $\langle \delta \rangle = 0$ or $\langle h_k \rangle = 0$. We now obtain:

$$\left\langle \delta_{kp}{}^2 \right\rangle = \left\langle (h_k - h_p)^2 \right\rangle = \left\langle h_k{}^2 \right\rangle + \left\langle h_p{}^2 \right\rangle - 2 \left\langle h_k h_p \right\rangle = s^2 + s^2 - 0 = 2s^2$$

where we have introduced $s$ as the standard variation of the height distribution.

The double summation includes $N^2$ terms. For $N$ of these we have $k = p$ and the exponential factors are equal to 1. We then obtain with $g = 2ks\cos\theta$

$$\langle E_s E_s{}^* \rangle = \left[ N + e^{-g^2} \sum_{1,k\neq p}^N \sum_1^N e^{-i2\mathbf{k}\cdot(\mathbf{R}_k - \mathbf{R}_p)} \right] \tag{36}$$

With $g = 0$ we have $\langle E_s E_s{}^* \rangle = N^2$ and consequently we know that the double summation must be equal to $N^2 - N$. We thus obtain:

$$\langle E_s E_s{}^* \rangle = N^2 e^{-g^2} + N(1 - e^{-g^2}) \tag{37}$$

The first term is the coherent part and the second is the incoherent part.

Thus we get two contributions with different meanings. In the first part all contributions are phase coherent and add up to a $N^2$ contribution, while in the incoherent part we have a power summation from the $N$ facets. The factor $g$ is related to the Rayleigh criterion. When $g < \pi/4$, the coherent part dominates. When $g > \pi/4$, the incoherent scattering dominates. When $g = \pi/4$ the two contributions are approximately equal in the above expression. The Rayleigh criterion for a smooth surface is defined as

$$2ks\cos\theta < \pi/8 \tag{38}$$

where $s$ is the rms height variations, see Figure 21(a). Figure 21(b) illustrates how the backward scattering is increasing with increasing roughness.

To characterise surface roughness, the standard deviation of its one-dimensional height profile $z(x)$, can be used (rms height).

$$s = \sqrt{E[z(x)^2] - (E[z(x)])^2} \tag{39}$$

In general the surface correlation function

$$\rho(\xi) = \frac{E[z(x + \xi)z(x)] - (E[z(x)])^2}{E[z(x)^2] - (E[z(x)])^2} \tag{40}$$

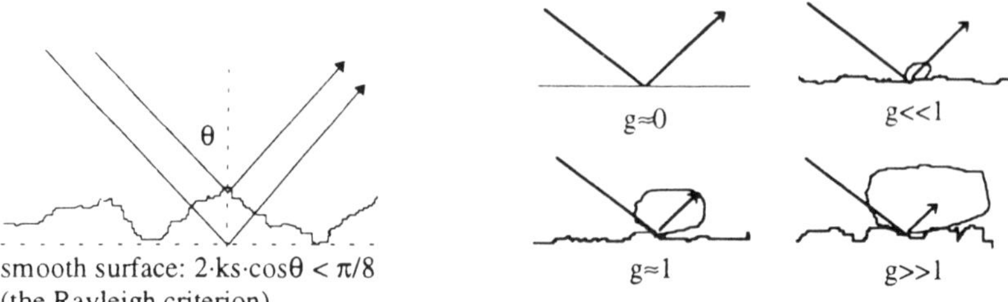



**Figure 21.** *(a) Characterisation of an uneven surface; (b) different cases of scattering illustrating the diffuse, non coherent component and the directly reflected, coherent component.*

also affects the radar backscattering, Often the correlation function can be expressed as an exponential or Gaussian function, characterised by a typical length, the correlation length $l$.

A number of theoretical models to predict the radar backscattering from rough surfaces have been developed, assuming that the rms height, the correlation length and the dielectric constant are known. The models are developed based on certain assumptions about the radar wavelength compared with the rms height and correlation length of the surface.

At present one of the most advanced surface scattering models is the integral equation model (IEM). The IEM covers the validity ranges of the Small Perturbation Method (SPM), the Physical Optics Model (PO) and partly the Geometrical Optics Model (GO). The validity of surface scattering models is illustrated in Figure 22 in comparison with data of surface roughness measured for sea ice during ARCTIC'91. For the IEM method, see (Fung *et al.* 1992), for the other methods, see (Ulaby *et al.* 1986).

## Volume scattering

If the radar wave penetrates into the volume below the rough surface we may obtain volume scattering from imbedded dielectric discontinuities. For volume scattering the radar cross section $\sigma_s$ is determined by the dielectric constant, $\varepsilon_s$, of the scatterer and its size, $r$, relative to the wavelength, $\lambda$. When the scatterers are much smaller than the wavelength we have Rayleigh scattering. In this case we obtain

$$\sigma_s = \frac{2^6 \pi^5 r^6}{\lambda^4} \left| \frac{\varepsilon_s - \varepsilon_b}{\varepsilon_s + 2\varepsilon_b} \right| \tag{41}$$

if $kr\left|\sqrt{(\varepsilon_s/\varepsilon_b)}\right| < 0.5$, where $r$ is the radius of the inclusions. $N = V/(4\pi r^3/3)$ is the number density of scatterers, and $\varepsilon_b$ is the dielectric constant of the background medium.

An assumption of independent scattering is valid only for volumetric fractions of scatterers less than 3–5%. Multiple scattering is a second order effect which often can be neglected in the case of like-polarized scattering but is important for explaining cross-polarization signatures.

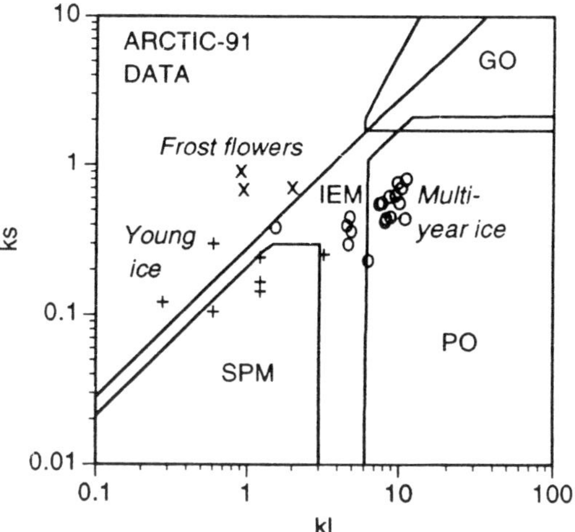


**Figure 22.** *Ranges of validity for the IEM in comparison to SPM, PO and GO. The magnitude of roughness parameters observed during ARCTIC'91 is also shown. IEM– Integral Equation Method, SPM–Small Perturbation Method, GO–Geometrical Optics, PO–Physical Optics, k–wavenumber, s–rms height and l–correlation length.*

## Surface and volume scattering

By the direct method we mean to compare observed radar signatures to model simulations which are based on *in-situ* data of field parameters. In *e.g.* the case of sea ice the semi-empirical model proposed by (Kim *et al.* 1985) which combines surface and volume scattering based on an intensity approach, is useful as a first step in the interpretation of radar data, because different physical effects are isolated easily. For a snow covered ice layer the Kim-model is given by

$$\sigma_0^0(\theta_0) = \sigma_{1s}^0(\theta_0) + \tau_{01}{}^2(\theta_0) \left\{ \sigma_{1v}^0(\theta_1) + \frac{1}{L_1{}^2(\theta_1)} \left[ \sigma_{2s}^0(\theta_1) + \tau_{12}{}^2 \sigma_{2v}^0(\theta_2) \right] \right\} \qquad (42)$$

where the contributing backscattering coefficients are denoted by $\sigma_{1s}^0$ for the surface, and $\sigma_{1v}^0$ for the volume of the first layer (snow) and similarly for the second layer (ice). $L_1$ is the loss factor of the first layer. The power transmission coefficients for the first and second interface are denoted $\tau_{01}$ and $\tau_{12}$, respectively. The incidence angle is $\theta_0$, and the refracted angles in the first and second volume are $\theta_1$ and $\theta_2$, respectively. Sometimes it may be relevant to add a third or more layers in the same manner. The layering is dependent on the estimated penetration depth and on the extent the scattering properties are varying with depth.

## 3.5   SAR image analysis

### Segmentation

The classification of SAR images is mainly based on the information obtained from the radar backscattering coefficient $\sigma_0$ and the standard deviation, $\sigma$, over a suitable window with a mean value $\mu$ for the radar cross section. Sometimes higher order statistics in the form of *e.g.* co-occurrence matrices are used. The speckle is a problem in radar images causing a minimum $\sigma/\mu$, and the texture standard deviation, $\sigma_T$, can be estimated (Ulaby *et al.* 1986) as

$$\sigma_T^2 = \frac{(\sigma/\mu)^2 - (1/N)}{1 + (1/N)} \tag{43}$$

where $N$ is the effective number of looks.

The radar cross section and possibly the texture standard deviation can be modelled (the direct method) from information on a number of surface properties. It is then possible to investigate the relative contribution from various parameters and sometimes isolate one phenomenon as more important than others.

We have at least two principally different SAR classification methods. One type is based on local considerations using some speckle filtering method and another using edge detection and region growing. A large number of articles have been produced in conjunction with different applications. For simplicity we well only refer to a very few articles related to the work in the Remote Sensing Group at Chalmers.

### Local Averaging Methods

Local averaging methods are based on speckle reduction by using information close to the point of interest. The simplest method is to determine the mean or the median value of $N$ pixels around the pixel of interest using a fixed $N$ over the entire image. A more refined method is to vary $N$ in an adaptive way over the image by first checking the variance over the area of interest. If the variance is larger than a certain value no averaging is performed and smoothing of borders avoided. One often-used adaptive method is Lee filtering (Lee 1986). For classification of the SAR images we need to determine the typical variation of the various classes and to determine $N$ in such a way that the class distributions do no overlap, i.e. increase the radiometric resolution at the expense of spatial resolution. Examples on sea ice classification are given in Kwok *et al.* (1992), Pettersson *et al.* (1994), Rignot *et al.* (1993), and Sun *et al.* (1992).

### Edge detection and region growing methods

Instead of adaptive local averaging we can first search for edges by some method such as the gamma detector and then determine if the edges are closed and if so determine the mean values inside the closed area. The segmentation method used here is described in Skriver (1989) and Sun *et al.* (1992). To use it we had to select the edge detector window size, the threshold value for the edge detected image and the thresholds in the segmented image to obtain a classified image. The gamma detector window is related to the size of the different areas and the speckle.

**Problems with the two classifications methods**

Classification with these two methods cause problems which are unique for each method.

In the SEG method it is very difficult to get an error estimation because of borders not beeing detected. This is in particular a problem when we have smooth borders between different ice areas. This is the case even if there is a large separation in backscatter between the two areas representing two different classes. Because the lack of border detection the two areas will be classified as belonging to the same class.

The LAM will not result in large homogeneous classified areas because we will get error pixels within a homogenous area. However, error pixels may not cause much error in the classification due to the maximum *a posteriori* estimation of the thresholds. This means that the error pixels of one class in another class are as many as the other way around and they should cancel. One problem with the LAM method is the border pixels between different areas, especially in complex border areas.

The best method for a certain application then depends on the risk for errors in each one of the methods.

**Concluding comments**

What has been described are examples on supervised classification where the classes have been determined by human interaction for identification of typical areas for the different classes. Unsupervised methods determine classes automatically, but normally have a problem in dividing the surface into natural classes. Neural networks have been given much attention recently for image classification, see *e.g.* Hara *et al.* (1994) and may offer interesting possibilities.

Often, however, we may have more than one single image available over a specific area. In particular with an all-weather sensor like the SAR we can count on images at the regular times of the satellite crossing and we can use repeated images over the same area to conclude about properties of changing areas, see Lichtenegger (1992).

GIS,(a Geographical Information System), is a method to combine information with its geographical coordinates in different information layers. In the SAR case it suffices to combine information from the SAR sensor with complementary information: images from optical sensors; topographic information (which can be used to take into account the geometric aspects in the image formation and in the calibration procedure); temperature information, (*e.g.* when surfaces are close to the melting point); wind information over the sea *etc.*

# 4   From Ocean to Land Application

## 4.1   Introduction

The SAR is in particular sensitive to dielectric changes primarily those caused by the water content and is also sensitive to various slope effects. Sometimes we have resonant effects such as Bragg scattering from capillary waves on the sea surface and generally we have a sensitivity to geomtric properties of the size of the wavelength. From the beginning the sensitivity for ocean features were realised as well as the sea ice sensitivity. Properties of the vast areas of oceans and of sea ice are very suitable for remote sensing however the applications are often specific and taken care of by meteorological and oceanographic institutes. The high resolution of the SAR also means a potential for land applications, which are more varied in their nature and not yet as well developed. In particular the new application of INSAR yields a possibility to determine topographic properties which recently has attracted high interest. Forestry properties are suitable for remote sensing due to the vast areas covered by forest, but the single channel SAR applications are relatively limited due to the saturation with bole volume or biomass. However change detection is a potential application in this case as well as in the case of agricultural applications where temporal changes are important to follow in order to deduce the surface properties, although very complex and so far relatively little studied in a qualitative manner. Soil moisture is a potential application, but the SAR sensitivity for surface roughness and vegetation must be taken account of, often in a refined manner.

## 4.2　SAR imaging of ocean waves

One of the principal applications of the SAR in oceanography is to study the ocean-surface waves to determine the wave spectrum. During the last few years there has also been a consensus amongst scientists about the mechanisms whereby waves are images. The problems have been the complexity of the problem in combination with the problems of *in situ* observations for validation. The SAR image is formed over a certain time and this is a fundamental problem in association with ocean waves since many of the phenomena occur on this time scale.

The basic phenomenon is radar scattering caused by resonant short-wavelength gravity or even capillary/gravity waves. These waves are modulated by the much longer wavelengths associated with the swell. The approach has been to study the scattering by using a two-scale model consisting of short waves causing Bragg resonance and swell waves of a scale large enough to be resolved by the pixel size of the radar. The spectrum of wavelengths which exist in between these scales is then ignored.

For the assumption that the short gravity waves by the Bragg mechanism is the prime scatterer it is fundamental to assume that the these waves do not substantially change within the integration time of the SAR, which in the ERS-1 case is of the order of 2 seconds. This means that the coherence time of the small scale ripple (*i.e.* the time that ripples should remain at a nearly constant amplitude) should persist during the SAR integration time. The viscous decay time of the ripples is, according to Tucker (1983), sufficient for Bragg resonance wavelengths larger than 4cm. Meanwhile the long swell waves beeing imaged have a period of typically 8–20s, and so it is reasonable to assume that their troughs and crests stay approximately in the same pixel during the integration time.

There are three distinct types of wave-imaging mechanism: hydrodynamic interactions, tilt modulation, and motion effects such as velocity bunching and azimuth image smear, of which the last one is specific for only the SAR due to the sensitivity for moving

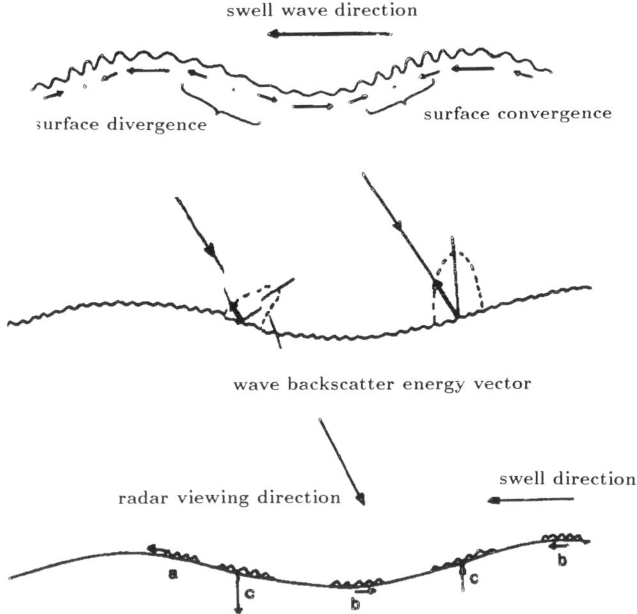


**Figure 23.** *Illustrating (a) hydrodynamic effect, (b) tilt modulation, and (c) motion effects. (From Robinson 1985)*

scatterers.

In imaging ocean waves we obtain a Doppler shift which is caused by the movement of the scatterers at the ocean surface due to ocean currents, orbital velocity due to the swell, and the phase velocity of the short resonant waves (see Figure 23). The part of the Doppler shift which is associated with the swell orbital velocity varies periodically in time as well as in space. This causes the velocity bunching effect and this mechanism contributes to the distortion of the image by smearing. The contrast in the SAR image can be increased by focusing which means shifting the matched filter by $\Delta V$ which is the difference between the platform velocity and the speed of the moving targets.

A good basic description of SAR imaging of ocean waves is given in Robinson (1985). Basic contributions to the theory have been given by Alpers and Rufenach (1979), Hasselman *et al.* (1985), Lyzenga (1988), Plant (1992), Raney and Vachon (1988), Shemdin (1988) and Swift and Wilson (1979).

## 4.3   Wave Spectra

A directional wave spectrum derived from the SAR image of ocean waves shows the direction and length of the dominant wave trains and, when properly calibrated, also shows the height and energy of waves.

Figure 24 illustrates three wave spectra derived at the ice edge: (a) illustrates the spectrum outside the ice, (b) between ice floes, and (c) the FFT of an ice flow showing

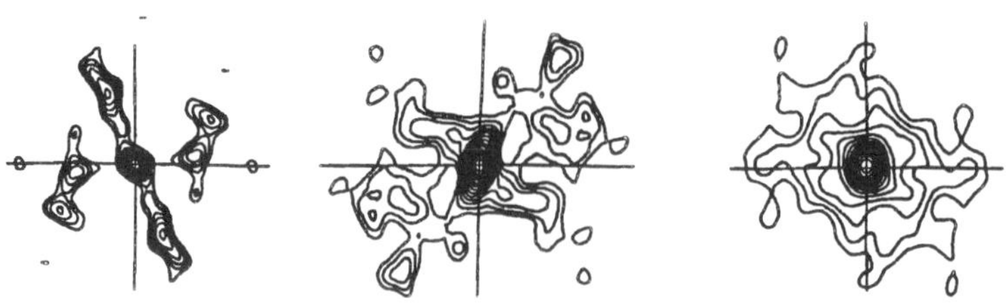

**Figure 24.** *Examples on SAR image derived directional wavenumber spectrum.*

just noise. The spectra have been derived by selecting sub scenes of 256×256 pixels, performing a 2-d Fourier transform and using a Gaussian filter over a 15×15 window for smoothing. The wavelength of the swell in Figure 24(a) is 523m .

## Oil spill

The damage caused by oil pollution is well known and it is important to use all available methods to detect oil spills in time to avoid large damage. For this purpose satellite-borne synthetic aperture radars can be used as well as airborne radar systems.

The oil on the surface can either form a thin film or have thicker parts. The thin parts may be as thin as micrometers while the thicker parts can be several centimetres. Investigations have shown that 90% of the oil can be found in 10% of the visually detected oil spill area. This means it is important to have equipment to detect the oil thickness such as microwave radiometers used together with airborne systems. Satellite imaging is useful for large area surveys while aircraft are useful to follow up an actual spill.

## Radar backscatter from oil spill

The reason why oil slicks are able to be detected on radar images is that even extremely thin oil films have a dampening effect on the capillary waves being responsible for the radar backscatter. The dominant backscatter mechanism from a sea surface is Bragg scattering from short wavelength capillary-gravity waves. The $n$th-order Bragg wavelength $\Lambda$ is given by

$$\Lambda = n\lambda/2\sin\theta_i$$

where $\lambda$ is the radar wavelength and $\theta_i$ is the incidence angle. The first order Bragg backscatter, $n = 1$, is normally the strongest, and for the C-band ERS-1 SAR with $\lambda = 5.3$cm. and $\theta_i = 23°$, this means a capillary-gravity wavelength of 6.8cm.

Radars are very sensitive to small changes in the surface roughness of the sea and the oil spill will cause such a change due to changes in the surface viscosity and surface tension.

A comparative study of ocean backscatter depression due to surface oil was carried out by Singh *et al.* (1986) using C-band and Ku-band scatterometers and using C- and X-band SAR imagery using HH-polarisation. The results show that ERS-1 is expected to detect a backscatter depression of the order 5 dB for light winds decreasing to 3–4 dB for winds of the order $10\text{ms}^{-1}$. Experimental results with ERS-1 show that for windspeeds between 2 and $6\text{ms}^{-1}$ oil spills are well recognised, that low resolution images can be used and that the seach for oilspills can be performed automatically.

The investigation (Singh *et al.* 1986) also demonstrated some interesting differences between the two radar frequencies over the angle of incidence interval. These differences may be caused by selective damping caused by a resonance effect, the Marangoni-effect, due to interaction with longitudinal waves in connection with visco-elastic surface films. Such effects can be obtained by monomolecular oleic alcohol film on the sea surface, causing very complex reaction between the surface film and the water molecules (Hühnerfuss 1986). Consequently thin slicks caused by natural organism can also cause radar backscatter depression of similar type as oil spills. The dark appearance of oil slicks on radar images is also similar to the appearance of areas of calm water. However, it is the distinctive shape of oil slicks which enables them to be identified with a high degree of confidence.

Controlled field experiments with oil spillage have been performed in the Haltenbanken area and reported by Bern *et al.* (1992).

## Detection of ships in SAR-images

Ship detection is a topic closely related to the question of oil pollution as well as other application, and the striking appearance of ship wakes in SAR images makes the detection of ships possible. Reports at the first ERS-1 Symposium in Cannes, November 1992, indicate that ship detection is most easy in far range at wind speeds up to $4\text{ms}^{-1}$ for ships larger than 50m and at wind speeds up to $10\text{ms}^{-1}$ if ships are larger than 100m. This means that fishing boats are not possible to detect at all and at wind speeds higher than $10\text{ms}^{-1}$ only very few ships can be detected. Ship wakes were reported to be seen with lengths up to 100km. Reports have been indicating that wakes are made by biologically produced chemicals, known as surfactants, that have an affinity for the ocean surface where they adsorb (hold molecules to their surface) in moderate to light winds.

Detection of ships from space have both civil and military applications and efforts are under way for fully automatic systems detecting ship signatures. To detect the ships at the same time as possible oil spills increases the possibility to determine who is causing the oil pollution.

A special feature in SAR images connected with moving ships is the displacement of the ship in the image due to the Doppler effect used by the SAR to locate objects along track. The SAR is 'fooled' by the extra Doppler shift caused by the ship velocity and shifts the boat in the image a distance related to its velocity. As the direction of the ship motion can be determined from the ship wake and the displacement of the ship relative the ship wake can be determined from the image, the ship velocity can be determined from SAR images.

## 4.4   Sea ice

### Introduction

Models of the earth as a system indicate an increase in global temperature due to the greenhouse effect. The effect is taking place over time scales of decades and early warning signs are important to identify. The models predict that effects due to a global temperature rise will be amplified in the Arctic (Stouffer *et al.* (1989)) due to the various feedback mechanisms associated with a strong coupling between the Arctic sea ice cover, the atmosphere and the ocean.

Modeling is a very complex process and each one of the different models for the atmosphere, sea ice, and ocean has to be further developed. Recently a model of the Arctic sea ice was analysed (Chapman *et al.* (1994)). Different parameters of the ice model were varied. Ice mass, ice area, mean ice drift, and winter-summer range of area were used as response parameters of the model. The strongest sensitivities arise from the minimum lead fraction, the sensible heat exchange coefficient and the atmospheric and oceanic drag coefficients.

Consequently there is a need to observe the various parameters of interest as indicators of a global temperature change and several remote sensing sensors are used to follow the ice properties and the cloud situation, (Gloersen and Campbell, (1991), Barry *et al.* (1993)). However, we cannot use a single remote sensing sensor to study the important properties of the Arctic ice cover but have to combine observations from various sensors and develop new techniques to observe some of the properties.

### Sea ice model for climate aspects

We will follow Chapman *et al.* (1994), who used a two-level dynamic formulation going back to Hibler (1979) and a thermodynamic formulation from Parkinson and Washington (1979) with various extensions. The dynamic formulation is based on momentum balance for a mass of ice which is affected by the air and water stresses prescribed by drag coefficients, Coriolis force, acceleration due to sea surface tilt and internal stress related to ice deformation and ice strength. The thermodynamic model is based on the surface energy budget, where the flux of the solar radiation, the downcoming flux of longwave radiation from the atmosphere, the transfer of sensible and latent heat between the ice and the atmosphere, the surface emission of Planck radiation, and the thermal conductivity to the ocean are balanced.

The ice mass is found to be most sensitive to the minimum lead fraction, while the ice area is most sensitive to the turbulent exchange of sensible heat, cloud depletion of longwave flux, the turbulent exchange of latent heat, and the snow albedo. The model demonstrates a negative trend in the wintertime maxima of ice covered area, which decrease by approximately 5% from the mid-1970s to the mid-1980s, consistent with observations by Gloersen and Campbell (1991). However it is warned that the model can easily be adjusted to produce trends of either sign in the simulated ice mass or ice area and the model was indeed tuned to give a temporally stable time series by varying the cloud depletion of solar flux. It is also stressed that the correlation between simulated and observed variations of ice coverage in various parts of the Arctic is such that much of the observed variance is unexplained by the model.

Although models of the kind described above are hampered by the complexity of the phenomena, the limitations of computer simulations and the uncertainties of parameter variations, the models can give information on important parameters to monitor, the need for accuracy, the need for sampling frequency *etc.* For example the total ice mass varies by nearly a factor of 2 as the minimum lead fraction varies from 1–4%.

The advance of the melt process during the summer should be important for global climate models. The ice floe melts from the surface, sides, and bottom (Steele (1992)). The lateral melt is significant only for small floes while the surface and bottom melt dominates and are nearly equal for large floes. One aspect of the melt process is the melt ponding. The variability of the melt pond area should be of interest to determine in order to characterise early stages in the degree of melting. The snow melt is typical for the Arctic (in contrast to the Antarctic) and the Arctic old ice may be covered in the summer by 10–60% of melt ponds (Gow and Tucker (1990)). Increasing melt pond coverage serves to accelerate melt rates as more heat is absorbed from incoming short-wave radiation.

The sensitivity analysis described above (Chapman *et al.* (1994)) illustrates the importance of a large number of parameters, only some of them may be possible to obtain by means of present day remote sensing techniques. Such properties include the ice area, and drift speed and it should be possible to relate these properties to *e.g.* the minimum lead fraction. However the modeling does not include a very detailed analysis of all possible phenomena and their variation by season, and there may be other properties as well of interest to determine by means of remote sensing.

## Remote sensing of the Arctic

The Arctic has been covered regularly independent of clouds and weather conditions by passive multichannel microwave radiometry since 1978, (Gloersen *et al.* 1992). Radar altimeters have also been used a long time and have been used for sea ice studies, (Laxon 1990). Synthetic aperture radars onboard Seasat, ERS-1, and JERS-1 provide high resolution information of Arctic ice. With the swath limited to 100km the sensor can be used to analyze test areas of the Arctic, while sensors like SSM/I are more appropriate for large coverage. RADARSAT and ASAR on Envisat will provide information over swaths of the order of 400/500km and the SAR will be more useful for coverage of large areas of the Arctic.

With the high resolution of the SAR we can observe other parameters than with low resolution sensors such as SSM/I, and we may observe the minimum lead fraction and the ice concentration with high accuracy. We can determine ice drift, we can observe properties of the wind and may relate these to the atmospheric drag coefficient of the sea ice. From high resolution SAR images we may also identify different ice classes and then indirectly various types of ice properties. By combination with information from the observed ice motion and possibly also with thermodynamic modeling, further constraints can be obtained in order to identify ice thickness and ice mass. However this is a complex procedure.

The SAR is mainly used for limited areas of the Arctic but by statistical sampling

                            *J Askne*

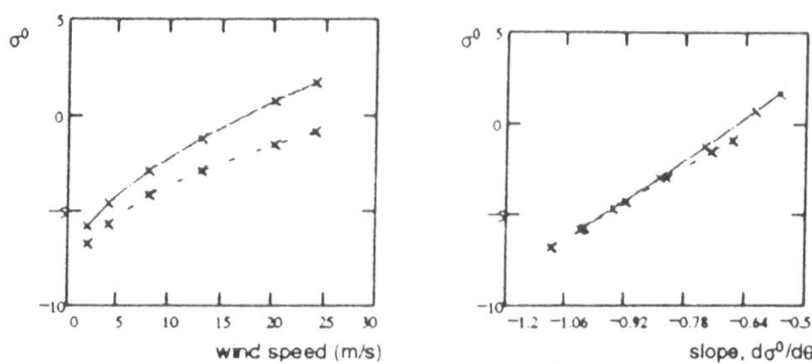


**Figure 25.** *Radar backscattering coefficient versus wind velocity and slope. The marked points correspond to 2, 4, 8, 13, 20, and 24 $s^{-1}$ wind speed. Solid line marks upwind and dotted line cross wind conditions. (From Long, (1992), Askne, (1993).)*

of properties over different parts of the Arctic the entire Arctic may be characterised. The SAR can also be used to derive local tie point values for the passive microwave sensors. The combined use of data from different sensors is necessary to overcome the limitations of each one with the goal to provide data for the Arctic sea ice.

## Open water signatures

The basis for ice concentration analysis is the contrast in the SAR image between the ice and the open leads. The open water leads are wind roughened and can show quite varying radar backscatter. The open water signature is not established completely for a SAR, but the signature for the ERS-1 scatterometer is illustrated by Figure 25.

## Advanced melt period

During this period we have the minimum ice extent and various properties contribute to this minimum value, including the maximum meltpond coverage.

## SAR signatures during the melt period

During the advanced melt period we only have old ice covered by melt ponds and open water. The old ice has a very low backscatter and, with a wind speed of only a few meters per second and a contrast relative to the open water, ice floes can be identified by their form and typical backscatter values over the image. An illustration is given in Figure 26 as function of incidence angle. If we have no systematic changes of the backscatter conditions along track we may fit a line to the identified values and these lines then represent the best estimate of the ice and open water signatures which may be compared to the along track mean backscatter value to determine the ice concentration, see Askne *et al.* (1994). The variability of the old ice illustrated in Figure 26 is caused by variability in melt pond properties and in the bare ice conditions, while the variability in the open water signature is mainly caused by wind variability but also by *e.g.* melt water runoff etc.

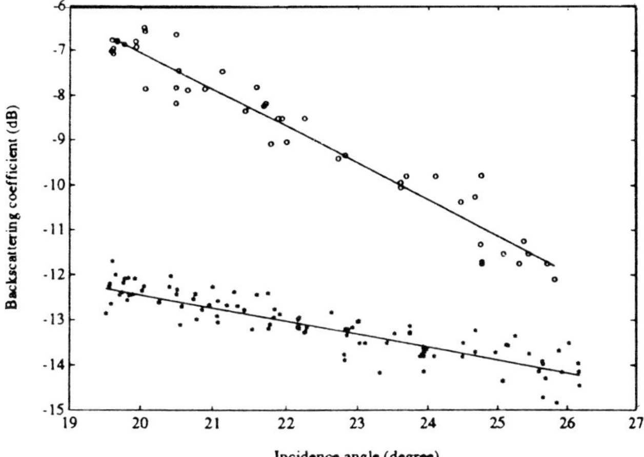



**Figure 26.** *Illustration of backscattering coefficient values obtained over the swath of ERS-1 SAR on 18 August 1991. Backscatter values over areas typical for open water or sea ice have been determined and the regression line is an estimate of the mean properties.*

## The SAR melt pond signature

The melt ponds generally are hard to detect with the resolution of the SAR, but under some conditions we may determine the meltpond coverage by means of the SAR. From meteorological charts we may determine that we have have melting conditions and that the meltponds have the same backscattering signature as open water. The backscattering coefficient would then be determined by an area average

$$10^{0.1\sigma_{ice}} = \mu 10^{0.1\sigma_{ow}} + (1 - \mu)10^{0.1\sigma_{bi}} \tag{44}$$

where $\sigma_{ice}$ is the backscatter value for ice including meltponds, $\sigma_{ow}$ is the value for open water and $\sigma_{bi}$ is the value for bare old ice, all values expressed in dB. In the high Arctic two consecutive orbits (after 100 minutes) cover partly the same area and we may then determine the two unknown, $\sigma_{bi}$ and the melt pond concentration $\mu$. This was described in more detail in Askne (1994) but the method should be futher tested before it can be applied in a reliable manner.

## Early freeze-up period

The early freeze-up period studied was studied by the ARCTIC-91 expedition from 21 September to 4 October 1991. During this period we have a decrease in temperature from zero to -15°C with some variability and the precipitation is in the form of snowfall. New ice is formed, sometimes combined with frostflower formation on the new ice surface, (Ulander *et al.* 1993). The backscatter of the multi-year ice stabilises at a level of $-5$ to $-8$dB while the new ice formation yields signatures at a lower level as long as the transient frostflower phenomenon does not take place. Areas covered by frost flower and saline wet snow may as well as the open water areas yield backscattering coefficients of the same order as multi year ice. Thin ice is important for the salt production and

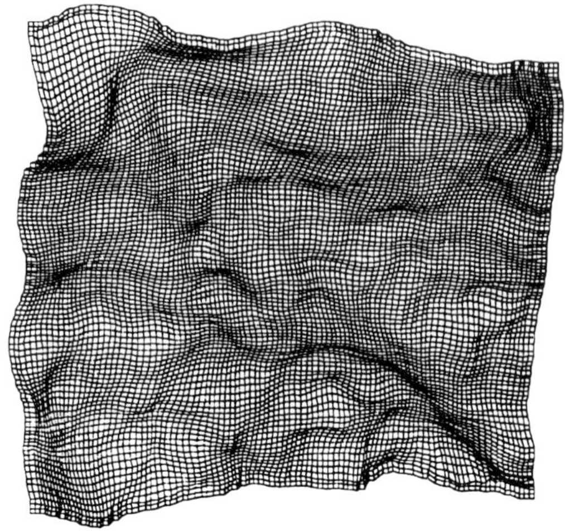

**Figure 27.** *Illustrating ice motion deformation from automatic routines Sun and Askne (1994) The area covers 80×80km² at the Yermak plateau in the Arctic Basin. (The ERS-1* SAR *images are from 28 Sept and 1 Oct. 1991.)*

for the thermal exchange between the atmosphere and ocean and can be recognised by its low signature. However the ice concentration, which is very high, can only be determined with high accuracy if the wind is sufficiently high, of the order of 5ms$^{-1}$ and higher. The variability of the signatures causes problems for automatic routines for ice class segementation, but are not sufficiently serious to be a great problem for ice motion recognition by identifying major similarities between the images, (Sun 1993). During this period the ice concentration is very low and we are approaching the period with the minimum lead extent. It is important to combine the information from the high resolution SAR sensor with the passive microwave sensors with high radiometric resolution in order to identify features of importance to characterise parameters of importance for the climate system description.

**Sea ice motion**

One important application of SAR imaging of Arctic sea ice is to determine ice motion. For that purpose repeated high resolution images over the same area within time spans of a few days as the ice may move several kilometres per day. High resolution microwave remote sensing is the only possible observation method. For this kind of image analysis automatic routines are necessary. Several algorithms for extracting ice motion information from satellite images have been developed since late 1980s. There are several approaches which are based on identifying similarities between features in two images over the same area, *e.g.* by using correlation technique (Fily and Rothrock 1987), matching edges of ice floes (Vesecky *et al.* 1988, McConnell *et al.* 1991, Banfield, 1991), using the optical flow method for changes in image intensities in space and time (Sun, 1993). Normally the algorithms have a pyramid structure in resolution, i.e. the images are averaged and large scale features are identified in the two images and the images rotated and translated to agree as much as possible, and then the approach is repeated at a higher resolution.The accuracy is dependent on the deformation of the ice field, but in the Arctic ocean it is of the order of 0.5km. An example of the deformation of the ice field between two images with a 3-day interval is shown in Figure 27. If the internal movements and deformation within an image is not large entire scenes may be matched by using Fourier transform techniques (Sun 1993). Ice motion analysis can be used to study images over the same area of the Arctic using images with some overlap taken between different orbits and then on time scales from hours to days using ERS-1 and to study the effect of wind as well as other phenomena on the ice motion (Sun and Askne, (1994)).

### Operational applications

Methods for fast transfer of the satellite information is the basis for real time application, *e.g.* to supply the ice breakers with ERS-1 images and other ice information, (see *e.g.* Askne *et al.* 1993). For that purpose either low resolution images or compressed images or subimages of full resolution images have normally to be used. Fast delivery of low resolution (100m ) images can be obtained from the Tromso receiving station by means of Internet and FD-images (fast delivery) images are also produced at the ESA receiving stations with full resolution which are transferred with larger delays to the national receiving stations using geostationary satellite. In the sea ice applications case the images were transferred to Mercator projection with coastlines and lighthouses included. SAR scenes averaged to 200m resolution and partial scenes with full resolution were transmitted using the Nordic Mobile Telephone (NMT) system to a presentation system onboard the ice breakers together with traditional ice data.

## 4.5   Forestry and agriculture[1]

### Introduction

Compared to optical systems, the use of microwave systems for forestry applications is fairly new and its potential is far from examined. However it is evident that radar data alone cannot provide all information needed for different types of forest surveys, but could be a good complement to data obtained from optical sensors. Potentially important forest applications of radar remote sensing include forest cover identification and mapping, discrimination of forest compartments, forest types, tree species and diseased areas as well as clear-cuts and areas affected by fire.

An application of great interest is estimation of forest biomass. This is important on a global scale as a crucial input parameter to global carbon models. Biomass estimation is of major importance also at a regional level in assessment of forest properties.

In certain areas of the world, cloudiness makes radar data the only feasible way of acquiring remote sensing data within a limited period of time. The probability of finding a cloud free scene over a specific area in Sweden, for example, may be less than once a year, taking into account the orbits of the Landsat and SPOT satellites. However these

---

[1]This section on forestry is adapted from (Israelsson and Askne, Israelson *et al.* 1994).

satellites have proven to give good results for classification of forest properties, and it is important to investigate the possibilities of the radar sensor to provide information in between acquisitions by optical sensors.

## SAR data survey

Radar forestry can be viewed from two starting points. First there is an analytical way based upon electromagnetic theory, where the radar targets are modelled as exactly as possible. Many natural targets, such as forested areas, have however a complexity that makes an exact treatment very difficult. The other extreme starting point is to use regression technique to fit a simple mathematical expression to the experimental data. The limitation of this method is that no general knowledge is gained about the significance of different scattering phenomena. Both approaches are considered below.

A forest in the Flevoland area in the Netherlands was chosen as a test area since data is available from the MAESTRO-1 campaign in 1989 and the MAC'91 campaign in 1991 as well as ERS-1 and JERS-1 data. Poplar forest planted in regular row patterns have in particular been studied below. The stand sizes vary between 5 and 10 hectares. The area is almost completely flat, hence no effect of topography is expected.

The bole volume represents the volume of the trunks without branches and is defined,

$$V_b = \frac{f d^2 \pi h N}{4} \left[ \frac{\mathrm{m}^3}{\mathrm{hectare}} \right] \tag{45}$$

where $d$ is the diameter at breast height, $h$ is the tree height, $N$ is the number of trees per hectare and $f$ is a formfactor taking into account that trunks are not perfect cylinders. The formfactor was in this case put to 0.45. Forty stands in the poplar forest were examined. In each stand, the bole volume was estimated at a number of points using a relascope and height measure instruments. The range of bole volumes that was found span from 10 to 315 m$^3$/hectare . The relative accuracy of the measured values was estimated to be around 10%.

Three different SAR systems were used with a frequency range from 440MHz to 5.3Ghz, see Table 4. The JPL/AIRSAR data (Held *et al.* 1985) were acquired during the MAESTRO-1, in August 1989, and MAC'91, four dates during the summer of 1991, campaigns. They have been polarimetrically and radiometrically calibrated using the responses from point-targets giving the radiometric accuracy of 1dB. The relatively wide swath of incidence angles shown in Table 3 corresponds to all flights during the

| System | Frequency | Polarisation | Inclination Angle |
|---|---|---|---|
| JPL/AIRSAR | 440MHz | fully | 30°–40° |
|  | 1.25GHz | polarimetric | 23° |
|  | 5.3GHz |  |  |
| ERS-1 | 5.3GHz | VV | 23° |
| JERS-1 | 1.25GHz | HH | 40° |

**Table 4.** *SAR systems used in the study*

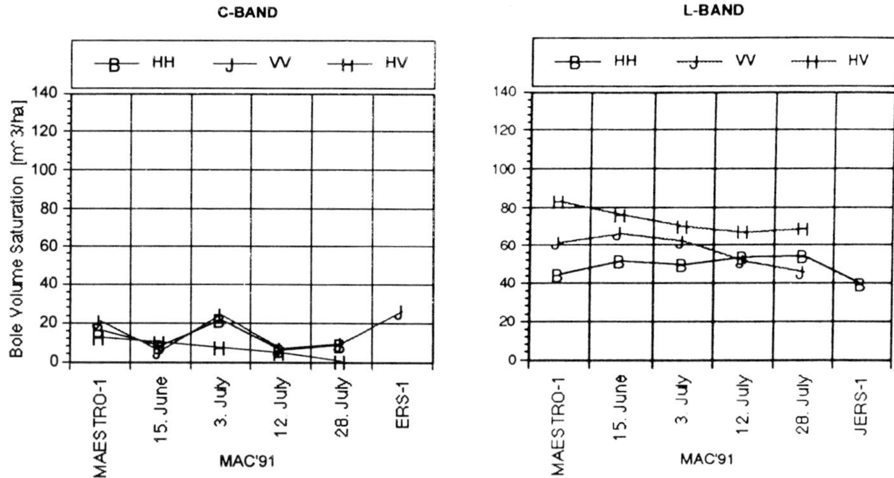


**Figure 28.** *(a) C-band and (b) L-band bole volume saturation levels.*

two campaigns. Within a single scene, the forest stands are located within a range of around 10° . The ERS-1 scene used is a FD product acquired 91-10-25, which is radiometrically calibrated with the accuracy of 0.3dB . The JERS-1 scene that is used is an ESA test image acquired 93-03-28. The accuracy and precision of that calibration factor is not known. Thus only relative analyses of the JERS-1 data were performed.

## Empirical results

It is a common phenomenon for types of forests that the radar backscattering becomes saturated at a constant level of $\sigma^0$ (Imhoff 1993, Dobson *et al.* 1992, Freeman *et al.* 1992) . This can be expected when the scattering mainly originates from the canopy. The measured radar backscattering coefficients have therefore been matched in the least square sense to an empirical expression,

$$\sigma^0 = \sigma^0{}_{\text{sat}} \left(1 - e^{-\alpha B}\right) \tag{46}$$

where $B$ is the bole volumes. Equation 46 is similar to the expression used by (Attema and Ulaby 1978), $a$ is a sensitivity factor, and the bole volume sensitivity saturation level is now defined as the bole volume level where the sensitivity is 0.1 [dB]/100[m$^3$/hectare]. The bole volume saturation levels at C-band, Figure 28(a), are well below 30m$^3$/hectare at all polarisations and registrations. 30m$^3$/hectare corresponds to a very young or intensely thinned forest. Hence C-band cannot be used to estimate forest biomass. It is in many cases even hard to distinguish a mature forest stand from an agricultural field. No significant variation among different polarisations and incidence angles was found.

At L-band, Figure 28(b), the bole volume saturation levels are higher than at C-band. The results indicate that it is possible at L-band to distinguish very young or thinned forest stands from mature stands, particularly at cross polarisation. The saturation levels are, however, too low to permit operational biomass assessment. A

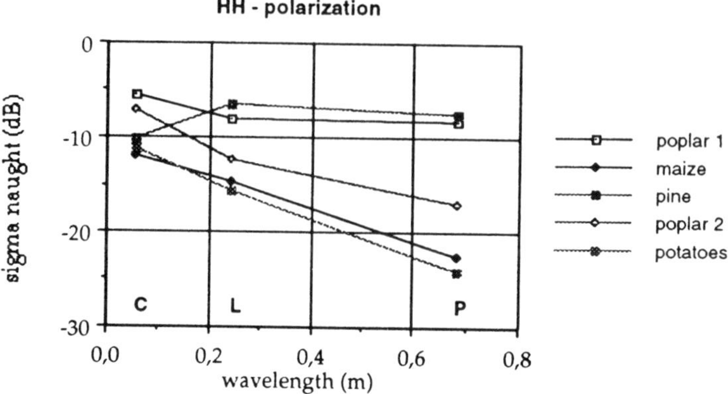



**Figure 29.** *Calibrated radar cross sections of different areas*

similar analysis of P-band shows saturation levels around 120m³/hectare indicating that a few bole volume classes can be determined. The saturation levels are however still too low to enable bole volume estimations of mature forests.

## Forestry and vegetation

The differentiation between forest types and vegetation, for some few species, are illustrated in Figure 29 and the small span of cross sections in the case of C-band increases with wave length. We see a different C/L-band behaviour for pine than for the rest indicating some possibilities combining C- and L-band combined.

## Radar Modelling of Forest Stands

The first and most obvious step in a modelling attempt is to try to identify the physical elements that most likely take part in the electromagnetic scattering process. The next step is to obtain analytical expressions for the different energy-matter interactions that take place. Some efforts to model a forest stand have been accomplished recently, and the most common approach is to decompose the radar scattering into discrete scattering components (Richards *et al.* 1987 and Ulaby *et al.* 1990).

The radiative transfer equation (47) can be used to examine the problem.

$$\cos\theta\frac{\mathrm{d}}{\mathrm{d}z}\mathbf{I}(\theta,\phi,z)=-\kappa_e(\theta,\phi)\mathbf{I}(\theta,\phi,z)+\int_0^{2\pi}d\phi'\int_0^{\pi}d\theta'\sin\theta\,\mathbf{P}(\theta,\phi;\theta',\phi')\mathbf{I}(\theta',\phi',z) \quad (47)$$

$$\mathbf{I}=\begin{pmatrix}I_h\\I_v\\U\\V\end{pmatrix}=\frac{1}{\eta}\begin{pmatrix}\langle|E_h|^2\rangle\\\langle|E_v|^2\rangle\\2\mathrm{Re}\,\langle E_h E_v{}^*\rangle\\2\mathrm{Im}\,\langle E_h E_v{}^*\rangle\end{pmatrix}$$

**I** represents the Stokes vector of the electromagnetic wave, $\kappa_e(\theta,\phi)$ contains the extinction cross sections of the canopy constituents in the forward direction and **P** is

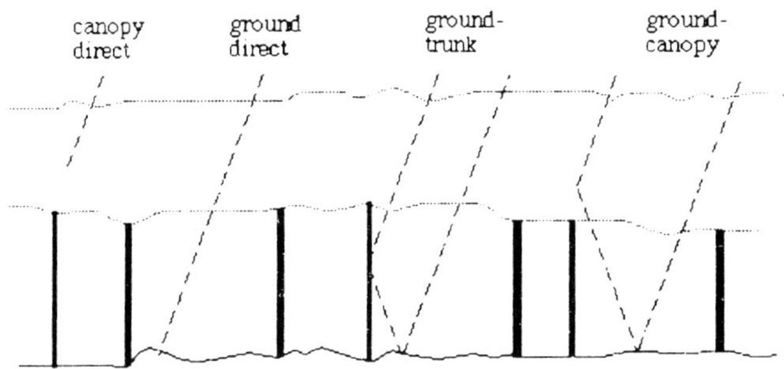


**Figure 30.** *Backscattering from a forest decomposed into four components*

the phase matrix which returns the scattered intensity from the $(\theta, \phi)$ direction onto the $(\theta', \phi')$ direction. The first order iterative solution (Kong, 1990) produces a sum of terms that can be related to different scattering paths. In Figure 30, the four most typical scattering paths have been distinguished. In this example, the forest canopy has been divided into two horizontal layers. No multiple scattering components, i.e. more than two interactions within one layer, have been taken into account, since they are introduced by solutions of higher orders. The energy loss of each scattering event makes multiple scattering irrelevant, except at cross polarisation where it might be of importance.

The MIMICS scattering model (Ulaby *et al.* 1990), which has shown good agreement between theory and experiments is based on the first order radiative transfer solution, with the forest canopy modelled as two homogenous layers. The different terms of the solution correspond to different scattering paths.

The backscattering coefficient was simulated as a function of increasing bole volumes. Using one single variable means that all variations of the model input parameters had to be related to the bole volume. Such relations were established from the ground data collection.

This procedure may introduce errors which effect the different frequencies differently. The moisture parameters were assumed independent on bole volumes and were given characteristic numbers.

Figure 31 shows the bole volume saturation levels according to MIMICS. According to MIMICS the C-band backscattering is totally predominated by backscattering from the tree crown, even at low bole volumes. This effect saturates the backscattered signal at very low bole volumes. At L and P-band the bole volume saturation levels are overestimated at co-polarisation, while they are fairly well predicted at cross polarisation. According to the model output, the tree crown dominates the cross polarised backscattering at all the three modelled frequencies. The higher bole volume sensitivity at the longer wavelengths is due to lower attenuation. At L-band and even more at P-band, in HH polarisation particularly, the trunk- ground backscattering also becomes important. This gives a theoretical sensitivity level that is higher than at cross polarisation. The overestimation compared to the measured values can be explained by the fact that the real forest is not as homogenous as the simulated one.

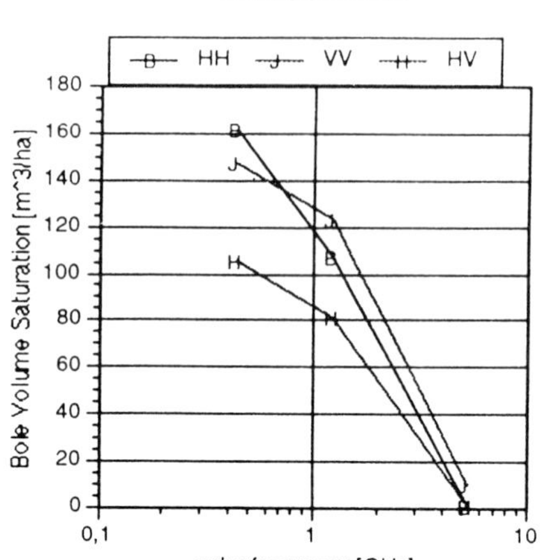



**Figure 31.** MIMICS *simulation of the bole volume saturation levels at PLC-band*

## Concluding comments

It should be mentioned that the resolutions of ERS-1and JERS-1 are a problem for small forest stands like those in Flevoland. However this does not affect the AIRSAR measurements and the common result is that none of the examined frequencies can be used to assess bole volumes in an operational way. The JERS-1 (L-band) seems somewhat more promising for forest investigations than ERS-1.

The analysis has been based on the radar cross section derived from a single measurement. Useful information about forestry and vegetation can be derived by combining temporal sequences of SAR images. Applications when cloud conditions is a problem are related to mapping fast changes such as clear cuts, forest fires and storm damages. In this cases although the biomass or bole volume sensitivity of the radar cross section is low we may determine changes by a combination of radar cross section and INSAR coherence maps (Askne and Hagberg, 1993).

## 4.6   Topography[1]

### Introduction

In a conventional SAR image, the image brightness is the primary measurable which is proportional to the backscattered echo amplitude or power in each resolution cell. It is obtained by taking the magnitude (and squaring to get power rather than amplitude) of the output complex signal which is output from the SAR signal processor. The phase of the signal is not used since it does not contain any useful information. However, the phase may be used to determine ground topography by combining two SAR images of the same area but acquired with slightly different viewing geometries (Graham 1974).

---

[1]This section on Topography is partly adapted from Ulander and Hagberg (1994)

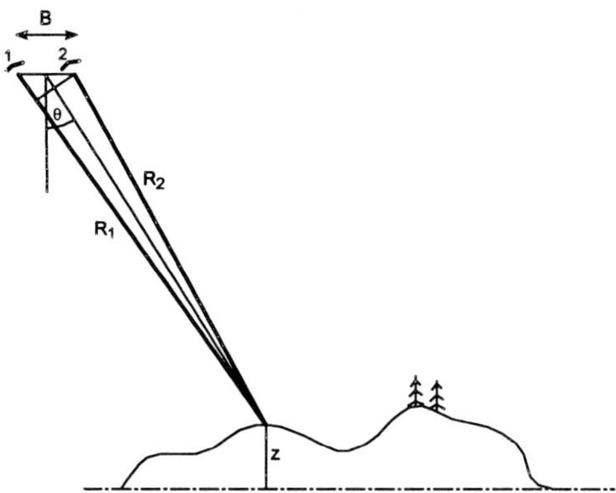


**Figure 32.** *Geometry of a SAR interferometer. The flight track is normal to the plane of the figure.*

The technique named INSAR, interferometric synthetic aperture radar technique, may also use satellites where the two SAR images are acquired on nearly repeating passes ('repeat-pass interferometry') (Li and Goldstein 1990).

The backscattered echo from a point-like object is essentially a very weak time-delayed version of the transmitted pulse. The phase angle of the echo is thus determined by the two- way electrical path length to the object and a phase shift associated with the scattering interaction between the radar wave and the object. If a second SAR image is acquired along a slightly offset flight track, the scattering phase shift is constant whereas the electric path length has changed. The phase difference $\phi$ between the same resolution cell in the two SAR images is thus determined by the angle $\theta$ and the baseline $B$ in Figure 32.

$$\phi = 2k(R_2 - R_1) + 2\pi n = B \sin \theta \tag{48}$$

where $k = 2\pi/\lambda$ is the wavenumber, $\lambda$ is the wavelength, $R_1$ and $R_2$ are the slant ranges to the object in the two SAR images, respectively: the term $2\pi n$ ($n$ integer) is included since phase is measured modulo $(2\pi)$. Conversely, we may determine $\theta$ by inverting this equation when $\phi$ and $B$ are measured and the $2\pi$-ambiguity is resolved. Together with the measurement of azimuth and slant range, it is now possible to determine the three-dimensional position of the object. Most often we do not have a single isolated point scatterer but instead many elementary scatterers within a resolution cell (*cf* discussion of speckle above). The backscattered echo is thus a sum a echoes with slightly different delay times given by the slant range to each scatterer.

It can be shown that the resulting phase may again be divided into two terms: Firstly, the two-way path length to the resolution cell and secondly, the interference phase between scatterers which produces speckle. The second term will change in a random fashion between the two observations unless $B = 0$. It can be shown that statistical correlation of the speckle term drops from one for $B = 0$ to zero for $B = B_{\mathrm{max}}$.

(For further mathematical details, see section 2.) After 'flat Earth compensation' a phase change in the image is directly related to the topographic height change besides phase noise and the $2\pi$-ambiguity. The latter is removed by special processing called phase unwrapping whereby the image phase is checked from different directions for internal consistency (Goldstein *et al.* 1988).

## Coherence properties for image classification

The coherence, which determines the visibility of the fringes in the interferogram, can be determined when the local topography has been eliminated. The coherence is defined (see Born and Wolf, 1965) as :

$$\gamma = \frac{E[g_1 g_2{}^*]}{\sqrt{E[g_1 g_1{}^*]\,E[g_2 g_2{}^*]}} \tag{49}$$

where $g_1$ and $g_2$ are the corresponding complex pixel values in image 1 and 2, (related to the radar cross section) and the expectation value is taken over a $4\times16$ window in the single look complex image, assuming the ensemble average can be determined by spatial averaging.

The incoherent movements within one resolution cell or the averaging window will decrease the coherence as the phase noise is directly related to the coherence, which decays like $e^{-\sigma_\phi^2/2}$, (Hagberg *et al.* ). From Equation 49 we obtain (assuming independent Gaussian random variables) for the phase noise, $\sigma_\phi$ , over the averaging area used when coherence is determined.

$$\sigma_\phi{}^2 = \left(\frac{4\pi B_n}{\lambda R \sin\theta}\sigma_h\right)^2 + \left(\frac{4\pi}{\lambda}\sigma_R\right)^2 + (\sigma_\phi)^2 \tag{50}$$

These terms describe possible phase variations within a pixel, $\sigma_{h1}$, caused by volume scattering, or between pixels, $\sigma_{h2}$, caused by varying scattering centres *e.g.* varying tree height, varying roughness etc. When we change from an ensemble average to a spatial average we introduce effects due to local roughness or topography left over when a tilted plane has been fitted to the local area. The second term is the variability of the location in the phase centre and the last term is caused by variations in the phase of the reflectivity function and other phase noise contributions. These two terms may not increase when spatial averaging is performed. The reflectivity function in the interferogram as well as the coherence are determined by the weighted mean of components differently affected by these terms,

$$\sum f_i e^{-\sigma_{\phi_i}{}^2/2} \tag{51}$$

From Equation 51 we conclude that the baseline should be small in order to obtain high coherence. On the other hand we are then more sensitive to $\sigma_R$ and would expect a higher variability in the coherence over the averaging area. $\sigma_R$ is the major term causing temporally decreasing coherence.

Intensity and texture together with intensity changes have been used for a long time for classification of SAR images. The coherence which is sensitive to small changes should be a valuable tool to identify effects like clear-cuts or forest fire, where the coherence properties are expected to change drastically but the radar cross section may not change outside its expected variability due to different phenomena. Hoever, not only for changes of this type coherence should be useful, but in order to characterise different surfaces in a different manner than the intensity and then increase the classification possibilities (Askne and Hagberg 1993)

## Temporal decorrelation

In repeat-pass interferometry, it is assumed that the imaged area is essentially unchanged between the two SAR overpasses. In practise, this means that temporal changes of the scatterers within a resolution cell may destroy the fringes. It can be shown that the rms displacement in the slant range direction should be less than about $\lambda/20$ to ensure high-quality fringes. This corresponds to 0.3cm for the ERS-1 SAR and is clearly a severe restriction for topographic mapping over natural terrain including forests, agricultural fields, grasslands, etc. It is clear that the repeat-pass technique will not work for lakes and other open water surfaces, and that wet precipitation will most likely destroy the fringes. It is also clear that the best potential is over bare and dry rock surfaces. For other surface types, *e.g.* forests and fields, we expect that fringe quality will depend on meteorological conditions, *e.g.* wind and temperature.

The high sensitivity to temporal changes also implies that correlated displacements over large areas may be measured with very high accuracy (of the order 1mm). This is called differential interferometry and has recently been used to map the displacement field before and after an earthquake (Massonnet *et al.* 1993). Limiting factors for this type of measurement are due to delays in the ionosphere and atmosphere, and satellite orbit stability.

## Topographic height accuracy

It is clear from (6) that the baseline $B_n$ should be sufficiently large to maximise the sensitivity of the retrieved height, but less than $B_{\mathrm{max}}$ according to (5) to ensure adequate speckle correlation. In fact, as $B_n$ increases so does the phase noise and there is an optimum baseline which minimises the height error. It has been found that surface slopes also affect the height error since steep slopes towards the radar increase the phase noise and make phase unwrapping more difficult (Hagberg and Ulander 1993). Simulation results are shown in Figure 33 for the ERS-1 SAR and three surfaces with increasing rms slopes. The middle curve is a simulation based on a real digital elevation model with altitude between 325 and 600m over a $10\times9$km area, whereas the top and bottom curves are based on scaling the height by 1.5 and 0.1, respectively. The bottom curve agrees with the expected limit for a flat surface, whereas the top curve gives a minimum height error of about 4m which probably is a realistic estimate of the achievable height accuracy. An experimental verification of the height accuracy which can be achieved using ERS-1 SAR interferometry has not yet been done.

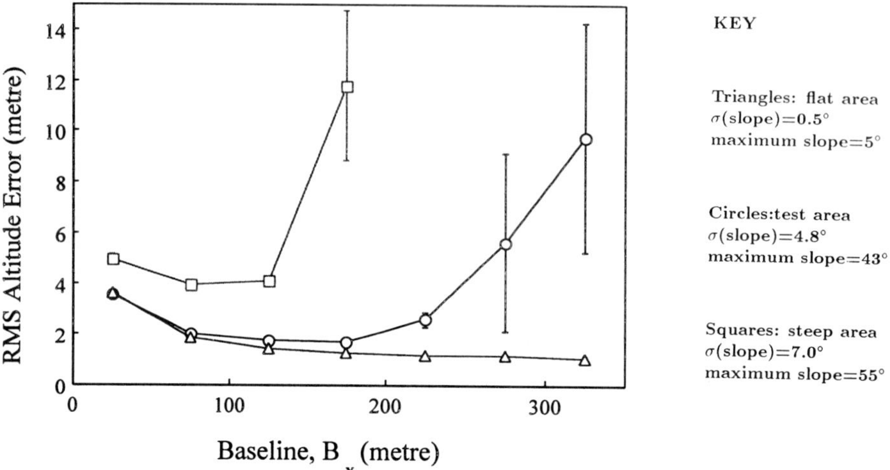



**Figure 33.** *Height error versus interferometer baseline. The results are based on simulations of three areas with different slope distributions. For the 'flat' area the rms. error is decreasing slowly until baseline decorrelation sets in at relatively large values. For the test area we have increasing error as the baseline decorrelation noise causes phase ambiguity problems. For the very steep areas we also have effects of layover. A good baseline for topography seems to be 150m .*

## Topographic mapping over a forested area

It is fundamental to investigate whether different surface types are sufficiently stable to allow interferometric measurements with the ERS-1 SAR. This is particularly important for forests which often cover topographic terrain and clearly are affected by the wind. As a starting point, we have investigated the potential of interferometry in a test area mainly covered by boreal forests in northern Sweden, about 100km north of Umeå (Hagberg 1994, Hagberg *et al.* submitted, Ulander *et al.* 1993).

## Atmospheric effects on SAR imaging

We normally say that microwaves are independent of clouds and atmospheric effects this is not entirely true. For example images of the ocean surface reveal not only surface waves, internal waves, eddies, oceanic fronts and underwater bottom topography but also effects of katabatic wind fields, internal gravity waves, atmospheric vortex rows behind islands, atmospheric boundary rolls, and convective cells above the sea (Alpers, 1994).

The troposphere and ionosphere affects the phase delay of waves and the accuracy of phase determination in connection with interferometry (Tarayre and Massonnet, 1994). For example, atmospheric delay can cause an error of half a fringe or in exceptional cases up to 3 fringes. This is a serious problem for digital elevation models where the phase error causes an altimetric error equal to the phase artifact×ambiguity level/ signal wavelength. The ambiguity altitude is dependent on the baseline (the altitude change causing a $2\pi$ phase shift in Equation 24 (section 2).

# Glossary

## Measuring with radars

**Radar cross section:** the basic concept describing the strength of the signal scattered from a certain object. To determine the radar cross section the system must be absolutely calibrated. With relative calibration the relative difference in radar cross section can be determined.

**Absolute radiometric accuracy:** the accuracy to which actual radar cross-section values can be associated with intensity levels in a SAR image.

**Precision or Relative radiometric accuracy:** the degree to which it is possible to discriminate between different radar cross-section values from the different intensity levels in a SAR image. The relative radiometric accuracy can be defined between two pixels at different locations in the same image (spatial r.r.a.) or between two images taken at different times (temporal r.r.a.) or with the same frequency but with different polarisation (polarisation r.r.a.) or different images obtained simultaneously with the same polarisation but with a different frequency (spectral r.r.a.).

**Radiometric resolution:** the minimum distinguishable difference in $s$ between two areas of distributed targets.

**System dynamic range:** the dynamic range of the system determines the radar cross section values (in dB) over which the stated absolute and relative radiometric performance is achieved.

**Noise-equivalent $\sigma_0$:** the normalised radar cross-section (in dB) which generates a signal equal to the system noise level for a single-look scene.

**Detectability:** the ability of a system to detect if an object is present. An object with a large radar cross section like a corner reflector is easy to detect even if it is much smaller than the resolution of the system. (However two radar reflectors within a resolution cell can not be detected as two different objects.)

**Speckle:** the scattered signal from a resolution cell is made up of a superposition of a large number of wavelets with varying amplitude and phase. Constructive and destructive interference yields a strongly varying signal variation between neighbouring resolution cells giving rise to the speckle noise in the image.

**Number of looks:** the number of sub apertures in the processing of the data. Pixel averaging in single look images can also be used in order to reduce the speckle noise.

## Wave properties

**Incidence angle:** the incidence angle is defined as the angle made by the radar look direction to the vertical, at the ground. ( Comment: the radar cross section is dependent on the incidence angle. The incidence angle for ERS-1 is $20°-26°$ and for future satellites intervals in the range $10°-60°$ are proposed).

**Frequency:** a number of frequency bands are traditionally used for microwaves. P and K bands are expected to cause difficulties for imaging from space due to the effects of the

ionosphere. Other bands and frequences in the bands of interest for space applications
are X-, C-, S-, and L-band. Future satellite SAR instruments may include more than
one frequency like present airborne systems do.

**Polarization:** the polarization of the electromagnetic wave is determined by the direction of the electric field component for the incident and reflected wave. The following
definitions are used:

      HH–horizontal transmit and receive
      VV–vertical transmit and receive
      HV–horizontal transmit and vertical receive
      VH–vertical transmit and horizontal receive

Future satellite SAR instruments may include more than one polarization combination
like present airborne systems do. If the phase component is measured as well as all the
amplitude components any polarization may be determined.


# Image properties and image products

**Resolution:** the resolution is the distance between two objects in order for the system
to recognise them as two different objects.

**Pixel distance:** this term is simply the ground dimension of the pixel and can be
chosen independent of the resolution (although normally there is a relation between the
two expression).

**Azimuth and range direction:** azimuth direction is along the satellite track and
range direction is across track.

**Swath width:** the design of a SAR system involves a trade-off between the width of
the imaged swath and spatial resolution (in azimuth) that can be obtained. Typical
performance figures using present generation technology might be 100km swath / 30m
resolution (6 azimuth looks) or 450km swath / 100m resolution (2–4 azimuth looks)
while in the future 30km swath / 10m resolution might be envisaged.

**Slant range:** the distance measured in a radial direction from the radar.

**Ground range:** the distance measured from the nearest sub satellite point on the
ground.

**Layover:** geometric artifact for side looking systems caused by the slant range distance
to the top of an object is smaller than the slant range to the ground base of the object.

**Foreshortening:** geometric artifact for side looking systems where the forefront of a
slope looks shorter than the slope of the backside.

**Radar shadow:** geometric artifact for side looking systems, causing a shadow behind
some objects.

**Data products:** SAR processed data are the outputs resulting from a double decon-
volution process applied to the SAR primary data, making use as appropriate of the
auxiliary data. By SAR analysed information or value-added products we mean information resulting from the interpretation of SAR processed data, inputs of data and
knowledge from other sources and which do not retain the original pattern of the primary or processed product(s) from which they were derived. Sometimes various levels

and names are used for low bit rate (LBR) data products: raw/primary data are defines as level 0/1; engineering products are level 1.5; geophysical products are level 2.

**Raw data:** SAR data not processed into image form.

**Quick look:** an image produced fast and simple with low spatial and radiometric resolution useful for preliminary checking of major image features.

**Image data:** image data can be in complex (in-phase and out of phase components) or detected (only amplitude) form and include a basic set of auxiliary data (i.e. location, timing details etc.)

**Georeferenced images:** geocoded image on a cartographic grid but without relief corrections.

**Geocoded images:** images on a cartographic grid with relief corrections using ground control points and a Digital Elevation Model (DEM).

# References


Alpers W R and Rufenach C L, 1979, The effect of orbital motions on synthetic aperture radar imagery of ocean waves, *IEEE Trans Antennas Propag*, **AP-27,** 685

Aplpers W, 1994, ERS-1 SAR views the ocean: An assessment, *Proc IGARSS'94, Pasadena, August 8 - 12, 1951*

Askne J, Thompson T, Ottersten H, and Fäst O, 1993, From research to operational ice service using ERS-1: The Swedish Sea Ice Service for the Baltic, *Proc EARSeL 13th Symposium, 28 June to 1 July 1993, Dundee, Scotland, UK*

Askne J, 1993, Wind derived from leads in ice covered areas as seen in SAR images, *Third Scandinavian SAR Symposium, Göteborg*

Askne J, Carlström A *et al.* , 1994, ERS-1 SAR backscatter modelling and interpretation of sea ice signatures. *IGARSS-94, Los Angeles, ?????*

Askne, J and Ulander, L M H, 1992, Remote sensing of Arctic sea ice using the ERS-1 SAR, *Central Symposium of the International Space Year*, **SP-341:** 129

Askne J, Ulander L M H, *et al.* , 1993, Accuracy of ice concentration derived from ERS-1 SAR images during the late melt period in the Arctic, *Microwave Remote Sensing of Ice, Lyngby, EARSeL*

Askne J, Ulander L M H *et al.* , 1994, Validation of ERS-1 SAR for Arctic ice properties from ARCTIC-91, *Second ERS-1 Symposium, Hamburg, ESA*

Askne J and Hagberg J, 1993, Potential of interferometric SAR for classification of land surfaces, *Proc. IGARSS'93*

Attema E P W and Ulaby F T, 1978, Vegetation modeled as a water cloud, *Radio Science* **13** 357

Banfield J, 1991, Automated tracking of ice floes: A stochastic approach, *IEEE Trans. Geosci. Remote Sensing*, **GE-29,** 905-911.

Barry R G, Serreze M C etal, 1993, The Arctic Sea Ice - Climate System: Observations and Modeling, *Reviews of Geophysics* **31** 397

Bern T I, T Wahl, T Anderssen, and R Olsen, 1992, Oil spill detection using satellite-based SAR: Experiences from a field experiment, *Proc First ERS-1 Symp, Cannes, France 4-6 Nov.*

Born M and Wolf E, 1965, Principles of Optics, Pergamon Press, Oxford.

Carlström A and Ulander L M H, 1993, C-band backscatter signatures of old sea ice in the central Arctic during freeze-up, *IEEE Trans Geoscience and Remote Sensing* **31** 819

Chapman W L, Welch W J *et al.* , 1994, Arctic sea ice variability: Model sensitivities and a multidecadal simulation, *J Geophysical Research* **99** 919

Comiso J C and Kwok R, 1992, Summer Arctic ice concentrations and characteristics from SAR and SSM/I data, *Proc first ERS-1 Symposium*

Dobson M C, Ulaby F T, LeToan T, Beaudoin A , Kaischke A S and Christensen N, 1992, Dependence of radar backscatter on coniferous forest biomass, *IEEE Transactions on Geoscience and Remote Sensing* **30** 412

Elachi C, 1987, Spaceborne Radar Remote Sensing: Applications and Techniques, *IEEE Press New York*

Fily M, and Rothrock D A, 1987, Sea ice tracking by nested correlations, *IEEE Trans. Geosci. Remote Sensing* **GE-25,** 570

Freeman A, Durden S and Zimmerman R, 1992, Mapping sub-tropical vegetation using multi-frequency, multi-polarization SAR data, *IGARSS'92 in Houston, Texas* 1686

Fung A K, Li Z and Chen K S, 1992, Backscattering from a randomly rough dielectric surface *IEEE Trans Geosci Remote Sens* **30** 356.

Gloersen P and Campbell W J, 1991, Recent Variations in Arctic and Antarctic Sea-Ice Cover, *Nature* **352** 33

Gloersen P, Campbell W J *et al.* , 1992 Arctic and Antarctic Sea ice, 1978-1987, Washington, DC, NASA.

Goldstein R M, Zebker H A and Werner C L, 1988, Satellite Radar Interferometry: Two-dimensional Phase Unwrapping, *Radio Science* **23** 713

Gow A J and Tucker W B, 1990, *Sea-ice in the polar regions* 47–122 (Polar Oceanography, San Diego, Academic Press.)

Graham L C, 1974, Synthetic Interferometer Radar for Topographic Mapping, *Proceedings of the IEEE* **62** 763

Hagberg J O, 1994, Repeat-Pass Satellite SAR Interferometry, Lic.Eng. Thesis, Technical Report No. 170L, School of Electrical and Computer Engineering, Chalmers University of Technology, Göteborg, Sweden.

Hagberg J O and Ulander L M H, 1993, On the Optimisation of Interferometric SAR for Topographic Mapping, *IEEE Transactions on Geoscience and Remote Sensing* **31** 303.

Hagberg J O, Ulander L M H and Askne J, Repeat-Pass SAR Interferometry over Forested Terrain, submitted to *IEEE Transactions on Geoscience and Remote Sensing.*

Hallikainen M T, Ulaby F T, Dobson M C, El-Rayes M A and Wu L K, 1985, Microwave Dielectric Spectrum of Vegetation-Part 1: Empirical Models and Experimental Observations, *IEEE Trans Geosc Remote Sensing* **23** 25.

Hara Y, Atkins R G, Yueh S H, Shin R T and Kong J A, 1994, Application of Neural Networks to Radar Image Classification, *IEEE Trans Geosc Remote Sensing* **32** 100.

Hasselman K *et al.* , 1985, Theory of synthetic aperture radar ocean imaging: a MARSEN view, *J. Geophys Res*, 90, 4659

Held D N *et al.* , 1988, The NASA/JPL Multifrequency, Multi-Polarisation Airborne SAR System, *IGARSS'88, Edinburgh, Scotland* 345

Imhoff M L, 1993, Radar backscatter/biomass saturation: Observations and implications for global biomass assessment, *IGARSS'93 in Tokyo, Japan* 43

Israelsson H and Askne J, 1994, Importance of dihedral reflection in radar remote sensing of forests, *Int J Rem Sensing* **15** 1585

Israelsson H, Askne J, 1994, Retrieval of forest biomass using SAR, *Proc of EARSeL 14th Symposium, Göteborg* to be published

Israelsson H, Askne J, Sylvander R, 1994, Potential of SAR for forest bole volume estimation, *MAESTRO-I special issue, Int J Remote Sensing,* to be published.

Hibler W D, 1979, A dynamic thermodynamic sea ice model, *J Phys Oceanogr* **9** 815

Hühnerfuss H, 1986, The molecular structure of the system water/monomolecular surface film and its influence on water and wave damping, *Habilatationsschrift vorgelagt dem Fachbereich Chemie der Universität Hamburg*

Kendell M G and Stuart A, 1958, The Advanced Theory of Statistics, *Charles Griffin & Company: London* 188-189.

Kim Y S, Moore R K, Onstott R G and Gogenini S, 1985, Towards identification of otimum radar parameters for sea ice monitoring, *J Glaciol* **31** 214

Kong J A (Ed), 1990, Polarimetric Remote Sensing, *Progress in Electromagnetic Research* **3,** Elsevier, New York, 520pp.

Kramer H J, 1994, Observation of the earth and its environment. Survey of missions and sensors, 2nd edition, Springer Verlag, Berlin

Kwok E, Cunninghan G and Holt B, 1992, An approach to identification of sea ice types from spaceborne SAR data, in *Microwave Remote Sensing of Sea Ice* edited by F. Carsey, AGU, Geophysical Monograph 68, 355-360

Laur H, 1992, ERS-1 SAR calibration, Derivation of backscattering coefficient s in ERS-1.SAR.PRI products, Issue 1 Rev. 0.

Laxon S, 1990, Seasonal and inter-annual variations in Antarctic sea ice extent as mapped by radar altimetry, *Geophysical Research Letters* **17** 1553

Laxon S. and Askne J, 1991, Comparison of Geosat and SSM/I mapping of sea ice in the Arctic. *IGARSS'91, Helsinki, ESA.*

Lee J S, 1986, Speckle Suppression and Analysis for Synthetic Aperture Radar, *Optical Engineering,* **25** 636

Li F K and Goldstein R M, 1990, Studies of Multibaseline Spaceborne Interferometric Synthetic Aperture Radars, *IEEE Transactions on Geoscience and Remote Sensing* **28** 88.

Lichtenegger J, 1992, ERS-1: Landuse mapping and crop monitoring: A first close look to SAR data, ESA earth observation quarterly, Nos 37-38, May-June 1992, pp 1-5.

Long A E, 1992, C-band V-Polarized Radar Sea-Echo Model from *ERS-1 Haltenbanken Campaign. ERS-1 Geophysical Validation, Penhors, ESA.*

Lyzenga D R, 1988, An analytic representation of the synthetic aperture radar image spectrum for ocean waves, *J Geophys Res,* **93,** 13859

Massonnet D, Rossi M, Carmona C, Adragna F, Peltzer G, Feigl K and Rabaute T, 1993, The Displacement Field of the Landers Earthquake Mapped by Radar Interferometry, *Nature* **364** 138

McConnell R, Kwok R, Curlander J C, Kober W and Pang S S, 19091, Y-S correlation and dynamic time warping: Two methods for tracking ice floes in SAR images, *IEEE Trans. Geosci. Remote Sensing* **GE-29** 1004

Parkinson C L and Washington W M, 1979, A large scale numerical model of sea ice, *J Geophys Res* **84** 311

Pettersson M, Askne J *et al.* , 1994, Classification of ice properties in the Arctic during freeze-up using ERS-1 SAR and SSM/I, *Proc EARSeL 14th Symposium. Balkema.*

Plant W J, 1992, Reconciliation of theories of synthetic aperture radar imagery of ocean waves, *J Geophys Res* **97,** 7493-7501

Raney R K, and Vachon P W, 1988, Synthetic aperture radar imaging of ocean waves from an airborne platform: focus and tracking issues, *J Geophys Res,* **93** 12475

Raney, R.K. T. Freeman, R.W. Hawkins, and R. Bamler, A plea for radar brightness,*Proc. IGARSS'94,* Pasadena, Cal August 8-12 1994, pp 1090-1092

Richards J A, Sun G Q and Simonett D S, 1987, L-Band Radar Backscattering Modelling of Forest Stands, *IEEE Trans Geosc Remote Sensing*, **GE-25** no 4.

Rignot E J M and Zyl J J v, 1993, Change Detection Techniques for ERS-1 SAR Data. *IEEE Transactions on Geoscience and Remote Sensing* **4** 896.

Shemdin O H, 1988, Tower ocean wave and radar dependence experiment: An overview, *J Geophys Res* **93** 13829

Singh K *et al.* , 1986, The influence of Surface Oil on C- and Ku-band Ocean Backscatter, IEEE *Trans on Geoscience and Remote Sensing*, **GE-24** 738

Skriver H, 1989, Extraction of sea ice parameteres from synthetic aperture radar images, Ph. D. thesis, Electromagnetics Institute, Tech. Univ. Denmark, Lyngby, LD 74

Steele M, 1992, Sea ice melting and floe geometry in a simple ice-ocean model, *J Geophys Res* **97** 17729.

Stouffer R J, Manabe S *et al.* , 1989, Inter hemispheric Asymmetry in Climate Response to a Gradual Increase of Atmospheric CO2, *Nature* **342** 660

Sun Y, 1993, SAR image processing for the study of sea ice properties, *Techn Rep No. 155L, School of Electrical and Computer Engineering, Chalmers University of Technology.*

Sun Y, 1993, A new correlation technique for the ice motion analysis. EARSeL workshop on sea ice, Lyngby, to be published in *EARSeL Advances of Remote Sensing*

Sun Y and Askne J, 1994, Ice motion in the high Arctic on different time periods by a fast algorithm of motion estimation, *EARSeL'94, Göteborg, Balkema*

Sun Y, Carlström A and Askne J, 1992, SAR image classification of ice in the Gulf of Bothnia *Int J Remote Sensing* **13** 2489.

Swift C T, and L R Wilson, 1979, Synthetic aperture radar imaging of moving ocean waves, *IEEE Trans Antennas Propag*, **AP-27** 725

Tarayre H and Massonnet D, 1994, Effects of a refractive atmosphere on interpherometric processing, *Proc IGARSS'94, Pasadena, August 8 - 12* 717

Ulaby F T, Moore K M and Fung A K, 1982, Microwave Remote Sensing: Active and Passive, **2** (Addison-Wesley Publishing Company)

Ulaby F T, Moore R K and Fung A K, 1986, Microwave remote sensing: Active and passive, 3 vols. Dedham, MA: Artech House.

Ulaby F T, Sarabandi K, McDonald K, Whitt M and Dobson M C, 1990, Michigan Microwave Canopy Scattering Model, *Int J Remote Sensing*, **11,** no 7, 1223

Ulander L M H, 1992. Calibration of the ERS-1 SAR fast-delivery images, *Proc. of First ERS-1 Symposium, Cannes, France, 4-6 Nov.*: 173-178. ESA SP-359.

Ulander L M H and Hagberg J O, 1994, Topographic mapping using satellite radar interferometry, *Bildteknik (Image Science)* **3** 10

Ulander L M H, Hagberg J O and Askne J, 1993, ERS-1 SAR Interferometry over Forested Terrain, *Proceedings of the Second ERS-1 Symposium held in Hamburg, 11-14 Oct 1993*, pp. 475-480

Ulander L, Carlström A *et al.* , 1992, *ARCTIC-91: Remote Sensing and Sea Ice Data Report, Dept of Radio and Space Science*

Ulander L M H, Carlström A *et al.* , 1993, ERS-1 backscatter from Nilas and Young Ice during Freeze-up, *Microwave Remote Sensing of Ice, Lyngby, EARSeL*

Vass P and Battrick B, 1992, ERS-1 System, European Space Agency, SP-1146

Vesecky J F, Samandi R, Smith M P and Daida J M, 1988, Observations of sea-ice dynamics using synthetic aperture radar images: Automated analysis, *IEEE Trans Geosci Remote Sensing* **GE-26,** 38

# An Automated Approach to the Detection of Oil Spills in Satellite-Based  SAR Imagery

Dave R Sloggett

Earth Observation Sciences Ltd,
Farnham, UK

## 1   Introduction

### 1.1   Background

Lean and Hinrichsen (1992) have shown that illegal discharges from tankers accounts for 22% of the oil that exists on the world's oceans. Other sources include municipal waste sites and natural seepage from oil deposits on the ocean floor. Environmental agencies, operating throughout Europe, are interested in obtaining reliable information on the scale of oil related pollution in maritime basins to monitor the effectiveness of regulatory measures that have been introduced. Of particular interest is the 1978 Barcelona Convention governing pollution in the Mediterranean Sea, where monitoring and research on the sources, levels and effects of pollution have been one of the cornerstones of the Mediterranean Action Plan (MAP)—see Jeftic (1993).  Given access to such information these agencies can work to establish an effective regulatory framework designed to control the levels of pollution. Today the agencies rely upon access to reports derived from aircraft, ships, local authorities and conservation groups to formulate a picture of the scale of the pollution problem. The perspective gained from these sources is, however, necessarily localised and not readily extrapolated to the basin scale as many factors influence the density of pollution in certain areas of the sea where the close correlation of detected oil spills with shipping lanes is clear (Pellemans *et al.* 1994). Some marine pollution also breaks up at sea under wind and wave action and never reaches shore. The total extent of the marine pollution problem is therefore not clear providing difficulties for the regulatory authorities tasked with its control. In essence contemporary marine pollution policy is set in a partial vacuum of knowledge.

In order to derive an accurate basin-scale assessment of the nature of the oil pollution problem a source of data is required that provides a routine source of information over large geographic areas. Satellites provide one means of accessing such a data source and offer the potential to enhance the effectiveness of the patrol patterns used to monitor the pollution by the agencies concerned. With missions such as ERS1, ERS2, Radarsat, ASAR and the Earth Watch missions, pollution monitoring agencies will have access to a routine supply of information over the oceans and coastal regions. Discussions with these agencies have revealed a need to develop decision support aids that are able to provide accurate and timely information and forecasts when oil pollution is detected. These decision support aids should be capable of carrying out a risk assessment where the trajectory of the slick is compared with known local assets such as beaches and marine aqua-culture sites. The decision support aids should provide assistance in formulating alternative response strategies based upon an assessment of the type of oil that has been found on the surface.

This paper reviews the potential use of satellite-based radar instruments, such as those deployed on the European Space Agency ERS1 satellite, to detect oil pollution automatically. The paper considers the requirements for information from the pollution monitoring authorities. It highlights the need to provide an assessment of the type of oil detected on the surface and the potential use of Model Based Image Analysis (MBIA) techniques to infer the type of oil from its dispersal under wind/wave action— see Mulder and Schutte (1992). The paper then presents the results of the first phase of the research and looks in detail at the potential sources of signatures that resemble an oil slick that occur in radar based images of the oceans surface. These are referred to as sources of false alarms. The paper presents the results of the research and describes an operational infrastructure in which an automated algorithm would feature as part of an overall service to users including a forecasting facility. It concludes that provided an automated approach to oil slick detection could be made reliable it would provide an important source of information to marine pollution agencies operating at both a national and regional level to help with monitoring the scale of pollution and formulating policies that seek to reduce its impact upon the environment.

## 1.2   Contemporary research activities

Research reported by many European scientists such as Bern *et al.* (1992), Wismann (1993), and Pellemans *et al.* (1993), has shown that oil slicks can be detected using the Synthetic Aperture Radar (SAR) system on ESA's ERS1 satellite. Wismann (1993) has shown that the detection of oil slicks is hampered in three ways. In conditions where the surface of the ocean is totally calm oil slicks will remain undetected as they will merge into the image background. In extreme weather conditions, were the wind speed is very high ($>15\text{ms}^{-1}$) and the sea surface is well developed, research has shown that the oil slicks become increasingly difficult to detect. Often they break up rapidly and disperse. The second problem arises due to the presence of other features in ERS1 imagery of oceans and coastal regions arising from natural phenomena, such as other types of surface slicks (*e.g.* algal blooms) or effects that arise due to being in the lee of coastal features. Having established the potential use of ERS1 in this field several agencies have started activities aimed at studying the operational potential of satellite-based SAR to enhance the effectiveness of aircraft operations and to collect improved assessments of the nature of the pollution problem that exists.

Bern *et al.* (1992) have described a Norwegian project based on the Tromso satellite receiving station. This project uses a near-real time SAR processing system to generate fast delivery low resolution images. These are passed over a satellite link to the Norwegian Defence Research Establishment (NDRE) in southern Norway where they are examined by an operator for oil slicks. Any slicks that are detected are then reported to the state pollution control agency by telephone. A similar approach has been described by Pellemans *et al.* (1994) where the European Space Agency (ESA) Broadband Data Dissemination Network (BDDN) has been used to quickly relay low resolution data from Tromso to NLR and onto the Rijkswaterstaat in The Netherlands. In their report Pellemans highlights the potential delays that arise in using the BDDN. Their approach requires access to trained image interpretation staff throughout the week to support a fully operational service that is able to link closely to aircraft operations. These conclusions clearly suggest that an automated oil slick detection service could, if it were reliable, be an important development.

The present research activities have focused upon the operational potential of a European oil slick monitoring network based upon automated methods of slick detection. This activity has been sponsored by a number of agencies. These are: the British National Space Centre (BNCS); the Defence Research Agency (DRA); the Advanced Techniques group of the Institute of Remote Sensing Applications (IRSA) of the Joint Research Centre at Ispra.

The research reported in this paper has resulted in the development of a pre-operational capability to detect oil slicks on the surface of the ocean automatically. An oil slick detection workstation has been developed and installed at the DRA West Freugh ground station in Scotland. An SAR processor, the Matra Marconi RAIDS system, is used to process ERS1 images within 1 hour of an overpass. The images are then passed into the workstation where potential oil slicks are detected and highlighted to the operator. This facility entered its pre-operational phase on the 1st of June 1994 and initially concentrated upon an area surrounding UK coastal waters. Within two months several important oil spills in the English Channel, off the Welsh coast and the Humber estuary and off the coastline of The Netherlands were logged and reported to pollution control agencies—the majority of these events being confirmed by the agencies concerned. In the latter part of 1994 this monitoring area was extended to include coverage in several European coastal regions including Portugal, the coast of Spain, the Ligurian and Tyrrhenian Seas off the coast of Italy, and the Baltic (Figure 1).

# 2   User agencies and their information requirements

Sloggett and Jory (1994a and 1994b) have suggested that it may be possible to envisage an operational network, operated throughout Europe, based upon information automatically derived from ERS1 SAR imagery. In Europe there are several potential agencies that would be interested in information on the location and size of oil spills. They would also be interested in a simple assessment of the type of the slick—if that could be derived from shape and texture measures of the oil spill—as this is the single

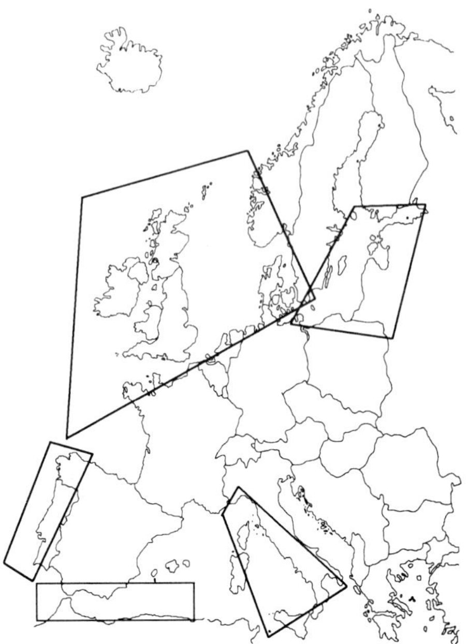

**Figure 1.** *Geographical area covered by the oil spill monitoring system*

most important piece of information used in deciding a response strategy. Coast guard agencies throughout Europe operate aircraft-based sensor systems and ships to monitor illegal discharges of oil. Their search areas and time on station are limited by a number of factors including: the duration of the flights; the speed of the ships; the area they can cover in a set period of time and budgetary constraints—which also serve to reduce the number of hours that aircraft can patrol. Given all of these restrictions an issue that arises is one of using limited resources in the most effective way possible to deter the potential polluters. This is where satellites may have a role to play. Given the principle of the polluter pays, access to data in near-real-time to assist in detection and prosecution is a vital element of an overall capability. Indeed it is possible to envisage that the income earned from such prosecutions could well be used to support the routine operation of the information services proposed.

It is envisaged that the information derived from ERS1 data could be integrated within an overall oil spill monitoring and nowcasting based service. This service would use information derived from the ERS-1 ground stations to supply reports on the location, size and type of slicks to local nowcasting centres based in each Member State. This would use local knowledge of coastal currents, tidal patterns, bathymetry and meteorological conditions to provide a nowcasting service to the coast guard authorities on the development of the spills. An important feature of the modelling facilities would be the integration of facilities to simulate the dispersal of the slick. This is where an assessment of the type of oil on the surface would be very important.

The nowcasting service could be used by the agencies concerned to carry out an assessment of the threat posed by the spill to a number of important assets, such as marine aqua-culture, coastal resorts, environmentally sensitive areas and harbour facilities. Given the results of this threat assessment the agencies concerned, such as conservation groups, fisheries managers and local authorities, can then make an informed decision as to how to respond most effectively to the spill.

Another agency that will be interested is the European Environmental Agency (EEA). One task of this organisation will be to collect information on the environment to provide accurate assessments to the European Parliament. It requires quantified assessments of many environmental variables in order to select the most appropriate policy response from one of a number of options. An important issue is the development of sustainable economic policies which take account of the environmental impact of each potential approach.

Before conducting socio-economic studies, policy makers require accurate assessments of the nature of the oil spill problems in their regions in order to draft legislation that will be effective. The satellite-based system outlined in this paper could provide one element of a system designed initially to collect the statistics that illustrate the nature of the problem and then monitor the effectiveness of any policy that was subsequently implemented.

# 3 The results of the pilot programme

## 3.1 Oil slick detection and sources of false alarms

Figure 2 shows a typical ERS1 scene off the coast of Ireland. An oil slick is shown in the bottom right hand side of the image. At the top of the image, and running down the coast, there are many other areas of low radar cross section. These are clearly not slicks. In viewing this image a human interpreter does a number of things to decide if there is a slick in the image. The first of these is to establish a context in which the dark areas are located. The second is to look at the morphological and spatial relationships between features detected in the image in an inter-linked reasoning process that incorporates model based reasoning as a key element of the overall process. These are the processes that an automated system has to replicate in order to eliminate potential sources of false alarms.

Figure 3 shows an image taken in the Dover Straits on the 16th of June 1994. The weather conditions were excellent and the sea was flat calm, with no wind ($<$1ms$^{-1}$). This image illustrates several vessels moving in the area. Figure 4 illustrates a section of the image where the cross channel ferries are crossing the English Channel. The ship targets stand out against the dark background of the calm sea and highlight how difficult it would be to detect oil slicks in totally calm weather conditions.

Figure 5 shows an image containing several natural phenomena. Detailed inspection of the scene shows several low radar cross sectional areas that have diffuse edges, a highly chaotic structure and persistent ship trails that indicate the surface film in those areas is not being disturbed by wind and wave effects. These are all sources of evidence that can be used in a reasoning process.

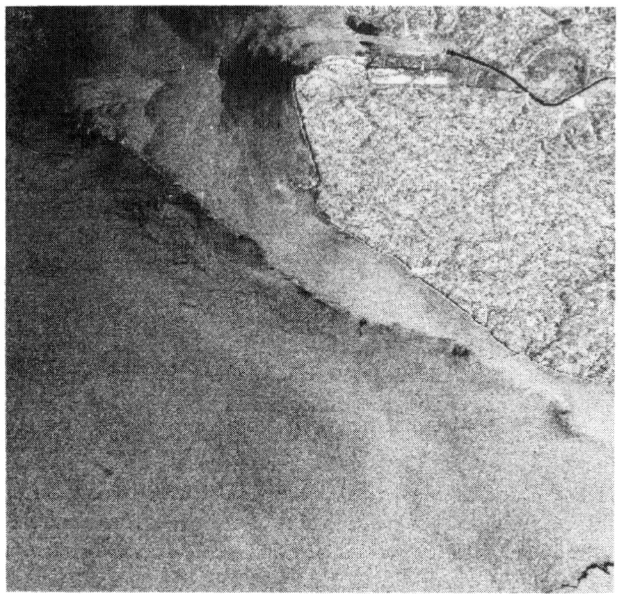

Figure 2. *An oil slick off the coast of Ireland*

Figure 3. *An ers1 image taken over the Dover Straits*

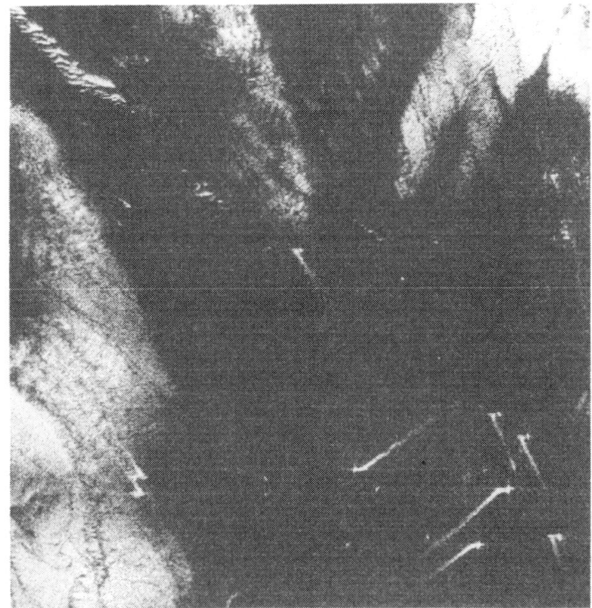

**Figure 4.** *An image of ferries moving in the Dover Straits*

**Figure 5.** *An ERS1 image showing complex areas of low radar cross section*

It is also clear from the image that knowledge of its location, the local meteorological conditions, bathymetry, and tidal patterns, would provide useful insights in an overall reasoning process designed to eliminate these low radar cross sectional areas as false alarms. At the bottom left hand side of the image a large dark area is seen to follow a ship that is heading north. This could be an oil spill or another phenomena known as a dark wake. Given that the sea in the local vicinity seems to be fairly well developed and that the feature is highly linear it would be possible to conclude that this feature is a dark wake. However on closer examination another highly linear white feature appears in the image that is linked to this object. This analysis suggests that the linear feature is an oil spill. This evidence is contradicted by the apparent increase in the width of the slick towards the bottom of the image, suggesting that the ship was in fact heading towards the top rather than the bottom. These individual sources of evidence have to be combined within a framework where the context in which the data is collected is fully understood.

Figure 6 provides us with another good example of typical surface features that exhibit low radar cross sectional characteristics. Close analysis of the image reveals a number of mesoscale features that are linked into an overall spiral shape. The long dark feature in the bottom right of the image is clearly linked into this image-scale feature and is not an isolated feature. This evidence suggests that this is not an oil slick but an oceanographic phenomena that is linked to local wind/wave interactions near to the shoreline.

Figure 7 shows a relatively straightforward image of a confirmed oil slick in the English Channel. The image was taken on the 27th of June 1994. It appears that the slick has spread from its release point under the influence of the wind and tide following the coast. The slick appears as two separate features one behind the other. Analysis suggests that the ship that discharged this oil was sailing from top right to bottom left of the scene as one of the two features is darker than the other. This suggests that the left hand slick is the more recent oil and the shape of the slick and its apparent thickness seem to confirm this hypothesis. Knowledge of the direction of the tide and wind also provides important insights into the likely shape of the oil; the rate of spreading towards the coast seemingly higher than that away from the shore, where the linearity of the edge is much clearer.

Figure 8 shows an image of the North Sea taken on the 20th of June 1994 with an oil slick on the left hand side of the image and several other natural phenomena with characteristics similar to slicks. This series of images highlight the problems faced by research groups whose interest is in the operational exploitation of oil slick information. An important issue is to reliably detect the oil spill whilst removing the false alarms that arise from other natural phenomena that create reduced radar cross section segments within the scene.

The natural phenomena illustrated in Figure 8 highlight the need in any automated processing algorithm to not only detect individual features in the water, such as the slick, but also to consider the relationship between features in the water to see if they are part of a larger scale feature. This type of reasoning is clear from the higher level structures in the top right hand corner of the image where the low radar cross sectional areas are clearly related to each other. The reasoning processes involved in this aspect of the processing use measures of similarity and dissimilarity to highlight differences

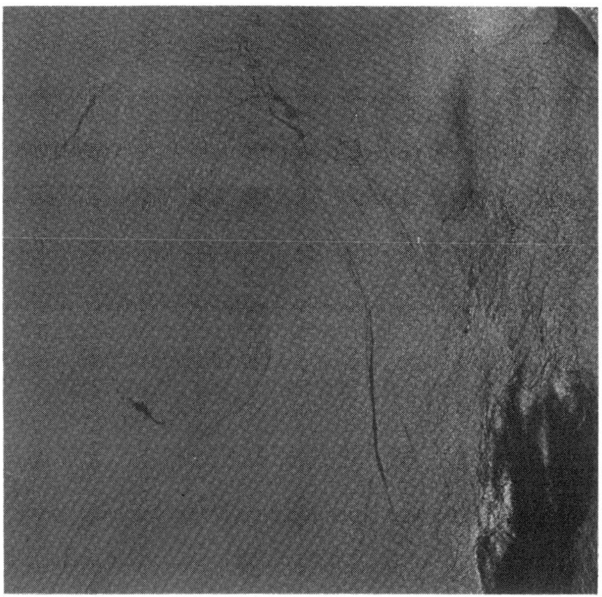

**Figure 6.** *An ERS1 Image containing linear features similar to oil spills*

**Figure 7.** *A confirmed oil spill in the English Channel*

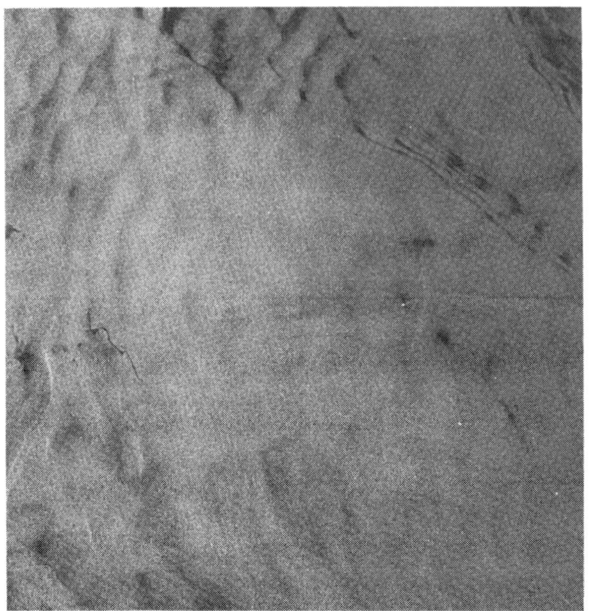

**Figure 8.** *A confirmed spill off the Humber Estuary*

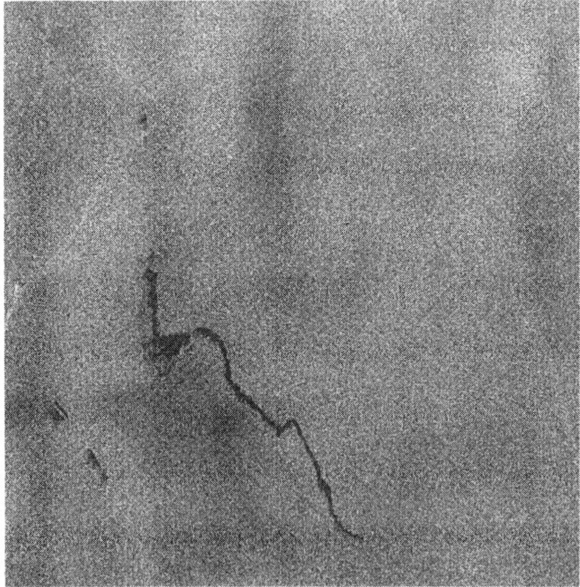

**Figure 9.** *An expansion of the Humber slick showing complex dispersal patterns indicative of the type of oil*

in individual features in the image and to cluster those features with similar spatial characteristics both individually and as part of a larger structure.

Figure 9 shows a detailed view of an oil slick detected off the Humber estuary. What is of immediate interest is the spreading that has occurred at the top of the slick in contrast to the lower part. It would be interesting to postulate if analysis of this spreading, under known wind/wave interaction, could give any indication of the type of oil on the water. This type of analysis could be possible for any slick detected provided their was some use of models in the reasoning process to postulate the size and shape of a slick, given an assumption of the oil type.

These practical examples of images highlight the ingredients of an automated approach to oil slick detection. These are:

- the ability to extract individual features from an image,

- the need to resolve relationships between features to assess if they belong to some higher level structure or are isolated,

- the need to establish the context of what is being observed from the perspective of the local geography and prevailing conditions,

- the use of models to predict the shape and texture of features with a view to carrying out assumption based reasoning.

## 3.2 The processing algorithm

The pre-operational version of the automated algorithm is currently undergoing validation with a number of collaborators in Europe. This algorithm, illustrated in Figure 10, shows that the image is subjected to a number of processing stages. The first of these reduces the resolution of the image prior to it being processed using an edge detection filter. The segmentation process then results in a set of initial regions in the image which are then contrasted with the background; see Figure 11. A series of potential oil spills is then identified, called targets, and these are individually processed to derive a series of features that can be used to discriminate natural targets from oil spills. Evidence is built up from each of the potential discriminants as to the probability that this is a slick before a final target region identification algorithm produces a definitive list of likely objects in the image that have characteristics similar to a spill.

## 3.3 Research results

Images collected at West Freugh have been processed and oil spills detected at the ground station have been reported, within a matter of hours, to the United Kingdom (UK) Marine Pollution Control Unit (MPCU) at Southampton who operate the oil spill detection service for the UK Department of Transport. In the period from the 14th of June when the West Freugh ground station started its pre-operational service, five slicks were detected in the first two weeks of operation in the test area. The first of these was detected on the 20th of June off the coast of the Humber estuary. The slick

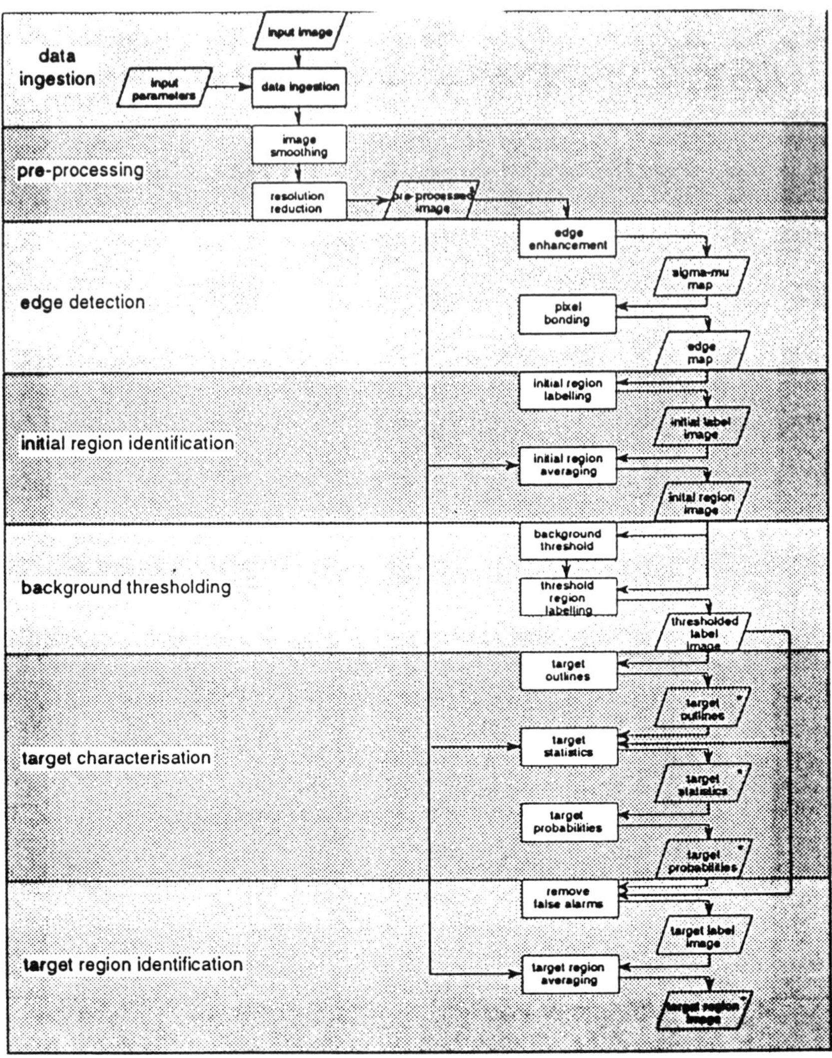



**Figure 10.** *The automated algorithm processing chain*

is shown in detail in Figure 9. It was detected and classified by the automated facility and the report was passed onto the MPCU. In another case, on the 27th of June, a slick was detected by the satellite after a midday pass over the Straits of Dover. The presence of the slick was reported and validated by the UK MPCU within an hour of the overpass. The slick detected off the coast of Eastbourne is shown in Figure 7. This had been reported slightly earlier by a ship operating in the area. Local observations

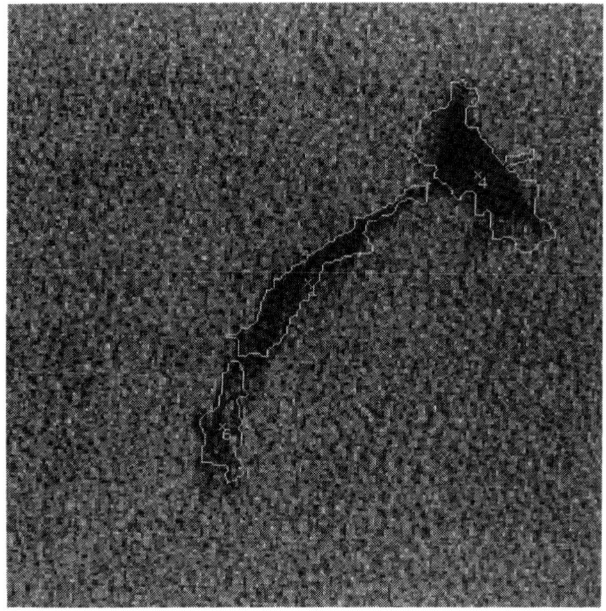

**Figure 11.** *A segmented oil spill*

of the wind at Dover showed its direction as south-westerly at 5–10 knots. At the time the image was taken it was 2–3 hours before high water on a spring tide which was also heading south-westerly at 2.6 knots. The combination of these forces of wind and tide seem to show up in the image in the way the oil has dispersed along the coastline. The dispersal pattern of the slick appears to support the inference that it developed from the top right hand side of the image with a combination of wind and tide spreading the slick in a south-westerly direction (towards the bottom left).

This research has provided some preliminary indications that it is possible to derive an automated approach to the detection of oil slicks. The key features identified in the images collected thus far are the fractal characteristics of the edges of the slick, its shape, measures of texture within the slick (which we believe can be used to infer the type of oil on the surface) and overall dimensions. Whilst these measures are important false alarms can still occur. Rejection of these is a key aspect of the ongoing research programme to ensure that an operational service is able to provide reliable information on the location and extent of slicks.

The current research programme is focusing upon the integration of more complex evidential reasoning techniques into the automated processor. The second generation of the processor is divided into three elements (Figure 12). The first element, the *direct* processing chain, is used to derive many parameters directly from the image and a segmented version of the image which is derived from a combination of an edge detection algorithm and a $\sigma$-$\mu$ filter. The direct parameters, which are derived for each segment, include statistical assessments of the radar backscatter, measures of the fractal nature of the edges, texture, shape and measures of the size of the slick. Assessments are

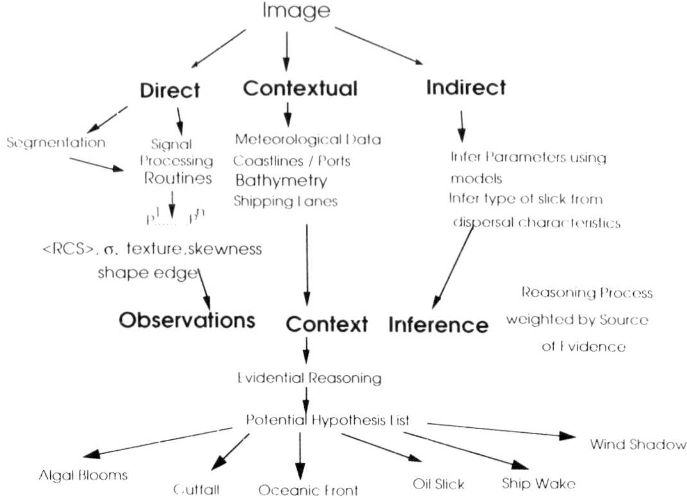



**Figure 12.** *An overview of the evidential reasoning process*

also carried out over the whole scene to derive background statistics and analysis of the wave spectra. These background measures allow the derivation of an assessment of effect of potential slicks on the underlying wave structure. This is potentially important for trying to classify the type of the slick.

The second processing element is the *contextual* processing chain. This element brings to the overall reasoning process information on the location of coastlines, meteorological data, tide and current movements and bathymetric data. These sources of information can be used in the reasoning process and in the third element of the oil slicks processor—the *indirect* element. This element uses information from the contextual element to produce inferences about what should be derived from the direct processing element. An example will illustrate this process. The indirect element can use models of the dispersal of a slick to forecast its development from a number of ways in which the slick is originated. Using a combination of wind, wave, tide and current information the shape of the slick can be inferred given a few assumptions about the type of oil present on the surface. The indirect processing chain can deduce several hypothesis about the likely dispersal pattern given that the oil is heavy crude, light crude, or any one of a number of other types of pollutant. This can then be correlated with the actual shapes derived from the image to determine which shapes most closely match those derived as potential slicks.

This use of models within an overall reasoning process has also been advocated by Mulder and Schutte (1992) using MBIA. Edwards (1993), in looking into the applicability of MBIA techniques in oil spill research, has stated that MBIA is a precursor to image understanding and uses hypotheses and physical models of interesting objects to conduct backward reasoning in an effort to find evidence that either supports or refutes a specific hypotheses about an object. This is the same approach reported by Sloggett and Jory (1994) at the ESA Pilot Projects workshop in Toledo, Spain.

The outputs from the three processing chains are brought together using a combination process based upon evidential reasoning. The outputs from the indirect processing element are compared with those derived from the direct processing element to see which of the inferences is most likely. In this combination process the contextual element provides the background to the reasoning to set the reasoning in context. The reasoning process seeks to weigh the evidence to support or refute one of several potential hypothesis as to the likely nature of the object detected in the water. These could be dark ship wakes, sewage outfalls, oceanographic features, wind effects, an algal bloom or a real slick. The evidence is weighed in the reasoning process to determine which of the possible solutions is most likely. If two outcomes are seen to be too close then the process is repeated with slightly different weighting applied to the reasoning to see if a clearer distinction emerges between the likely outcomes.

## 3.4   A nowcasting system

If our research proves successful we hope to establish a European-wide nowcasting service based upon the West Freugh ground station. This would use the automated facilities at the ground station to extract the location, size and type of oil and feed this directly to the pollution control agencies within 1–2 hours of a satellite overpass. At the same time the information on the slick would be passed to a modelling centre where this information would be combined with meteorological and tidal/current data and a hydrodynamic model to predict the movement of the oil on the oceans surface; ·see Edwards (1993). This information and the postulated trajectory would then be passed to those authorities mandated to monitor the coastline and carry out clean-up operations. The aim of such a system would be to provide accurate information that enabled effective resource planning and a situation assessment capability that allowed the agencies concerned to manage their response strategies correctly.

# 4   Conclusions

This research is showing some encouraging results in finding ways in increasing the reliability of detecting oil slicks in ERS1 SAR imagery and in reducing the probability of false alarms that originate from a variety of different sources. The research has shown that it is possible to envisage an automated service providing information to end user agencies on the location and size of oil spills in the North Sea. We envisage two forms of service being established. The first provides near-real-time access to information, as part of an operational structure designed to assist pollution monitoring agencies in their day-to-day operations and to provide an early warning system to coastal authorities charged with managing pollution incidents. The second is a service that provides a statistical summary of the location, extent and volumes of oil, on a geographic basis. This second element will be of great interest to agencies such as the EEA and to pollution control authorities who have a need to ensure that they effectively manage their limited resources to optimise their chances of successfully prosecuting polluters.

The automated process at the heart of this service is based upon the use of evidential reasoning techniques. Parameters, or *sources of evidence*, are derived directly, indirectly (through the use of models) and contextually using a generic concept for a geophysical processing system. The research is still in the pre-operational stage as we are continuing the validation of the existing processing system and considering improvements that will improve the quality of the information generated. In time we hope to demonstrate that it may be possible to make some form of assessment of the type of the oil on the surface as a result of analysis of the texture within a slick and the way in which it has formed on the surface under the influence of wind and tidal forces. This information is of particular interest to the user community as knowledge of the type of oil is a crucial parameter that is used in planning their response to specific incidents.

# 5   Acknowledgements

EOSL would like to acknowledge the support it has received from ESA (especially the help of Dr Guy Duchossois the ERS1 Mission Manager), BNCS, DRA and the Advanced Techniques Group of the Institute for Remote Sensing Applications (IRSA) at the Joint Research Centre (JRC) Ispra in carrying out this research activity. Dr Sloggett would like to also acknowledge the contribution made by Dr Alois Sieber, Mr Mike Boswell, Professor Werner Alpers, Dr Volkmar Vissman and Dr Eric van Halsema in the course of this programme.

# References

Bern T, Wahl T, Anderssen T and Olsen R, 1992, Oil Spill Detection using Satellite Based SAR: Experience from a field experiment, *Proc First ERS-1 Symposium - Space at the Service of our Environment*, Cannes, France, ESA SP-359

Edwards W W, 1993, Dynamic-Model-Based image analysis using minimum cost parameter estimation: An application to oil pollution flows in the marine environment, M.Sc Degree Thesis, ITC The Netherlands

Jeftic L 1993, Evaluation of the Med Pol Programme, *Proc of Clean Seas* O93, Valletta, Malta

Lean G and Hinrichsen D, 1992, Atlas of the Environment, World Wildlife Fund Publication

Mulder N J and Schutte K, 1992, Model based Image Analysis Using Solid Model and Ray Tracing, Symposium, *Remote Sensing and Space*, Hat Yai, Thailand, January

Pellemans A, Bos W G, van Swol R W, Tacoma A and Konings H, 1993, Operational use of real-time ERS-1 SAR data for oil spill detection on the North Sea, First Results, *Proc Second ERS-1 Symposium - Space at the Service of our Environment*, Hamburg, Germany, ESA SP-361

Pellemans A, Bos W G, Konings H, van Swol R, 1994, Oil spill detection on the North Sea using ERS-1 SAR Data, Report of the Directorate-General for Public Works and Water Management

Sloggett D R and Jory I, 1994, Oil slicks detection and information dissemination, *Proc 14th EARSeL Conference*, Goteborg, Sweden

Sloggett D R and Jory I, 1994, Automatic satellite based oil slicks detection and monitoring system, ESA ERS-1 Pilot Programmes Workshop in Toledo, Spain

Wismann V, 1993, Oil spill detection and monitoring with the ERS-1 SAR, *Proc Second ERS-1 Symposium - Space at the Service of our Environment*, Hamburg, Germany, ESA SP-361

# Radar Altimetry

## David Mantripp

Mullard Space Science Laboratory
Guildford, England

# 1 Basics of radar altimetry

## 1.1 Introduction

Satellite radar altimetry uses the ranging capability of radar sensors to measure surface topography. Spaceborne altimeters have principally been flown as part of oceanographic or ocean geodesy missions, but have increasingly been used to study other aspects of Earth geophysics. The first spaceborne radar altimeter was carried onboard Apollo 17. Since then, the sophistication of the instruments has increased. The Seasat radar altimeter (1978) was a significant step forward, and a similar instrument was flown aboard Geosat (1985-1989). The first European radar altimeter is currently in operation on board ERS-1, launched in 1991, soon to be joined by ERS-2 (January 1995). In parallel the joint NASA/CNES mission Topex-Poseidon is currently in orbit carrying two different instruments. Several future missions including radar altimeters are currently being planned, for example ESA's ENVISAT 1. Such missions will provide continuous altimeter coverage from the launch of ERS-1 onwards, enabling the accumulation of the long time series of data required for the study of many geophysical features, a number of which are thought to be sensitive to both natural and anthropogenically forced climate change. Satellite radar altimeters are simple in concept. A pulsed, vertical (nadir) pointing radar is used to measure the time delay between the transmission of each pulse, and the reception of the return echo. Since we know the velocity of the electromagnetic waves, it is a simple matter to convert this time measurement to range, and thus obtain the distance between the satellite and the ground. Current designs allow this measurement to be made with a resolution of just a few centimetres. By recording the strength of the return signal over time, we can infer other properties of the mean scattering surface.

In practice, however, things are not quite so simple. We are not really interested in range. The geophysical parameter that we need to extract is the surface elevation, and for this we need to know the precise position of the satellite. We need to understand how to interpret the return signal as a function of the radar altimeter's electronics, the antenna, and the satellite platform. The assumption that the propagation velocity of electromagnetic waves *in vacuo* can be applied to signals travelling through the atmosphere leads to errors which must be accounted for. Finally, we need to consider exactly what we mean by 'the surface'.

## 1.2 Principles of operation

### The range measurement

The fundamental measurement made by a radar altimeter is the range to some target. For a satellite in Earth orbit, this is simply given by:

$$h = \frac{ct}{2} \tag{1}$$

where $h$ is the range, $c$ the speed of light and $t$ the two-way travel time of the radar pulse. The satellite operating altitude, referenced to Earth ellipsoid, is generally around 800km. This allows a reasonable compromise between the requirements for limiting the effects of atmospheric drag and the high-order gravity field components on the one hand, and those on altimeter transmitter power and signal to noise ratio on the other.

The range resolution is given by:

$$\Delta h = \frac{c\tau}{2} = \frac{c}{2B} \tag{2}$$

where $t$ is the pulse length and $B$ the bandwidth.

In order to resolve individual echoes, we require:

$$\tau \leq \frac{2\Delta h}{c} \tag{3}$$

Thus for the ERS-1 altimeter operating in 'ocean mode', which has a pulse length $\tau$ of 3.03ns, the resolution of one discrete measurement of range is approximately 0.45m.

A major application of radar altimetry is the measurement of ocean dynamic signals. This requires the measurement of $h$ to a precision of a few centimetres, requiring a timing precision of less than 0.2ns. This is achieved by averaging the estimates from many individual pulses, and by fitting the echo waveform data to a known instrument surface response.

In order to understand the practical limits on accuracy of range measurement, we need to consider the nature of the transmitted pulse, its interaction with the surface topography and fabric, and the propagation delays introduced along the signal path. Finally, when converting the range measurement to surface height, we must consider the geodetic reference frame and the determination of the satellite's position.

### Operating method

All satellite radar altimeters use the same basic design, although variations and enhancements have been introduced over the years.

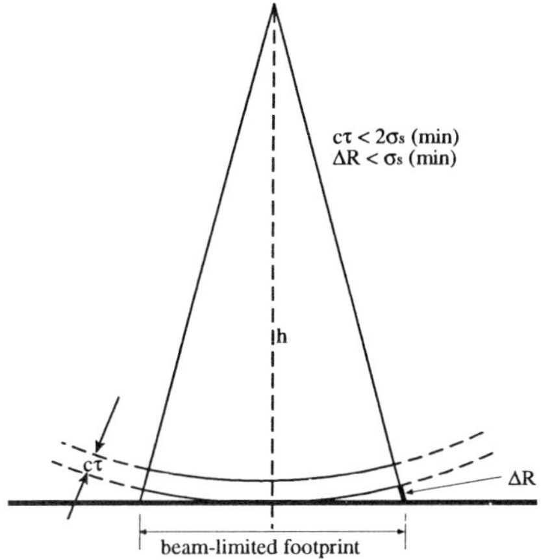


**Figure 1.** *The geometry of a beam-limited radar altimeter*

An altimeter may be designed for operation in either beam-limited or pulse-limited mode. The mode defines the diameter of the measurement footprint, the area of the surface which is 'seen' by the altimeter. It is important to understand that the range measured by the altimeter is not to a single point target, but to a mean surface corresponding to an integration of echoes from discrete scatterers over the area of the footprint.

Figure 1 describes the geometry of beam-limited operation. The pulse length $\tau$ is chosen to resolve the minimum surface roughness of interest, $\sigma_s[min]$, from Equation 3. In order to ensure that the entire area of the footprint is illuminated at the same time by the transmitted pulse, it is necessary that the difference $\Delta R$ between the nadir range and the slant range at the edge of the beam-limited footprint be less than $\sigma_s[min]$. The advantage of a beam-limited altimeter lies in the simplicity of the interpretation of the signal. However, it is operationally difficult to fly in space: from geometry and antenna theory considerations, it can be shown that in order to meet the resolution requirements for the ERS-1 mission, a beam-limited radar altimeter would require an antenna diameter approaching 10m, which would be very expensive and difficult to accommodate. Furthermore, for beam-limited operation it is necessary to know the pointing direction of the antenna boresight extremely accurately, which requires complex and costly attitude control systems.

Pulse-limited operation, which has been used by all spaceborne radar altimeters flown so far, is described in Figure 2. In this mode, a broad antenna beam is used, requiring a much smaller diameter antenna (1.2m on ERS-1), and requirements on pointing are less stringent. For pulse-limited operation, the excess slant range at the edge of the beam-limited footprint is $\Delta R$. Each pulse of microwave energy propagates away from the antenna as part of an expanding spherical shell, the thickness of which is given by the pulse length. As this shell intersects the ground, it illuminates a growing disk (assuming a flat surface) which spreads out across the beam-limited footprint to a maximum diameter defined by the pulse length. Return echoes are recorded over the

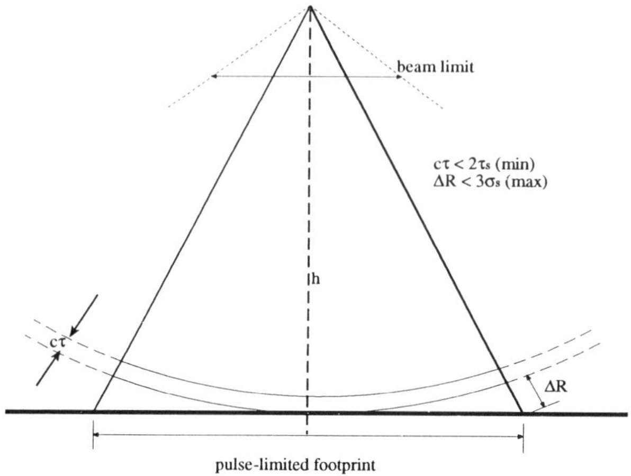


**Figure 2.** *The geometry of a pulse-limited radar altimeter for a flat reflecting surface*

duration of the pulse, building up an echo waveform with a rapid rise and long decay. The basic range measurement is made by timing between the transmit pulse and the leading edge of the echo waveform.

## 1.3   The echo waveform

The build up of the echo waveform is illustrated in Figure 3. Initially, as the pulse intersects the nearest point on the ground, and the illuminated disc spreads out rapidly, the power of the return echoes builds up quickly to a maximum. This maximum is reached when the maximum area is illuminated, that is to say just as the rear of the pulse intersects the ground. Following this point, the illuminated area remains constant, but the received power slowly decays due to the effect of the antenna angular response. Echoes are received from scatterers throughout the pulse-limited footprint. In order to reduce speckle noise arising from coherency in individual waveforms, telemetered waveforms are actually an average of a number of pulses, 50 in the case of ERS-1.

The geometry for deriving the pulse-limited footprint size is illustrated in Figure 4. From this figure we can see that at the altimeter echoes are received from the nearest point N from time $t = 2t_0$ to time $t = 2t_0 + \tau$, where $t_0 = h/c$. Note that the effective pulse duration $\tau$ is used to estimate the pulse-limited footprint size. This assumption is only valid for targets exhibiting small surface roughness, for example a calm ocean surface. At any instant $t = (2t_0 + \Delta t)$ power is just beginning to be received from the point P$'$ such that:

$$R'(\Delta t) = \frac{(2t_0 + \Delta t)c}{2} = h + 0.5c\Delta t \tag{4}$$

Since:

$$l' = \sqrt{(R')^2 - h^2} \tag{5}$$

We have:

$$A_f(\Delta t) = \pi hc\Delta t \tag{6}$$

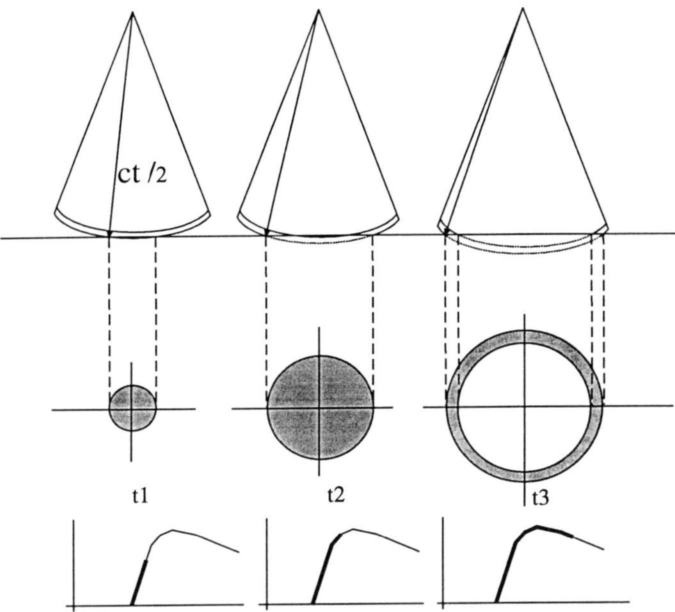



**Figure 3.** *The build up of an echo waveform over the duration of a pulse. As the leading edge of the pulse intersects the surface, it illuminates a small disk. Echoes are received from this area, and the power in the waveform builds up rapidly as the illuminated area grows. It reaches a peak just as the rear of the pulse intersects the surface, and then decays slightly towards a cut-off defined by the pulse length or the range window, whichever is the shortest. The decay is due to off nadir characteristics of the antenna pattern. Note that, for clarity, an average waveform is depicted; a single waveform would be more noisy.*

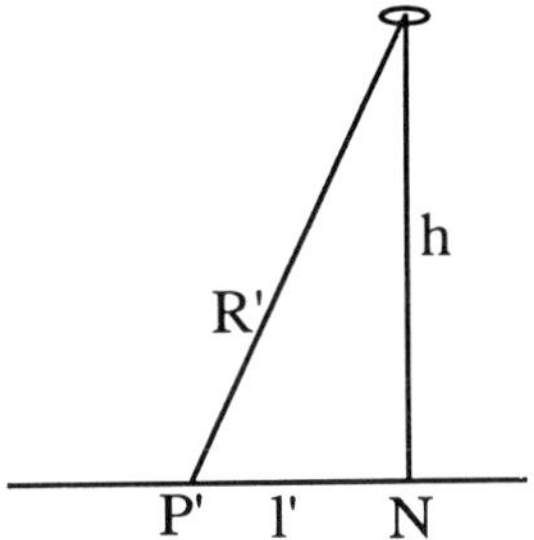



**Figure 4.** *The geometry for deriving the footprint size for pulse limited operation*

where $A_f(\Delta t)$ is the area of flat surface perceived by the altimeter to be illuminated at time $t = (2t_0 + \Delta t)$. This area grows linearly with time, according to equation (6), until $\Delta t = \tau$, after which the disk changes to an annulus with constant area (Figure 3). Thus the area of the pulse-limited footprint for a flat surface without significant roughness is:

$$A_f(\Delta t) = \pi h c \tau \tag{7}$$

and the pulse-limited footprint diameter is:

$$D_f = 2\sqrt{hc\tau} \tag{8}$$

For the ERS-1 altimeter, $D_f = 1.7$km.

When significant surface roughness is present, that is to say that the surface within the beam-limited footprint exhibits variations in elevation away from the mean of a magnitude similar to or larger than the effective pulse length, the situation becomes more complex. In the presence of ocean waves, the echo waveform rise time is longer, hence the leading edge is less steep, and the area within which surface facets appear to be illuminated simultaneously is larger, since the wave crests are illuminated earlier than the wave troughs. Therefore over waves, the pulse-limited footprint diameter is given by:

$$D_f = 2\sqrt{hc\tau''} \tag{9}$$

where:

$$\tau'' = \sqrt{\tau^2 + (1/c)(\ln 2)H_{1/3}} \tag{10}$$

where $H_{1/3}$ is the crest-to-trough height of the 1/3 largest waves, and is related to the RMS wave height $\sigma$ by:

$$H_{1/3} = 4\sigma \tag{11}$$

The maximum pulse-limited footprint possible over ocean surfaces, where the distribution of scattering facets can always be considered Gaussian, is 7km, and the maximum value of $H_{1/3}$ is 20m. Over non-ocean surfaces, the distribution of facets contributing to the waveform can be wider, but footprint then becomes fragmented.

## 1.4   On-board electronics

In order to understand how to carry out signal processing, calibration and correction of radar altimeter data, it is useful to have a general understanding of how the instrument functions. A simplified block diagram of the ERS-1 radar altimeter is given in Figure 5.

Radar altimeters need to be capable of both transmitting very high power signals and receiving very faint echoes. A typical value for the receive/transmit ratio is $-138$dB. It can be shown that the power required to achieve this performance using a conventional constant frequency pulse would be around 900W, which is barely possible when considering other requirements and current technologies. In order to meet the requirement, pulse compression techniques are used. This allows a long- duration, low power but high energy pulse to be transmitted in such a way that after compression the effect is the same as having transmitted a short duration high power pulse with the same energy. The compression method used is linear frequency modulation, where a linear frequency shift is applied to the transmitted signal, resulting in a 'chirped' pulse.

When the first echo is expected to return, after a time determined by an automatic control loop known as the Height Tracking Loop (HTL), a second chirp is generated

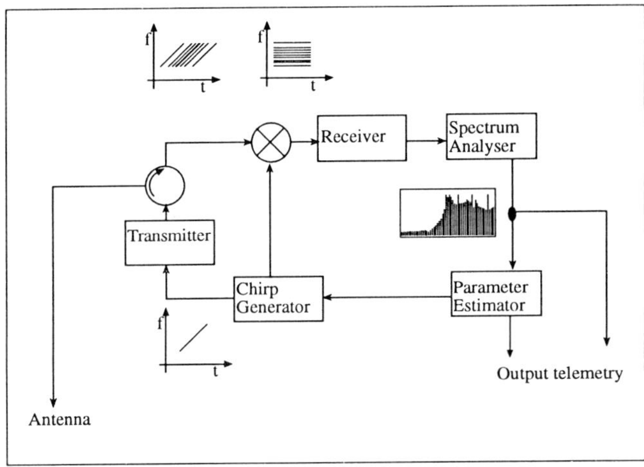



**Figure 5.** *Simplified schematic diagram of the ERS-1 radar altimeter electronics*

at an offset frequency, and mixed with the received signal. The resultant set of tones, one corresponding to each received echo, is converted from the time domain to the frequency domain using the chirp slope as a conversion factor. The frequency spectrum is analysed by a hardware digital signal processor, which extracts a series of evenly spaced spectral components (64 for ERS-1) using a Fast Fourier Transform. For ERS-1's 'ocean' mode, the spacing of these spectral components is 50kHz, which corresponds to 3.03ns separation. The timing of the triggering of the receive chirp for the next transmit pulse has a resolution of 12.5ns. However, this can be fine-tuned by applying a phase rotation to the complex signal.

The range between the satellite and the Earth's surface varies around the orbit by several kilometres. However, in the vertical dimension the altimeter can receive echoes only within a 'range window' specified by the width of the frequency spectrum, which for ERS-1 is about 60m in ocean mode. In order to receive echoes all around the orbit, the range window offset must be continuously adjusted to keep the leading edge of the echo waveform at a specified position, usually the centre of the window. This position is known as the tracking point, and is range bin 34 for ERS-1 in ocean mode. The tracking point is adjusted to the current surface range estimated by the HTL, in coarse and fine steps (Figure 6). The value of the coarse tuning is used to update the receiver trigger timing; the fine tuning value is relayed via telemetry, and is added to the on-board estimate of range during data processing. The HTL range estimate is smoothed by a second order low-pass '$\alpha\beta$' filter to damp down oscillations due to speckle in individual waveforms. The HTL also produces an error signal based on the difference between the received echo waveform and an idealised echo which is used as the basis for parameter estimation. Similar tracking Loops are also implemented for Automatic Gain Control (AGC), and for Leading Edge Slope (STL).

When the surface elevation varies too rapidly for the tracker to respond, the altimeter is said to have 'lost lock'. When this occurs, the instrument automatically goes into

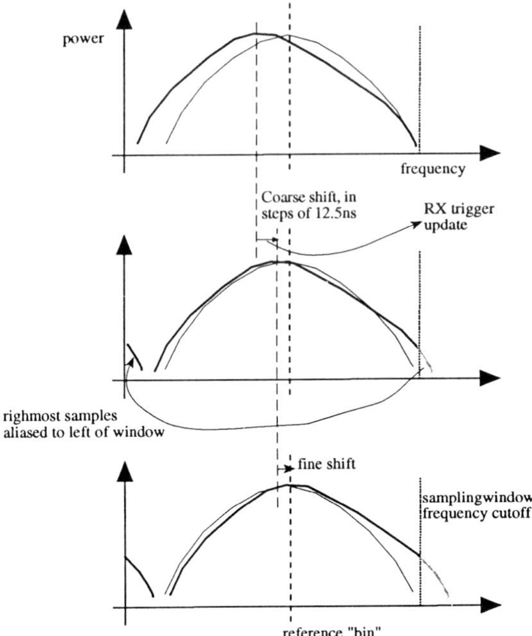


**Figure 6.** *Illustration of the adjustment of the tracking point by the altimeter Height Tracking Loop (*HTL*), based on fitting the echo frequency spectrum to a predicted shape*

acquisition mode, and hunts for a surface. Alternatively, a dynamic range window may be implemented, where the width of the window is adjusted rather than its centre, by increasing the width of the individual filter bins and therefore the effective area of the pulse-limited footprint. Thus resolution is traded off against the ability to track rough terrain. ERS-1 has two fixed window widths, 'Ocean' mode, where the filter bins are 3.03ns wide in the time domain, and 'Ice' mode (so called because it is designed to permit mapping of the polar ice caps) where the width is 12.12ns. The ERS-1 altimeter cannot track terrain where the mean slope is greater than about 1°.

## 1.5  Height corrections

In the introduction we stated that range was not of great geophysical interest, since what we really need to know is surface elevation. The preceding few sections have discussed the method by which time delay is measured very accurately, and, when range is referred to, it has been with the implicit assumption that the radar pulses are propagating in a vacuum. It is therefore necessary to make allowances for the transmission delays and path length fluctuations introduced by the atmosphere. Once these corrections have been made, we can infer a mean elevation for the scattering surface. The accuracy of this elevation, as well as its location, however, is limited by the accuracy to which we know the satellite's position, which can be improved by applying higher precision orbit

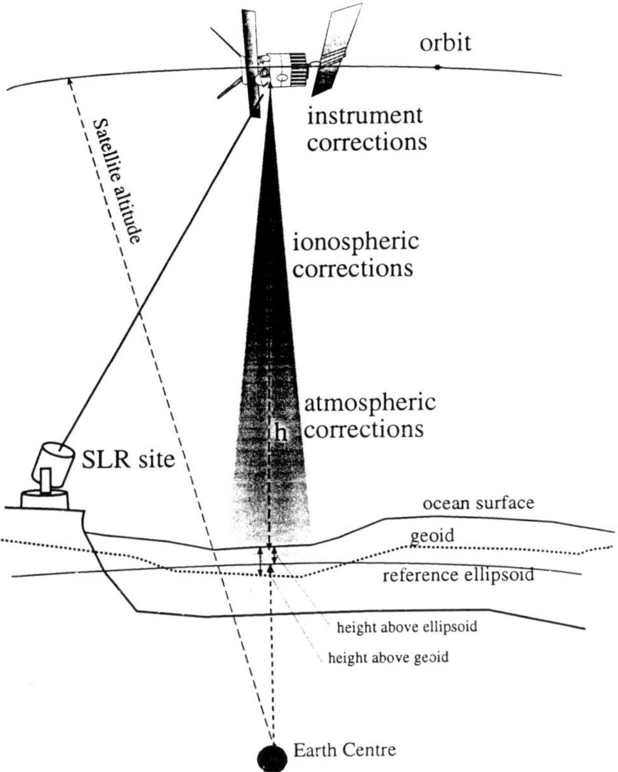


**Figure 7.** *The contributers to uncertainties in surface elevation estimates derived from radar altimeter data*

ephemerides. Figure 7 illustrates all of the components which we need to address in order to obtain a fully corrected height. Each of these is dealt with in turn.

## Satellite location

First of all we address the problem of satellite location. Tracking by ground-based Satellite Laser Rangers (SLRs) can determine the centre of gravity to around 5cm over approximately 2000km arcs, providing that sufficient tracking data are acquired. Passive laser retroflectors are carried on board the satellite to enable precise location of the laser reflection within the satellite frame of reference. The position of the altimeter antenna plane is known with reference to the centre of gravity from pre-launch measurements and corrections to this due to fuel burn are computed during the mission. Other methods of location include microwave rangers such as the PRARE system, although the PRARE included in the ERS-1 payload failed shortly after launch. GPS also is used on Topex/Poseidon. A model is then used to propagate the location of the centre of gravity between measurements, taking into account spatial variations in the Earth's gravity field, solar radiation pressure and atmospheric drag. The very best accuracies obtained from such models are currently around 50cm. However, these are improving as more satellite tracking data are incorporated into gravity field descriptions. This remains a major source of uncertainty for non-terrestrial radar altimeter mapping.

## Instrument corrections

These include fixed and variable components. The radar altimeter electronics introduce certain fixed biases, many of which can be determined during ground calibration. Fluctuating biases due to differential heating around the orbit, and long term drift during the mission, are estimated using data from periodic on-board calibrations. Clock frequency drift is accounted for by periodic correlation with reference clocks. A remaining bias which cannot be calibrated either on the ground on internally during flight is due to antenna effects and to the delay within electronic components lying outside of the internal calibration loop. For ERS-1, ESA designed and carried out a sophisticated activity to determine this bias, known as the Venice Calibration Campaign. The result of this campaign gave the external range bias as $-40.9 \pm 5$cm.

| Error Source | Correction | | Error | |
|---|---|---|---|---|
| **Range Corrections** | | | | |
| Internal calibration | $\pm 10$ | cm | $\pm 2$ | cm |
| External calibration | $\approx 40$ | cm | $\approx 5$ | cm |
| Doppler shift | $\pm 33$ | cm | $\pm 0.3$ | cm |
| Centre of gravity offset | 0–90 | cm | $\pm 0.1$ | cm |
| Ionospheric delay | 0–100 | cm | $\pm 5$ | cm |
| Tropospheric delay – dry | 1.7–2.5 | m | $\pm 1$ | cm |
| Tropospheric delay – wet | 0–50 | cm | $\pm 5$ | cm |
| Liquid water content | 0–3 | cm | $\pm 0.1$ | cm |
| Ocean Tide | $\pm 50$ | cm | $\pm 10$ | cm |
| Earth Tide | $\pm 30$ | cm | $\pm 1$ | cm |
| Ocean loading Tide | $\pm 10$ | cm | $< 1$ | cm |
| Geoid | $\pm 100$ | m | $\pm 1$ | m |
| **Backscatter measurements** | | | | |
| Internal calibration | $\pm 0.5$ | db | $\pm 0.1$ | db |
| Range | 0–0.33 | db | $\pm 0.01$ | db |
| Liquid water | 0–0.12 | db | $\pm 0.1$ | db |

**Table 1.** *Corrections applied to ERS-1 Waveform Products during off-line processing. The last column gives an estimate of the error in the correction.*

## Propagation corrections

The propagation velocity of electromagnetic waves depends on the refractive index of the medium through which they travel. Thus a correction $\Delta h$ must be applied to altimeter height estimates in order to account for the slight reduction in velocity within both the ionosphere and the neutral atmosphere.

For the ionosphere:

$$\Delta h_i = 40.3 \frac{N_t}{f^2} \tag{12}$$

where $\Delta h_i$ is in metres, and $N_t$ is the Total Electron Content (TEC) per m$^2$ along the propagation path. $N_t$ varies by day and night, with latitude and longitude, with season and with solar activity, in the range $10^{16}$ to $10^{19}$, giving:

$$0.002 \leq \Delta h_i \leq 0.02$$

Ionospheric models may be used to derive $\Delta h_i$ to within a factor two. Alternatively, a dual frequency altimeter, such as Topex, can take advantage of the dispersive nature of the delay to measure $N_t$ directly. For calibration and verification activities, estimates of $N_t$ from (ground based) dual frequency GPS can also be used.

For the neutral atmosphere, we have, to a first approximation:

$$\Delta h_a = 2.27 \times 10^{-5} P_s + \frac{1.723W}{T_a} \tag{13}$$

where $P_s$ is the surface atmospheric pressure, $W$ is the liquid water content, and $T_a$ is the average temperature of the lower atmosphere. The two terms are often referred to as the 'dry' and 'wet' atmospheric corrections. Typical magnitudes are 2m for the 'dry' and 6–30cm for the 'wet'. The 'dry' correction can be estimated to within $\pm 0.7$cm using a meteorological model, although this becomes more difficult in areas where the model suffers from a lack of observational data. The 'wet' component can be directly measured through multi-channel passive microwave detection (over ocean), by radiosonde ascent data, or estimated using a meteorological model. The best results are obtained by in-situ passive microwave radiometers, which can give results accurate to a few cm.

| Satellite | Operator | Start | Finish |
|---|---|---|---|
| GEOS-3 | NASA | 1974 | 1978 |
| Seasat | NASA | 1978 | 1978 |
| Geosat | US Navy | 1985 | 1989 |
| ERS-1 | ESA | 1991 | 1996? |
| Topex/Poseidon | NASA | 1992 | 1997? |
| ERS-2 | ESA | 1995 | 1998? |
| Geosat/TP | NASA | 1996? | 2001? |
| ENVISAT-1 | ESA | 1997? | 2002? |

**Table 2.** *Altimetry missions; past, present and future*

## 1.6 Satellite altimeter missions

**Past, present and future radar altimeter satellites**

Table 2 illustrates the life span of all satellite missions launched to date and carrying radar altimeters. Expected lifetimes of currently flying and future missions are depicted as dashed lines, but these should only be taken as estimates. Future mission planning is a volatile business due to conflicting requirements, policy pressures and expense. For example, the issue of a US Navy follow on mission to Geosat and/or Topex-Poseidon is not yet settled, although it seems likely that something will emerge. However, barring accidents it seems that the goal of continual coverage required for all applications concerned with temporal fluctuations in geophysical parameters will be met for at least the next decade.

The major difference between various radar altimeter designs is whether they are single or dual frequency. Single frequency designs have evolved through from the GEOS-3 instrument, to the similar Seasat and Geosat instruments, to ESA's 13.8GHz RA which is currently flying on ERS-1, and on ERS-2, which incorporates fully fledged dual mode operation. The CNES designed SSALT on Topex-Poseidon is a experimental lightweight single frequency instrument, which is of interest to those advocating cheap, multiple satellite programmes for widespread, short repeat period coverage. The NASA NRA and ESA RA-2 instruments are more complex designs incorporating dual frequency operation for in-situ ionospheric correction, and variable range window width allowing a trade-off of bandwidth against resolution for more robust tracking of severe terrain.

| Satellite | Inst. | Freq GHz | Pulse Hz | Beam width | Height Km | Incl | La | Ge | Gl | Oc |
|---|---|---|---|---|---|---|---|---|---|---|
| | | | | | | | Applications | | | |
| GEOS-3[1] | ALT | 13.9 | 100 | 2.6 | 839–853 | 115.00 | 3 | 3 | | |
| Seasat[2] | ALT | 13.5 | 1020 | 1.6 | 776–800 | 108.00 | 3 | 3 | 5 | 5 |
| Geosat | ALT | 13.5 | 1020 | 2.1 | 760–817 | 108.10 | 5 | 3 | 5 | 5 |
| ERS-1[3] | RA | 13.8 | 1020 | 1.3 | 780–800 | 98.52 | 3 | 3 | 4 | 4 |
| Topex/Poseidon | NRA[4] | 5.3/13.5 | 1000/4000 | 2.8/1.1 | 1335 | 66.02 | 3 | 3 | 5 | |
| | SSALT[5] | 13.65 | 1700 | 1.1 | | | 3 | 3 | 5 | |
| ERS-2[6] | RA | 13.8 | 1020 | 1.3 | 780–800 | 108.10 | 3 | 3 | 3 | 3 |
| Geosat/TP | NRA? | | | | | | 3 | 3 | 5 | |
| ENVISAT-1[7] | RA-2 | | | | | | 3 | 3 | 3 | 3 |

**Table 3.** *Mission parameters. The application subheadings refer to Land, Geology, Glaciology and Oceanography. The comments are as follows: (1) no on-board data recorder, (2) failed after 3 months, (3) near-polar orbit, (4) dual frequency altimeter, (5) experimental compact altimeter, (6) identical to ERS1, (7) RA2 dual frequency altimeter.*

## Mission scenarios and objectives

The wide variety of applications making use of radar altimeter data inevitably lead to conflicting requirements which must be dealt with by mission planners. Table 3 lists some relevant characteristics of past, present and future missions. The main trade off is between repeat period, ground track density and latitudinal extent. In general, the oceanographic community want a satellite in a medium repeat period orbit, and a relatively low inclination, typified by Topex-Poseidon. Mesoscale ocean circulation ideally requires an observation sampling rate of the order of 10 days. Oceanographic satellites must also list microwave radiometers for in-situ tropospheric correction high on the priority list. The second largest community is probably made up of those interested in aspects of polar science. Here the need is for dense coverage to high latitudes, with short term repeat a lower priority. Supporting instruments such as microwave radiometers are less important, but demands on altimeter tracking capability and data rate are higher. No dedicated polar mission has been flown to date. ERS-1 and ERS-2 represent a compromise between the various requirements, with the result that both missions have several successive orbit scenarios, and the risk that nobody is entirely satisfied! The forthcoming ENVISAT 1 mission is a very complex platform carrying a wide variety of instruments, all of which place various demands on mission planners. There have been calls for a return to simpler, dedicated radar altimeter missions, such as the mooted Geosat Follow-On (GFO), and proposals for constellations of small altimeter satellites. However, to date nothing of this nature exists as a definite plan.

## User products

Radar altimeter user products divide into several types. The first division concerns delivery time: some missions make fast turnaround data available, such as the ERS-1 Fast Delivery Product (FD). These are compact transcriptions of on-board estimates of basic parameters, with a limited set of corrections and other ancillary data. They are typically used in medium range weather forecasting, where high precision is less important than rapid availability. ERS-1 FD products are available through the Meteorological GTS telecommunications network around 3 hours post-acquisition. Although such data can be of use for other applications, they are generally compromised by low resolution orbit ephemerides. Fast delivery products are also termed 'on-line'. On the other hand, off-line products may take several months to produce, the delay being due to the time required to process high resolution orbits and atmospheric corrections, as well as repackaging data into contiguous segments and including a wider range of ancillary data. Off-line products divide down again into two main sub-categories, waveform and non-waveform products. Non-waveform products are generally termed GDRs (Geophysical Data Records), and include re-estimated fully calibrated basic parameters, various quality assessment data, atmospheric and engineering corrections and precision orbits. GDR products are generally used for looking at areas over which the user is confident that the echo waveform conforms to a well understood model, in other words the oceans. Waveform products, known as SDRs (Sensor Data Records) or WDRs (Waveform Data Records) usually contain all of the GDR parameters, with the addition of the full waveform data and other engineering level parameters. These are required when a more detailed analysis of the waveforms is required to be able to extract range and backscatter accurately. Note that the ESA products for the ERS series depart from the usual practice somewhat. The analogue to the GDR for these missions is the OPR (Ocean Processed Record), and to the SDR the WAP (WAveform Product). Some hybrid products also exist, for example the ERS-1 IGDR product produced in the USA by NOAA, which consists of FD products updated with high precision orbits. Clearly as the complexity of the product increases, so does the size. The ERS-1 FD product record length is 80 bytes (per second) whilst WAP records 5100 bytes long (per second).

# 2 Estimation of geophysical parameters

## 2.1 Introduction

In Section 1 we discussed the operation of a pulse-limited radar altimeter. This section covers the extraction of parameters from the echo waveform built up of all the replicas of the transmitted chirped pulse returned within the pulse-limited window. This topic is very clearly divided into two areas, extraction of parameters over ocean and over non-ocean surfaces. It is possible to statistically describe the distribution of scattering facets within the pulse-limited footprint over an ocean surface, but over non-ocean surfaces, the situation is both more complex, less predictable, and the nature of the return echoes varies very much as a function of surface type. The following covers the basic principles behind the use of the Brown Model for interpretation of ocean surface echoes, and describes a few of the approaches used to interpret data from non-ocean surfaces. Whilst not an exhaustive catalogue of methodologies, it should give a broad illustration of the principles underlying the extraction of geophysical data from radar altimetry.

## 2.2 Ocean surfaces

Satellite radar altimeters have principally been flown to study ocean variability and the ocean geoid. Three principal parameters can be obtained by analysis of echo waveforms over ocean surfaces: surface height, significant wave height, and wind speed. We have seen that the range resolution of a typical altimeter is approximately 0.5m, whilst the requirement on height precision is approximately 3cm. We can achieve this precision by fitting a model of waveform return to the data, which also enables the determination of wave height and wind speed. The model used for this purpose is known as the Brown Model, after its originator, and is applied both in a simplified form in on-board parameter estimation, and in more sophisticated post-processing on the ground.

**The Brown model**

The Brown Model describes the average impulse response of a rough surface to radar pulse as a convolution of three functions:

- the point target impulse response

- the flat surface impulse response

- the vertical distribution of surface scatters.

The average impulse response is described by the following expressions:

$$
\begin{aligned}
P_{fs}(\tau) &= \frac{G_0{}^2\lambda^2 c\sigma_0(\Psi_0)}{4(4\pi)^2 L_p h^3} \exp\left[-\frac{4}{\gamma}\sin^2\xi - \frac{4c}{\gamma h}\tau\cos 2\xi\right] * I_0\left(\frac{4}{\gamma}\sqrt{\frac{c\tau}{h}}\sin 2\xi\right) \\
P_r(\tau) &= \eta P_t P_{fs}(0)\sqrt{\frac{\pi}{2}}\sigma_p\left[1 + \mathrm{erf}\left(\frac{\tau}{\sqrt{2}\sigma_c}\right)\right] \quad \tau < 0
\end{aligned}
\tag{14}
$$

$$P_r(\tau) \;=\; \eta P_t P_{fs}(\tau) \sqrt{\frac{\pi}{2}} \sigma_p \left[ 1 + \mathrm{erf}\left( \frac{\tau}{\sqrt{2}\sigma_c} \right) \right] \quad \tau \geq 0$$

where:

| | |
|---|---|
| $P_r$ | average return echo power |
| $\tau$ | two-way time delay |
| $\eta$ | pulse compression ratio |
| $P_t$ | transmitted power |
| $P_{fs}$ | flat-surface impulse response |
| $\sigma_p$ | related to the width of the radar's point target response |
| $\sigma_c$ | scale-size of the distribution of surface scatterers |
| $G_0$ | on-axis antenna gain |
| $\lambda$ | radar wavelength |
| $c$ | velocity of light |
| $\sigma_0(\Psi_0)$ | surface backscatter coefficient at incidence $\Psi_0$ |
| $L_p$ | propagation loss |
| $h$ | satellite height |
| $\gamma$ | parameter related to the antenna beamwidth |
| $\xi$ | the antenna mispointing |

This model can only be applied where certain assumptions are reasonable. Brown quotes the most important of these, which relate to the convolutional model, as being the following:

1. The scattering surface may be considered to comprise a sufficiently large number of random independent scattering elements.

2. The surface height statistics are assumed to be constant over the total area illuminated by the radar during construction of the mean return.

3. The scattering is a scalar process with no polarisation effects and is frequency independent.

4. The variation of the scattering process with angle of incidence (relative to the nominal mean surface) is only dependent upon the backscattering cross section per unit scattering area $\sigma_0$ and the antenna pattern.

5. The total Doppler frequency spread $(4v_r/\lambda)$ due to a radial velocity $v_r$ between the radar and any scattering element on the illuminated surface is small relative to the frequency spread of the envelope of the transmitted pulse $(2/T$, where $T$ is the width of the transmitted pulse).

These assumptions are generally tenable over open ocean surfaces. However, care must be taken in selecting the averaging time to ensure that the surface statistical homogeneity is satisfied. Care must also be taken to avoid including any echoes returned from coastal areas which lie within the slant range of the altimeter. For non- ocean surfaces, it is not generally the case that these assumptions are valid, and hence an alternative approach must be sought.

## Applications of the Brown model

Since the Brown Model fully defines the shape of the echo waveform from an ocean surface, it can be implemented in the on-board tracking software to directly output estimates of the three principal parameters which are of geophysical interest:

- $\sigma_s$, the standard deviation of the Gaussian distribution used to represent the probability distribution of the height of surface scatterers (which is effectively the same $\sigma_c$ this case)

- $\sigma_{\text{eff}}^0$, the effective $\sigma_0$, equal to $\sigma_0(\Psi)/L_p$

- $\tau$, the time delay.

We can also extract mispointing (the angle of the boresight to mean surface normal), and noise, but this is computationally intensive, and not normally carried out within the on-board parameter estimation software.

The tracking software fits the return echoes to the model by adjusting these three parameters using a Sub-optimal Maximum Likelihood Estimator (SMLE) algorithm. The parameter values are telemetered to the ground and can be used directly, after processing has taken the various corrections described previously into account.

The SMLE uses simplified gates approximating to weighting functions to examine just the significant portion of the return signal when evaluating each parameter. The results can be improved during post-processing by implementing a true MLE algorithm using the exact weighting functions, but this is a computationally intensive task which is only justified in special cases.

## Physical interpretation of Brown model parameters

The section describes the derivation of the 'on-board' estimates of range, significant wave height, and effective backscatter obtained through the implementation of Brown Model based tracking software. Note that some details of these derivations are mission-specific, depending on the philosophy and logic of the on-board software. The following discussion strictly relates to the ERS series of altimeters only.

The geometry of a radar altimeter is such that it receives echoes only as specular returns from scattering facets normal to the incidence angle of the pulse. Over a rough ocean surface, such facets are assumed to be randomly distributed throughout the footprint, and at any instant only a certain proportion will be oriented in such a way to reflect energy back to the altimeter. Therefore the echo wave form from single transmitted pulse from such a surface will tend to be rather noisy. The signal to noise ratio can be improved (by a factor $1/\sqrt{n}$) by summing a number $n$ of discrete wave forms, each made up of echoes received from different facets throughout the footprint (Figure 8). It is important that $n$ is large enough to satisfy the assumptions constraining the use of the Brown Model. For ERS-1, 50 individual wave forms are summed for each telemetered 20Hz waveform. Data products designed for oceanographic applications generally average the 20Hz data to 1Hz, so that $n = 1000$.

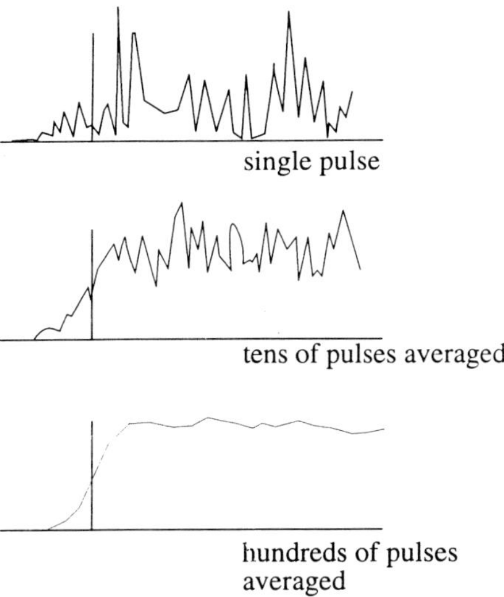



**Figure 8.** *A single radar altimeter echo wave form shows a lot of noise. This can be removed by averaging. ERS-1 averages 50 pulses on board prior to telemetering the data, so individual pulses are not retrievable.*

Of the three SMLE parameters, $\tau$ is the most straightforward, and its conversion to surface elevation has already been covered. However, some supplementary corrections need to be taken into account due to the topography of the sea surface. Since the sea surface profile is not sinusoidal, but made up of flattened troughs and steeper crests, more echoes are received from the troughs than the crests, leading to the so-called electromagnetic, or em bias, which is an overestimate of the range to the mean sea surface (Figure 9). Methods for correcting this effect are a topic of on-going research, and no consensus has yet been reached. The expression used in ERS-1 off-line product processing for the 'OPR' product is:

$$h_{\mathrm{em}} = -\frac{1000\lambda}{8} \times (SWH)$$

where $\lambda$ is the skewness coefficient, calculated from:

$$\lambda = 0.25 \times (SWH)^{-0.28}$$

The magnitude of the correction is generally approximately 5% of significant wave height ($SWH$). The presence of waves on the sea surface increases the area of the effective footprint, which introduces a further 'tracker bias' which increases with roughness. The overall effect of sea surface roughness is sometimes expressed as 'sea state bias', encompassing these two.

The $\sigma_s$ and $\sigma_0$ parameters respectively allow the derivation of wave height and wind speed. We have seen that $\sigma_s$ , a measure of wave form leading edge slope, is related to the probability distribution function of the height of surface scatterers, which in this

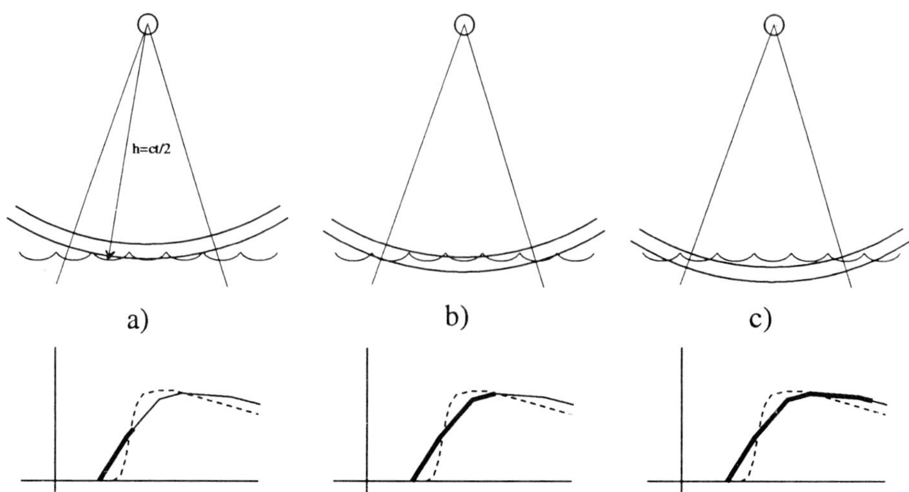

**Figure 9.** *The slope of the leading edge of the echo wave form is related to wave height, due to the time interval between the leading edge of the pulse intersecting the wave crests (a), and fully illuminating the troughs (b to c)*

case are facets distributed across the ocean waves. Since these facets are distributed in height, the incident pulse does not illuminate them all simultaneously, but rather over a period of time as the pulse leading edge travels down from the mean height of the crests to that of the troughs. The echo wave form shape is thus modified as it takes longer for the peak return power to build up. The slope of the leading edge is thus a function of wave height (Figure 10). Conventionally significant wave height ($H_s$) is defined as four times the standard deviation of sea surface elevation, which is approximately equal to the mean height of the highest third of the waves. It is related to leading edge slope as $H_s = 4\sigma_s$.

Backscattering energy from the sea is strongly related to the speed and direction of the wind. The effect of wind on the ocean surface is to generate very small capillary waves. Microwave wavelengths are resonant to these capillary waves, which are therefore much more dominant in the backscattering process than the large underlying ocean waves. Since surface roughness causes facets to tilt away from the normal to the incident pulse, increasing capillary wave activity reduces the mean received signal strength. There is therefore an inverse relationship between surface roughness and signal strength. Radar altimeters include a automatic gain control (AGC) which is used to keep the sum of the powers received across all bins in the range window constant. In ocean-mode operation on ERS-1, the AGC is governed by the Brown Model, so that the parameter produced by the tracking loop is $\sigma_0$. Expressions have been developed to relate $\sigma_0$ to the mean square surface slope $\bar{s}^2$, for example:

$$\sigma_0 = \frac{|R_0|^2}{\bar{s}^2} \sec^4 \theta \exp\left[-\frac{\tan^2 \theta}{\bar{s}^2}\right] \tag{15}$$

used by several authors and quoted by Robinson (1985), where $R_0$ is the Fresnel reflectance of the air-sea interface at normal incidence, which is a function of frequency,

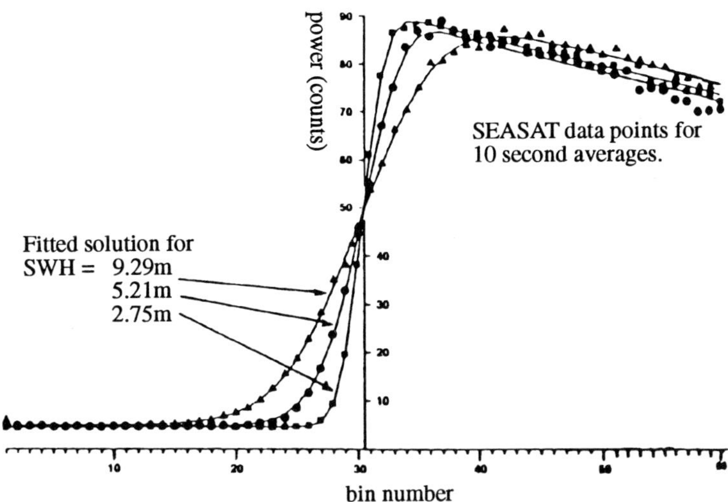


**Figure 10.** *Echo wave form leading edge as a function of increasing wave height. Adapted from I S Robinson, Satellite Oceanography.*

water temperature and water salinity. The relationship between mean-square slope and wind speed has been known for some time, and is given as $\bar{s}^2 = 0.003 + 0.00512U_{12.5}$, where $U_{12.5}$ is the averaged wind speed at 12.5m above sea level. However, recent research has favoured more complex models, where the influence of different mechanisms can be accounted for as a function of wind speed. Most off-line processing algorithms use look-up tables relating $\sigma_0$ to $U_{12.5}$, based on such models. Accurate determination of windspeed requires that any variation in $\sigma_0$ is a function of surface roughness only, so it is important to correct for mispointing prior to making any calculations. It should be noted that the value of $\sigma_0$ per unit area is entirely dependent on the surface characteristics and the radar equation. The parameter actually estimated by the altimeter includes some uncertainty due to the true antenna pattern, and care should be taken when comparing values of $\sigma_0$ or windspeed derived from different missions.

## Limitations of Brown model in ocean applications

Although the above parameters are widely applicable for oceanography, certain sea states are less well estimated by Brown Model based tracking. Where is sea surface is very calm, much of the reflected energy will be received from near-nadir, producing a more narrow, high peaked wave form which departs from the model. Another problem is the presence of sea ice, which seriously contaminates the wave form, both due to it's relative smoothness and it's much lower backscatter. In coastal regions areas of land can enter into the slant range, again disrupting the wave form. In these cases, the tracker cannot be expected to behave normally, and errors occur in the estimates. When these becomes acute, such a tracker will often loose lock. User products are generally quality controlled, and such data are filtered out. This can lead to problems in some areas, for example the Mediterranean Sea, where a large amount of data is lost to the filtering

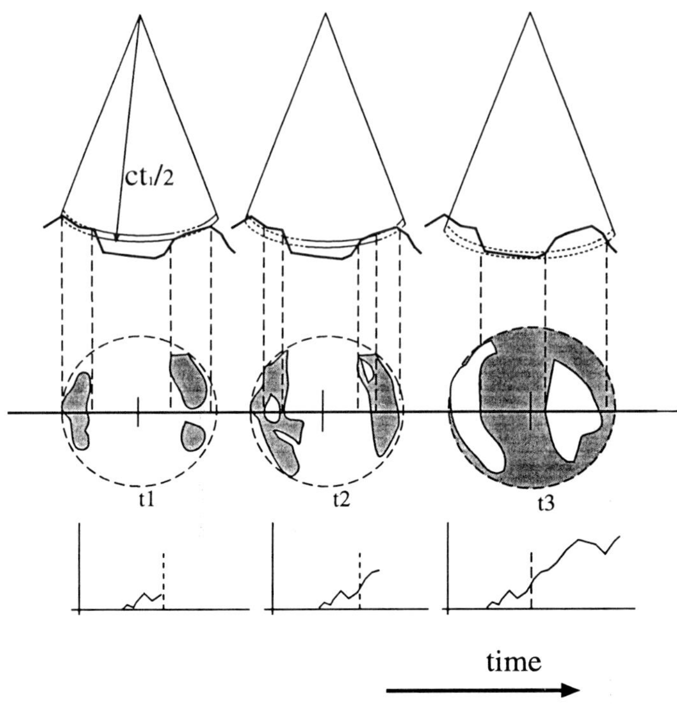



**Figure 11.** *An example of echo wave form build-up over a surface with significant topography. Three snapshots illustrate the progression of the pulse as it intersects various parts of the surface (top row), the area progressively illuminated (second row) and the echo wave form(bottom row). The beam-limited footprint is illustrated by a dotted circle. Note that the topographic profile in the top row corresponds to the topography at nadir, but we include the influence of some across-track topography in the second row.*

process. Specialised reprocessing of the raw wave form data is necessary for studies of such areas.

## 2.3   Non-ocean surfaces

We have seen that for ocean surfaces a well understood analytical model of radar altimeter echoes can be constructed. This is based on assumptions about the homogeneity of the scattering surface, which do not hold for non-ocean surfaces. The echo wave form from a radar altimeter pulse over land is comprised of reflections from an essentially static, non-homogeneous surface, with various types of cover within the beam-limited footprint, each exhibiting different reflection characteristics. The concept of a mean surface, which is quite straightforward over ocean, then becomes more tenuous. However, radar altimeters do record large amounts of data over non-ocean surfaces, and various methods of extracting surface elevation have been used. Historically these have been heuristic, but new techniques allowing the determination of analytical solutions for at least some surfaces are being developed.

It is often the case over non-ocean surfaces that the height distribution of scatterers is such that echoes are received from targets spread over a wide range, limited by the beam. The concept of the pulse-limited footprint becomes less straightforward when the height distribution is large in relation to the pulse width. The total energy received at the altimeter is the same as for a flat surface with identical mean reflectivity, but the area illuminated by the pulse is not necessarily contiguous (Figure 11).

A second complication is that for non-ocean surfaces the radar pulse can sometimes penetrate the surface, giving rise to a volume scattering component which is convolved with the surface scatter.

## Retracking

Retracking is the generic term usually given to the heuristic techniques used to retrieve information from the wave form. In general, the objective of retracking is to relate some point in the range window to mean surface. The particular approach taken depends on whether *a priori* assumptions can be made about the surface characteristics. The output of a retracking algorithm is a figure to be added to the onboard estimate, derived either by a Brown Model fit as described above (and hence inappropriate) or by some other form of tracking. The simplest form of retracking is to assume that the first return represents the surface. For a very specular surface this may be reasonable, but generally one has to assume that the shortest slant range is not necessarily at nadir. A simple retracker, the offset centre-of-gravity (OCOG) algorithm, is illustrated in Figure 12. In this algorithm, the wave form is characterised using two parameters, amplitude and width, where the amplitude is defined as twice the height of the centre of gravity of the wave form, and the width to be that of a rectangle whose height is the amplitude and whose area equals the area under the wave form, as follows:

$$\text{Amp} = \sqrt{\frac{2\sum_{i=1}^{n}(0.5r_i^4)}{\sum_{i=1}^{n} r_i^2}}, \qquad \text{Width} = \frac{\left(\sum_{i=1}^{n} r_i^2\right)^2}{\sum_{i=1}^{n} r_i^4}, \qquad \text{CoG} = \frac{\sum_{i=1}^{n} i r_i^2}{\sum_{i=1}^{n} r_i^2}$$

where $n$ is the number of bins.

Using squares of the bin values reduces the effect of pre-leading edge noise, and reducing the retracking bias. The retracked range is then defined as the position of the waveform leading edge at some threshold defined by some percentage height (or power) of the rectangle, often 50%, or alternatively as the leading edge of the rectangle. This type of retracking has the advantages of being robust and consistent for noisy, non-ocean like wave forms, and for making no *a priori* assumptions about the surface.

More complex retracking algorithms are used for more specific cases. A recent example is that developed by Curt Davis, which is designed to retrieve 6 parameters of snow covered surfaces, for example the Greenland and Antarctic ice caps, from the Geosat radar altimeter waveform data. These parameters are:

| | |
|---|---|
| $(DC)$ | Altimeter DC bias |
| $\sigma_s$ | r.m.s. surface roughness |

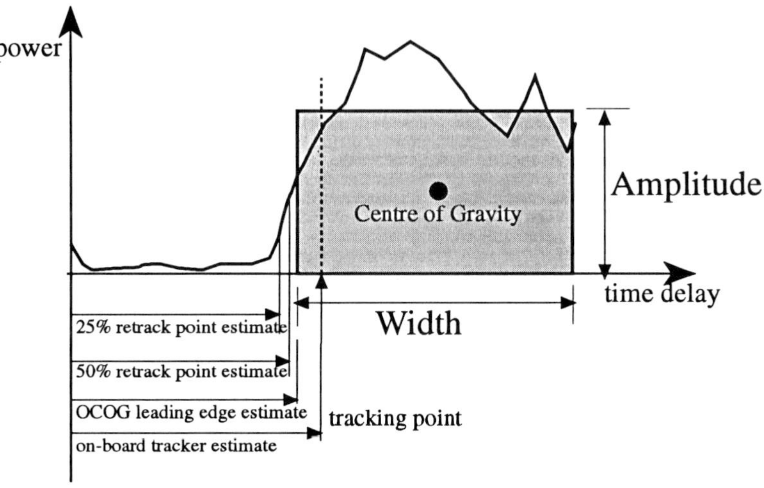



**Figure 12.** *Schematic diagram of the* OCOG *retracking technique. The various alternative tracking points, defined at the time at which a certain percentage of the integrated echo power has been received, are selected by the user based on assumptions about the properties of the scattering surface.*

| | |
|---|---|
| $n'$ | mean surface location |
| $A_m$ | maximum waveform amplitude |
| $K$ | volume-scattering coefficient |
| $k_e$ | extinction coefficient of snow |

The algorithm attempts to fit waveforms to a model which includes components both for surface scattering, and for volume scattering from radiation penetrating the surface. The combined model is given as:

$$SV(n) = (DC) + \frac{A_m}{s_2}\left\{ S(n) + \frac{K}{s_1}V(n)\right\}$$

where $s_1$ and $s_2$ are normalising factors. The expressions for the two components are given as:
Surface Scattering:

$$S(n) = \begin{cases} \frac{1}{2}\left[1 + \mathrm{erf}\left(\dfrac{g_{2s}(n - n')}{\sqrt{2}\sigma_c}\right)\right] & \text{for} \quad n < n' \\[4ex] \frac{1}{2}\exp\left[\dfrac{-4c}{\gamma h}g_{2s}(n - n')\right]\left\{1 + \mathrm{erf}\left(\dfrac{g_{2s}(n - n')}{\sqrt{2}\sigma_c}\right)\right\} & \text{for} \quad n \geq n' \end{cases}$$

Volume Scattering

$$V(n) = \exp\left[\frac{\beta_\tau^2}{t^2{}_0\beta^4} - \frac{2g_{2s}(n - n')}{t_0\beta^2}\right] - \exp\left[\beta_\tau^2 k_e^2 c^2{}_s - 2k_e c_s g_{2s}(n - n')\right]$$

where $n$ is the range bin number and $g_{2s}$ is a constant that converts from range bin number to time. This combined model is non-linear, and must be solved by an iterative method.

### Migration

Retracking, whilst simple to apply, has some serious drawbacks. An echo wave form is built up from many echoes from individual targets, and the closest of these is not necessarily at nadir. Therefore height measurements based either on an on-board or post-processing analysis of the wave form are biased towards the topographic maxima within the beam. It can be shown that a lower limit below which topographic cannot be resolved exists, which is a function of the beam width, and is typically of the order of 5km over undulating terrain. There is also a problem of non-uniqueness: several targets may exist within the footprint which are at the same slant range. It is not possible to resolve azimuth from radar altimetry, so it is not possible to locate the source of the echo to a better precision that the area of the footprint. This is not a problem for ocean altimetry, but for non-ocean applications, for example topographic mapping, it is a serious limitation. A method of overcoming this obstacle is to analyse sequences of wave forms so that individual targets can be tracked and correlated over a longer distance. Such techniques are commonly used in the analysis of seismic sounding data, and are referred to as 'migration'. Figures 13(a) and (b) show the advantages that can be gained from migration. The migrated data in Figure 13(b) show much more small-scale detail than the retracked profile in Figure 13(a). It has been shown that the theoretical lower limit horizontal resolution that can be achieved from migration is equal to the along track spacing of the wave forms, which for ERS-1 is approximately 330m.

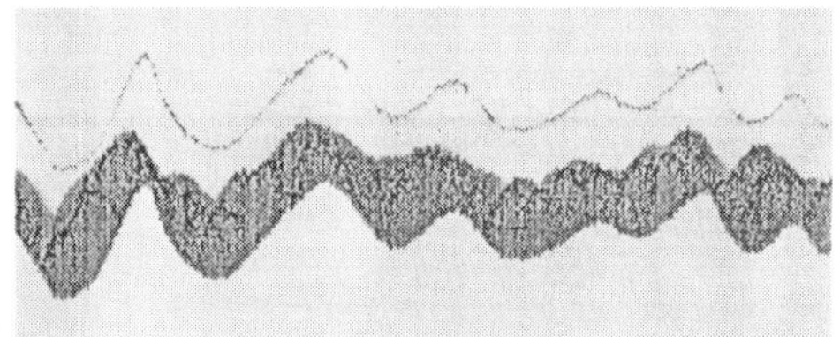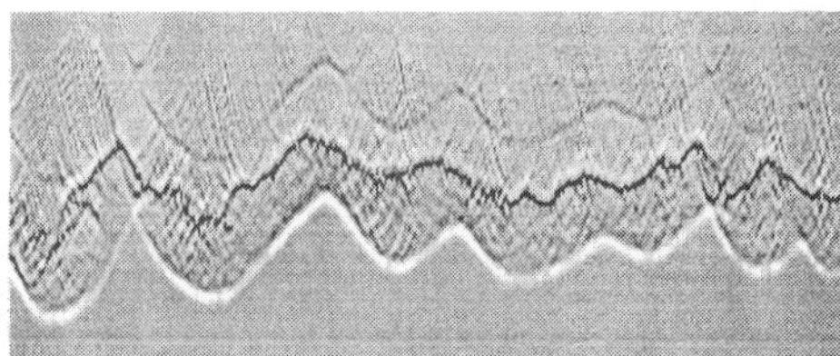

**Figure 13.** *(a) A 'Z-Scope' representation of a series of echoes received from a profile of Antartica. Note the lack of Short-wavelength resolution. (b) The same profile derived from a 2D migration technique. Note the much better short wavelength definition. The poor definition of some sections, for example at the left of the profile results from significant across track topography*

However, migrating a single profile is not sufficient to determine the surface uniquely, since echoes are also received from topography lying out of the plane of the section. Across track surface gradients also produce height biases. These problems can be reduced by including neighbouring profiles in the migration. This does require a densely spaced set of profiles, which can only be acquired by a satellite in a very long repeat period orbit. The ERS-1 'geodetic phase', with a 168 day repeat period, should satisfy this requirement for large areas of the polar ice caps, and the development of suitable three-dimensional methods is a topic of ongoing research.

## 2.4  Imaging altimeters

The pulse-limited radar altimeter was originally designed for use over ocean, and, as we have seen, extending it for use over land can be problematic due to the relatively wide beam widths used. The concept of imaging altimeters has been put forward, which use scanning beam-limited or synthetic aperture technology, with the objective of improving resolution.

# 3  Oceanographic and geodetic parameters

## 3.1  Introduction

The principal scientific purpose of radar altimeters has been to provide information about ocean dynamics and the marine geoid. Here we discuss the detailed processing necessary to optimise results in these areas, and provide some examples of recent applications. When using radar altimetry to study ocean surfaces, it is generally assumed that the Brown model for ocean returns provides an accurate description of the echo wave form. Since this is implemented in on-board tracking software, on-board estimates of range and related quality flagging are taken to be reliable, and only in special cases where the surface departs from the assumptions is any wave form reprocessing applied. Accurate determination of the satellite's radial position is extremely important, due to the small scale of the features of interest. The reduction of orbit error is carried out by statistical analysis of repeat orbits.

One type of area where on-board estimates cannot be used is ice-covered ocean. Special techniques have been developed to extract surface elevation from sea ice covered areas, and also to estimate the extent of this ice. More recently, these techniques have been extended to enable gravity field mapping of ice covered polar oceans. We also briefly discuss applications of radar altimetry to studying inland seas and lakes.

## 3.2  Oceanography from radar altimetry

Radar altimetry allows the determination of three types of data over ocean surfaces. The first relates to static ocean topography, or the marine geoid. The marine geoid is defined as the surface along which the Earth's combined gravitational and rotational potential is constant, in other words the surface at which the ocean would settle if all effects other than gravity field were removed. The second type relates to dynamic sea surface topography, that is the departure of the mean sea surface from the marine geoid due to the action of ocean currents. The third type exploits the altimeter's backscatter measurement to provide information on wind fields. The detection of marine geoid and dynamic sea surface topography depends on the range measurement.

## Ocean observables

When the marine geoid or ocean currents are studied, the ocean surface elevation deviations caused by solid-Earth and ocean tides, and by changes due to variations in barometric pressure must be modelled and removed. Having done this, and applied all other corrections to range, the surface we obtain is called the mean sea surface (MSS). This surface agrees closely with the marine geoid, and undulates by up to 100m relative to a best fitting Earth reference ellipsoid, with undulation wavelengths from a few to several thousand kilometres. The differences between the marine geoid and the MSS are of the order of a few metres, and are the result of the surface expression of ocean currents. The difference between the two surfaces is termed the dynamic sea surface topography. Since this is of the order of a few metres only, superimposed on marine geoid undulations of up to 100m in amplitude, it is clear that the altimeter range measurements principally observe the marine geoid, and that the dynamic ocean topography is secondary.

In global ocean circulation, two principal geostrophic components may be identified. A large-scale component associated with the quasi-permanent mean circulation which may be considered time-independent over time scales of the order of several months, and a meso-scale component (of the order of 100–1000km) associated with eddy circulation, which is time-dependent (although often seasonal). Quasi-permanent circulation features such as the Gulf Stream cause local sea surface height to depart from the marine geoid by only 1–1.5m over distances of 100–150km. In order to derive parameters of interest to oceanographers, such as actual water velocity, we need a very accurate measurement of these slopes, which in turn requires an extremely accurate model of the marine geoid. On the other hand, mesoscale circulation introduces variability in sea surface height due to meandering, changes in current velocity or direction, vortex shedding or eddy migration. Because the geoid is time invariant, studies of variations in these systems can be made without recourse to highly accurate geoid models.

Undulations in the Earth's geoid are due to density variations in the planetary fabric. Undulations are defined with reference to an ellipsoid which approximates to the Earth's shape. Present knowledge of the marine geoid is best on long scales and worst on short scales, but is constantly being improved. At long wavelengths (>10000km), geoid undulations are known to better than 5cm, at medium wavelengths ($\approx$2000km) to $\approx$40cm, and at short wavelengths ($\approx$100 km) to no better than 1.5km. At long wavelengths, various types of satellite tracking data have been used to improve the models. At shorter wavelengths, *in-situ* gravitational potential measurements from ships and aircraft are used, as are satellite altimeter measurements themselves.

## Orbit determination

Uncertainty in locating the position of the satellite is a primary limitation on radar altimetry. A satellite's position in space is calculated using a computed set of ephemeris data, which are derived from tracking data from satellite laser ranging stations (SLR), microwave beacon systems such as the French DORIS system deployed on, amongst others, Topex/Poseidon and SPOT-2, or the ESA PRARE system deployed (but failed) on ERS-1, on ERS-2 and one of the COSMOS series, or the US Navy's Global Positioning

Service (GPS). Orbits are modelled using these data, to produce ephemerides which then allow propagation of the observations to any point in the orbit. However, comparing the modelled orbit location at points for which tracking data are available reveals errors which arise from imperfect knowledge of the forces determining the orbit. The satellite orbit depends on the slope of the gravitational field, on air drag, on solar pressure and on Earth reflected radiation pressure. Instabilities are also introduced after orbit manoeuvres, carried out either to return the satellite to its predetermined orbit, or to shift it into a new orbit. Application of the best general purpose gravity model available at the time of the launch of Seasat in 1978, GEM-9, led to radial orbit errors of 3–5m (Wakker *et al.* 1985). By the end of 1993, orbit determination of the ERS-1 mission had improved to such an extent that Scharroo *et al.* (1993) quote a radial error of 13cm using orbits generated with the JGM-2 gravity model.

A number of techniques have been developed to analyse orbit radial accuracy, the detail of which is too complex to cover here. Altimeter data products are supplied with height parameters referenced to specific orbit and mean sea surface models. These are usually referred to by the name of the centre which computed them, for example 'Delft orbit' refers to high precision orbits computed by Delft University of Technology's Section Space Research and Technology. These orbits have quoted RMS residuals, which are typically determined using a combination of techniques. Range residuals from SLR measurements can be calculated by fitting the modelled orbit to the observed data. However, this will only give a good representation of global orbit error when the tracking data are regularly distributed in time and provide a good global coverage. This is never the case in fact, as SLR stations are concentrated in the Northern hemisphere, and coverage has to be shared between various satellites. In these circumstances, SLR residuals cannot give decisive information about the radial accuracy of the computed orbit, nor any information away from the SLR stations.

Several statistical tests can be applied to the altimeter height data themselves to improve this information. The altimeter height residuals with respect to the mean sea surface reference model include frequency components due to orbit error, to imperfections in the mean sea surface model, and to inadequacies in propagation and calibration corrections. However, orbit errors will be concentrated around the lower frequencies, particularly around 1 cycle per revolution, and those due to the reference model are expected to be found at higher frequencies (around 10 cycles per revolution). Secondly, crossover differences can be analysed. This involves computing the altimeter height at the crossover points of ascending and descending orbits (Figure 14). Typically, for each orbit a short arc is taken encompassing about 20 seconds of data centred on the crossover. A third-order polynomial is fitted to this arc, and the normal point sea height is given by the polynomial fit at the crossover point. Note that although we are principally interested in the radial accuracy, the horizontal (along-track and across-track) accuracy limits the precision to which the actual crossover can be determined. The resultant crossover height difference is a good indication of radial orbit error, although, again, part of it is made up from actual sea level change and calibration and correction errors. If orbit errors along ascending and descending passes were completely uncorrelated, then the crossover difference, $\delta h$, could be equally divided between the two passes. However, errors in the gravity field model lead to errors in the orbit which are constant from cycle to cycle, and possibly common to ascending and descending

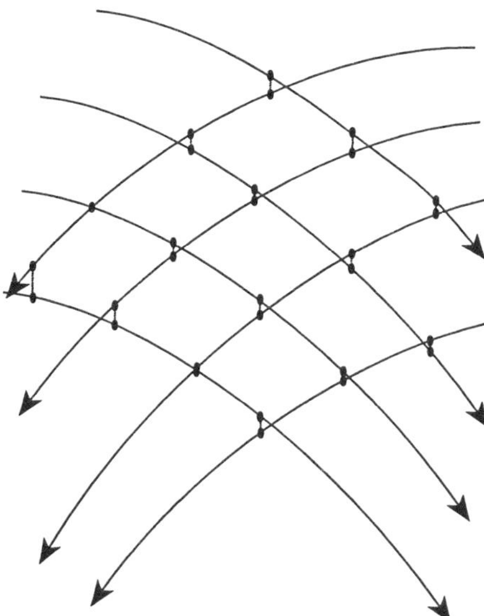

**Figure 14.** *Crossover analysis is the technique used to minimise orbit error by assigning it to the height residuals observed between ascending and descending tracks, then minimising those differences using some technique, e.g. least squares fitting*

passes. Therefore a considerable part of the gravity-influenced radial orbit error is geographically correlated, and cannot be observed by cross-over analysis or any other statistical method. One method of estimating the magnitude of geographically-related orbit error is to use dual crossover analysis, where two simultaneously orbit satellites in different inclinations are used. This technique has been successfully applied to ERS-1 and Topex-Poseidon.

## Sea surface parameters

Several integrated methods for extracting mean sea surface and sea surface variability, whilst correcting for radial orbit error have been put forward, one of which is described by Wakker *et al.* (1993). The procedure is illustrated schematically in Figure 15. The first step in the procedure is to filter the calibrated and corrected altimeter data using limits on selected parameters, such as pulse peakiness and wind speed. Height residuals are then calculated by subtracting geoid heights (using an appropriate model) and the filtered altimeter data from the orbit heights. Next, a crossover minimisation procedure is applied to the whole data set. The crossover residual differences are themselves minimised and the resultant Fourier functions representing the main part of the radial orbit error are subtracted from the sea height residuals.

The next step is to re-sample the corrected sea height residuals on to a reference grid, using distance weighting. The sea heights at the grid nodes are called the altimeter normal heights. The mean sea height is calculated at each point, so that a mean sea surface topography is produced. A topography for the North Atlantic produced from

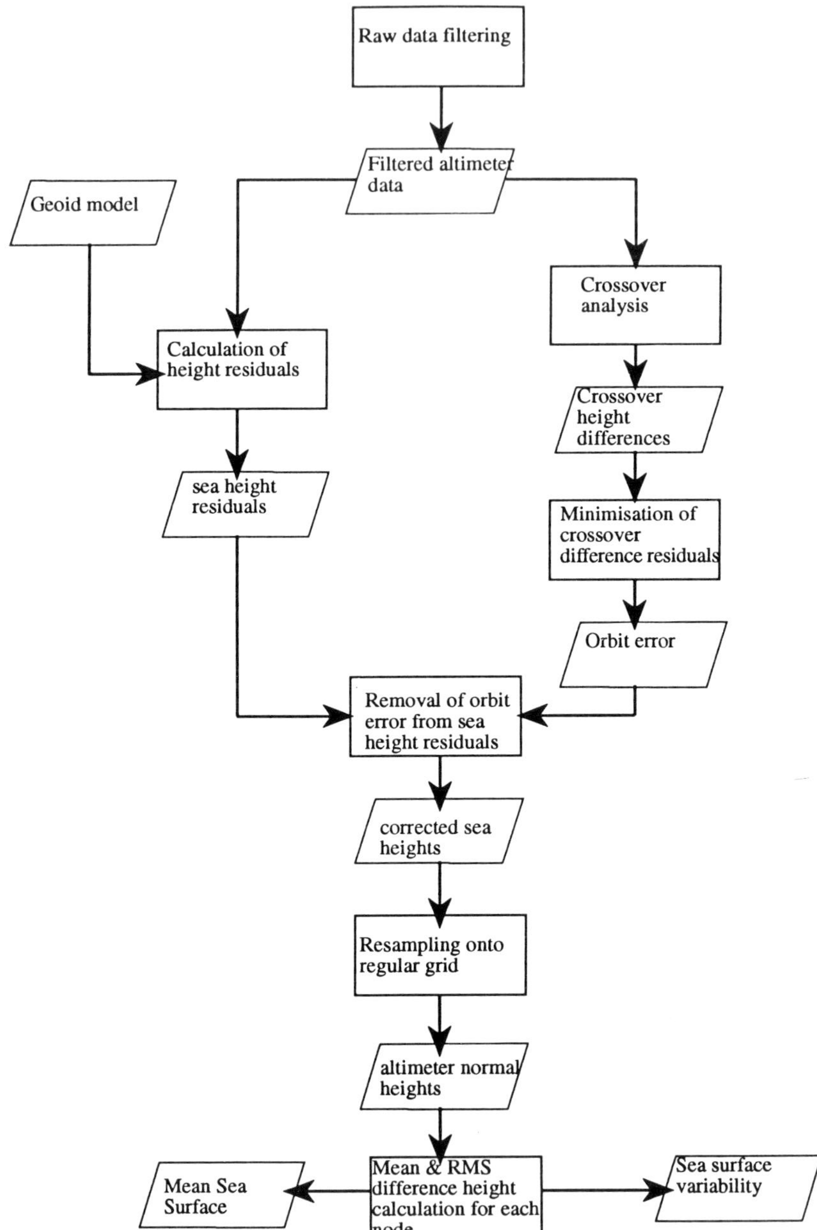


**Figure 15.** *Block diagram schematic of an integrated sea surface height/ orbit error minimisation algorithm*

ERS-1 by Wakker *et al.* is given in Figure 16. The differences from the mean value at each point can be calculated by subtracting the observed value at that point from the mean, and by gridding the RMS of these differences at each grid point, a diagram of sea surface variability can be produced (Figure 17). Finally, the time series of the differences in each grid node can be interpolated in time and space to create grids describing the relative ocean topography at various epochs, revealing the evolution of mesoscale ocean currents. An example of this is shown in Figure 18, which reveals the evolution of the El Nino Southern Oscillation event in the Pacific Ocean, from a combined Geosat/ERS-1 analysis. A remaining problem with the technique of crossover residual difference minimisation is that the solution is non-unique. This is because the surface can be lowered, raised or tilted as a whole without affecting the crossover difference residuals. Mean sea surface models with fine detail can be produced, but they may show long wavelength errors which directly or indirectly result from errors in the gravity model.

The study of specific mesoscale features can also be carried out using co-linear profile analysis. An early example of this is given in Figure 19. Nine co-linear tracks from Seasat were analysed across the Emperor seamount chain and the Kuroshio Extension in the North Pacific. A 2470km Seasat ground track yielded these 9 passes during the 3 day repeat cycle. Sea surface elevations were computed at regular intervals along track (top graph). The sharp peak in these indicates the crossing of the Emperor seamount ridge, which produces a local disturbance of the gravity field, and in turn a short-wavelength sea surface undulation. The constant part of the signal was eliminated by subtracting the mean from each profile, giving the results shown in the second graph. From these a linear trend was removed to absorb orbit and other long wavelength errors. The third graph shows the outcome of this, along with each profile separated by 1m vertically for clarity. Finally the RMS variability between each track is calculated. The Kuroshio Current appears as a 20cm RMS signal at 37.5°N, and a migrating cold ring indicated by the high variabilities at 33° and 35°N appears further South.

## Wave height estimation

Satellite remote sensing has had a major influence on wave research, and has drawn attention to certain areas where knowledge of the basic understanding of the hydrodynamics of waves has been lacking. Until the introduction of SAR, scatterometer and radar altimeter derived data, the widespread assumption was that the sea consists of a uniform train of long-crested waves, as computed by Stokes in 1847. Measuring wave data in situ using waverider buoys or shipborne wave recorders is expensive and restricted to a few sites.

The basic theory for measuring waves from radar altimetry was covered in the discussion of the Brown Model. The theory depends on three assumptions, that the sea surface is Gaussian, that the point target response is Gaussian, and that the specified flat surface response is valid. Work has been carried out to examine the implications of departures from these assumptions. In particular, sea surface skewness, which produces a non-Gaussian surface, has been the subject of considerable work (*e.g.* Srokosz 1985). Much work can be done using the 'Hs' parameter provided in user products, but caution is advised when using on-board estimates rather than ground processed data, as the

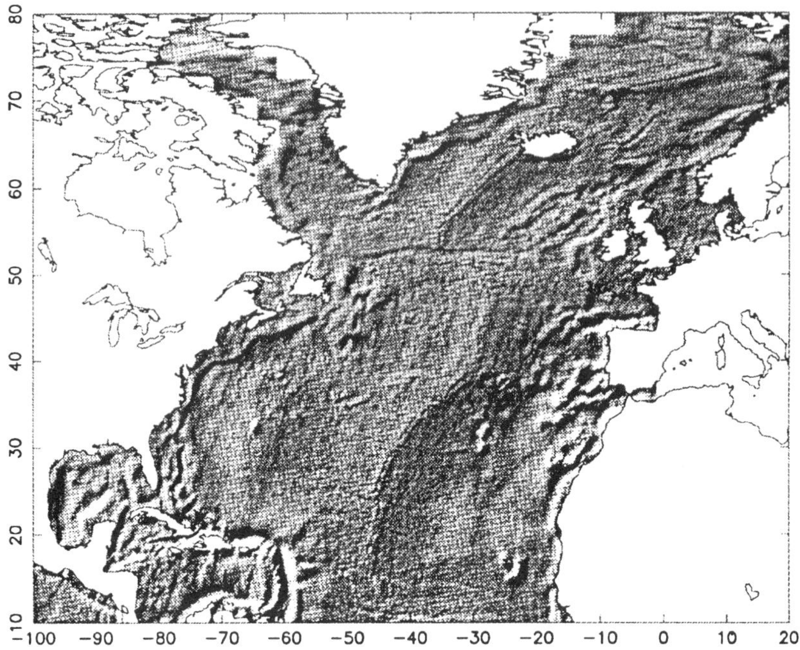

**Figure 16.** *The ERS-1 derived high order mean sea surface for the North Atlantic (from Walker et al. 1993)*

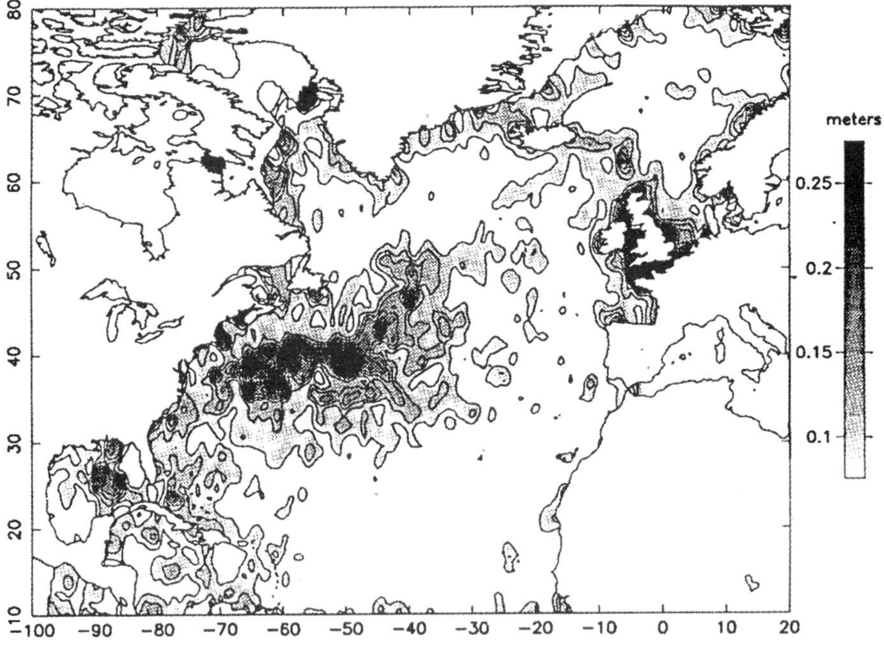



**Figure 17.** *The ERS-1 Sea Surface variability for the North Atlantic, contoured at 2.5m intervals (Walker et al. 1993)*

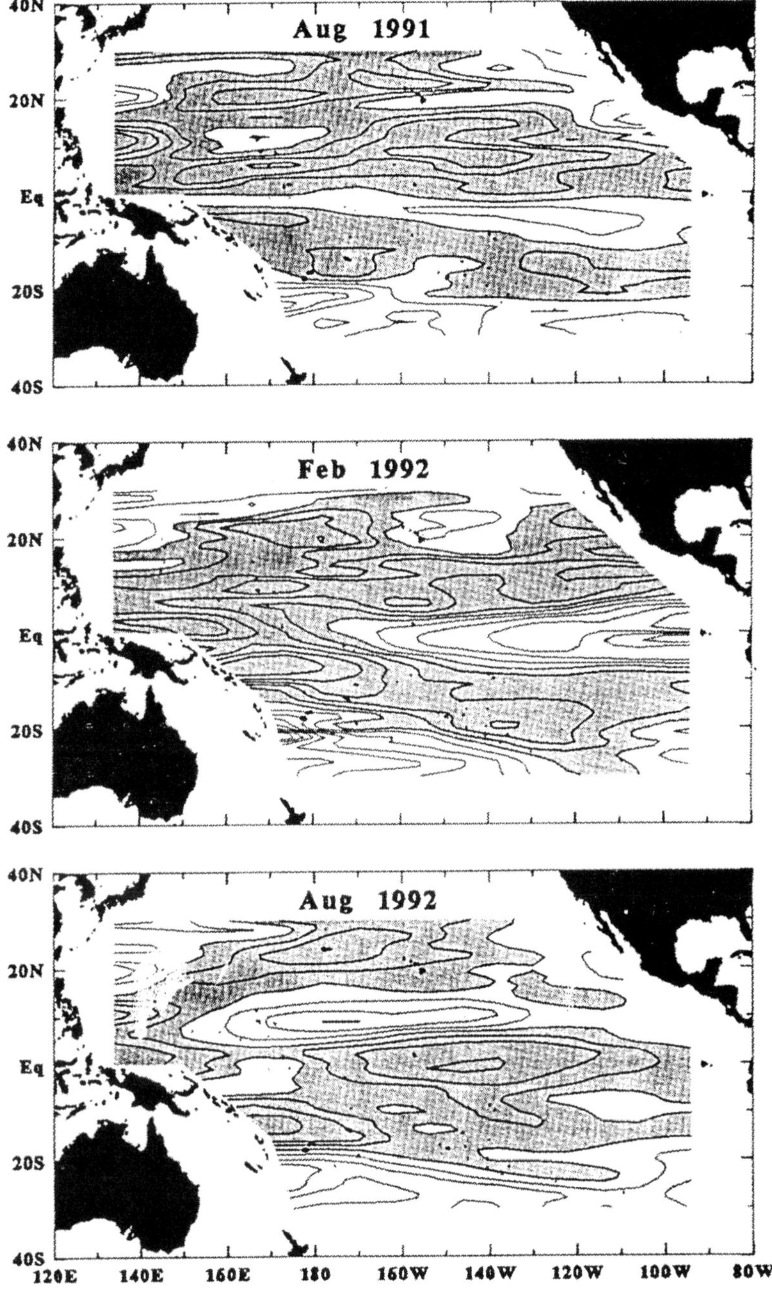



**Figure 18.** *Evolution of the El Nino Southern Oscillation as detected by the Geosat & ERS-1 radar altimeters between August 1991 and August 1992*

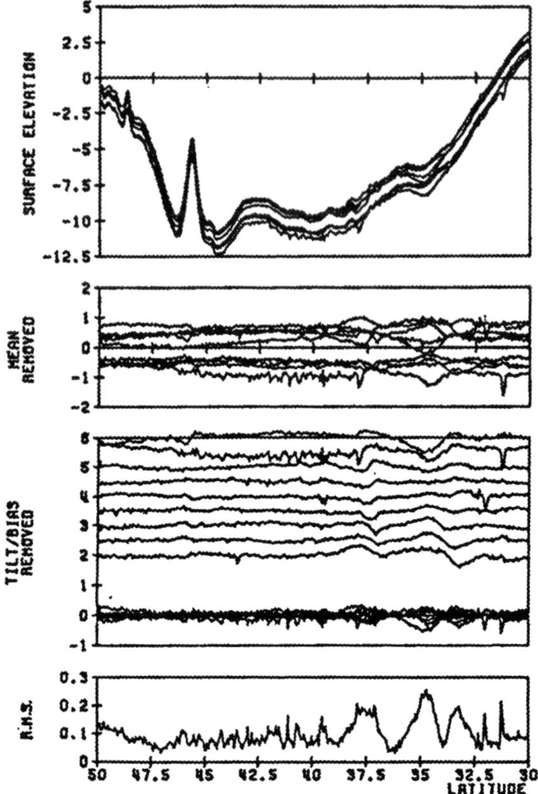


**Figure 19.** *An analysis of nine co-linear Seasat altimeter profiles across the Emperorchain and the Kurushio Extension in the North East Pacific Ocean (Walker et al. 1998)*

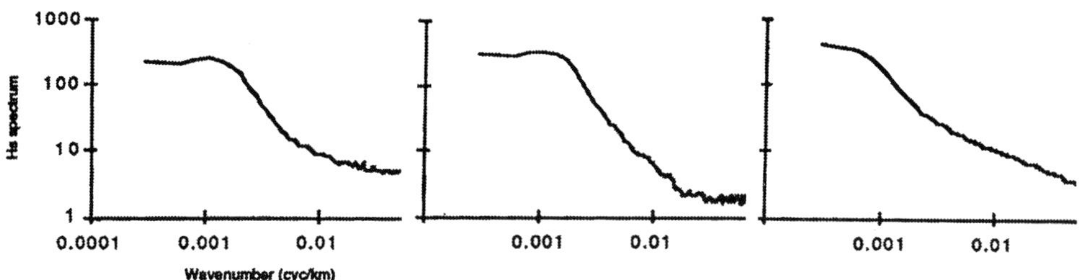


**Figure 20.** *Three average 'Hs' spectra of wave height for the North Atlantic, derived from (left) ERS-1 data, (middle) ERS-1 Fast Delivery data and (right) Geostat data. Note that the Geosat spectrum's relative smoothness is due to a longer time series. Assuming that the average wave spectrum for the region is constant, the differences must be due to instrumental effects (after Challenor 1993)*

former may be compromised by the side-effects of the tracking software.

The radar altimeter data are applied to wave climatology, in which the study of sea surface variability is used to attempt to characterise behaviour in different ocean basins, and to detect links between temporal variability and climate. Sea state variability encompasses wave height magnitude and wave spectra statistics. An example of wave spectrum extraction is given by Challenor (1993) for both Geosat and ERS-1 (Figure 20). Looking at the Geosat data, it is found that the along track spectrum of significant wave height is invariant across the world's oceans. Comparison between the on-board estimate based spectra computed from Geosat and ERS-1 does reveal significant differences, especially at low frequencies. This is thought to be due to instrumental effects, and demonstrates the importance of using re-processed data where possible.

## Wind speed estimation

Wind speed estimation from active microwave instruments is based on the principle that for near nadir incidence angles, backscattered power depends almost exclusively on short-wavelength roughness of the sea surface, which is itself dependent on wind speed. The radar altimeter is only capable of giving wind speed, not direction. Most oceanographic and atmospheric applications of sea surface winds require knowledge of the vector wind field, however for studies of latent heat flux, wind speed is sufficient. Wind direction is provided by wind scatterometers, which have been carried aboard several remote sensing satellites, including Seasat and ERS-1. Altimeter wind speed retrieval is a two step process: first of all, the normalised radar backscatter $\sigma_0$ is calculated from receiver gain (AGC), satellite attitude and satellite height, and corrected for atmospheric attenuation and internal calibration. The precise method of determination of $\sigma_0$ from AGC varies from instrument to instrument, but the general principle holds. The second step is relate wind speed to $\sigma_0$. Although $\sigma_0$ is strongly dependent on wind speed, the relationship is not linear, and the range of wind speeds encountered over ocean is concentrated in a relatively narrow range of $\sigma_0$. Small changes in $\sigma_0$ can be caused by large changes in wind speed. Wind speed retrieval algorithms need to be calibrated against either coincident and contemporaneous surface measurements, or against some other simultaneously operated remote sensing instrument for which a good calibration has been obtained. Ground data is difficult to obtain. Wind speeds given by anemometers on buoys or ships are local instantaneous measurements of the turbulent boundary layer, whereas altimeter wind speeds are a measure of the average over the footprint. Anemometer measurements must be averaged over time to remove high frequency fluctuations. A thorough investigation of wind speed estimation was carried out by Chelton and McCabe (1985) in advance of the Geosat mission. They found several inadequacies in the algorithm used for Seasat, and proposed a new method based on a cross-calibration with the Seasat wind scatterometer (SASS). This was used to relate wind speed at a reference height above mean sea level of 19.5m to $\sigma_0$ (in dB):

$$u_{19.5} = 10^{(0 \cdot 1\sigma_0 - G)/H}$$

where $G$ and $H$ are constants determined through direct comparison with *in-situ* datasets. The histogram of 19.5m wind speeds computed from the full 96 day Seasat mission is shown in Figure 21.

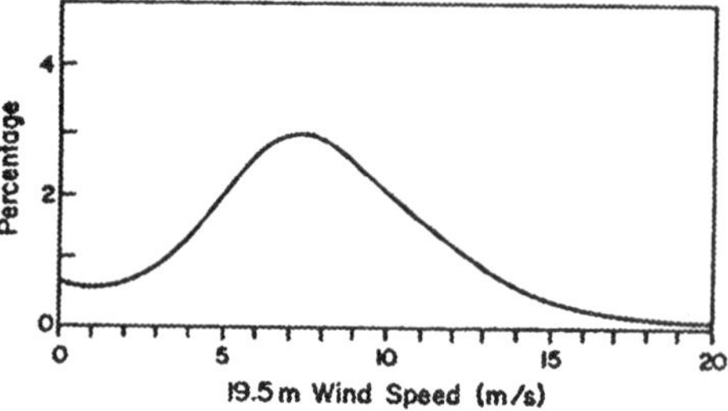



**Figure 21.** *Histogram of global ocean wind speeds referenced at 19.5m above mean sea level calculated by Chelton and McCabe for the Seasat mission (1978)*

## 3.3 Polar ocean applications

### Sea ice detection

The distribution and extent of sea ice is of great importance to climate research, to polar oceanography, and to operators of polar ships. Sea ice forms in high latitudes when the air temperature drops to sea water freezing temperatures. During the winter months, ice builds up rapidly. Some of it survives the winter, leading to multi-year sea ice with a different structure and microwave signature to new ice. Some areas of the world's oceans are never ice free, such as the high Arctic Ocean, the Weddell Sea and various other areas around the coast of Antarctica. Seasonal melt opens up important shipping routes in the Northern Hemisphere, and it is of great economic value to be able to accurately forecast the ice extent through these channels. Early results from GEOS-3 and Seasat demonstrated the feasibility of using radar altimetry to map sea ice; since then Geosat and ERS-1 data have been used both for further research and to contribute to operational monitoring. Data from other remote sensing instruments, such as visible and infra-red imagers and passive microwave detectors have made important contributions, but experience problems with ice-type classification (for example multi year ice represents a greater obstacle to shipping than new ice), with coarse resolution, with cloud contamination and with ice/water discrimination during melt/freeze periods. Radar altimetry, whilst not a universal panacea, can make a big contribution to our knowledge, being particularly suitable for ice edge mapping. Research has also been carried out to develop sea ice type indexing from analysis of radar altimeter wave form shape. Echoes from sea ice have a significantly different shape to those from open ocean. Ice tends to produce specular reflection, so that much less energy is scattered away from the antenna. Echo wave forms from sea ice tend to exhibit high AGC values when compared to ocean wave forms, and to have a sharp, single peaked shape with a rapid decay. A semi-empirical algorithm based on a simple characterisation of the wave form shape was developed for Geosat which produces a number known as the sea ice index (Hawking and Lebanon, 1989).

ERS-1, with it's high inclination, is well suited to sea ice mapping. A series of ERS-1 wave forms showing a series of surface type transitions in the Weddell Sea area is given in Figure 21, including a sequence of typically specular sea ice echoes. A plot of monthly ice edge means computed from a complete year's data is shown in Figure 22.

Certain questions have yet to be resolved. In climate studies, sea ice thickness is as important as extent, and measuring the small difference between sea ice elevation and local ocean surface elevation is a problem which so far has not been solved using radar altimetry or any other space remote sensing instrument.

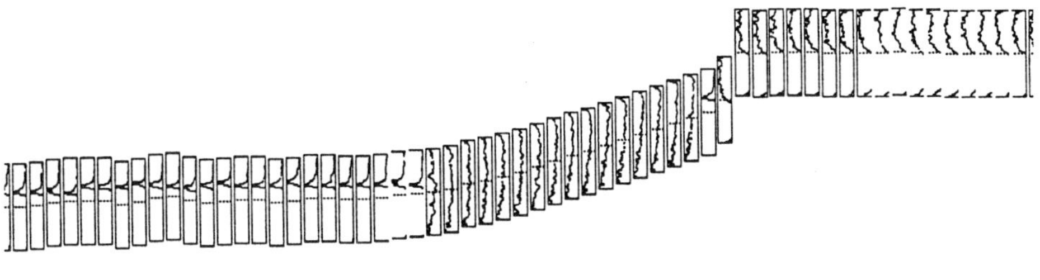

**Figure 22.** *A sequence of ERS-1 wave forms showing a transition from specular sea ice echoes (bordered in light grey), through a chaotic area, probably open water contaminated with small icebergs (dark grey border) and onto an ice shelf where echoes are near ocean-like (black border)*

## Polar marine gravity field mapping

Gravity field mapping using crossover difference residuals has been discussed above. In the polar oceans, as latitude increases, two complications arise. As the across track spacing decreases, it becomes increasingly difficult to minimise the height difference residuals without introducing large cross-track gradients, and also the number of crossover points is such that the computational task would be enormous (Sandwell, 1992). Furthermore, increasing sea ice coverage makes the use of on-board estimates or ocean type off-line processed products inappropriate. Recent work on solving these problems has produced spectacular results. If one is only interested in the horizontal and vertical derivatives of the gravitational potential, rather than the geoid height itself, profile adjustment is unnecessary. The radial orbit error spectrum for Geosat has a large peak at 1 cycle per orbit, and a much smaller one at 2 cycles per orbit, which correspond to 0.8 and $0.15\mu$rad of slope error at once and twice per orbit respectively, which are suppressed well below the noise level of the altimeter (which corresponds to approximately $6\mu$rad of slope error). Therefore taking the along track height derivatives provides a good estimate of the along track geoid slope. These are interpolated onto ascending and descending orbit grids, which are then summed and differenced to produce grids of east ($\xi$) and north ($\eta$) deflection.

The derivative of geoid height $H$ with respect to time $t$ along the ascending profile is:

$$\dot{H}_a = \frac{\partial H}{\partial \theta}\dot{\theta}_a + \frac{\partial H}{\partial \phi}\dot{\phi}_a$$

$$\dot{H}_d = \frac{\partial H}{\partial \theta}\dot{\theta}_d + \frac{\partial H}{\partial \phi}\dot{\phi}_d$$

where $q$ is geodetic latitude and $\phi$ is longitude, and it can be assumed that at crossovers. From this we can calculate the east and north geoid slope:

$$\frac{\partial H}{\partial \phi} = \frac{1}{2\dot{\phi}}(\dot{H}_a + \dot{H}_d)$$

 David Mantripp

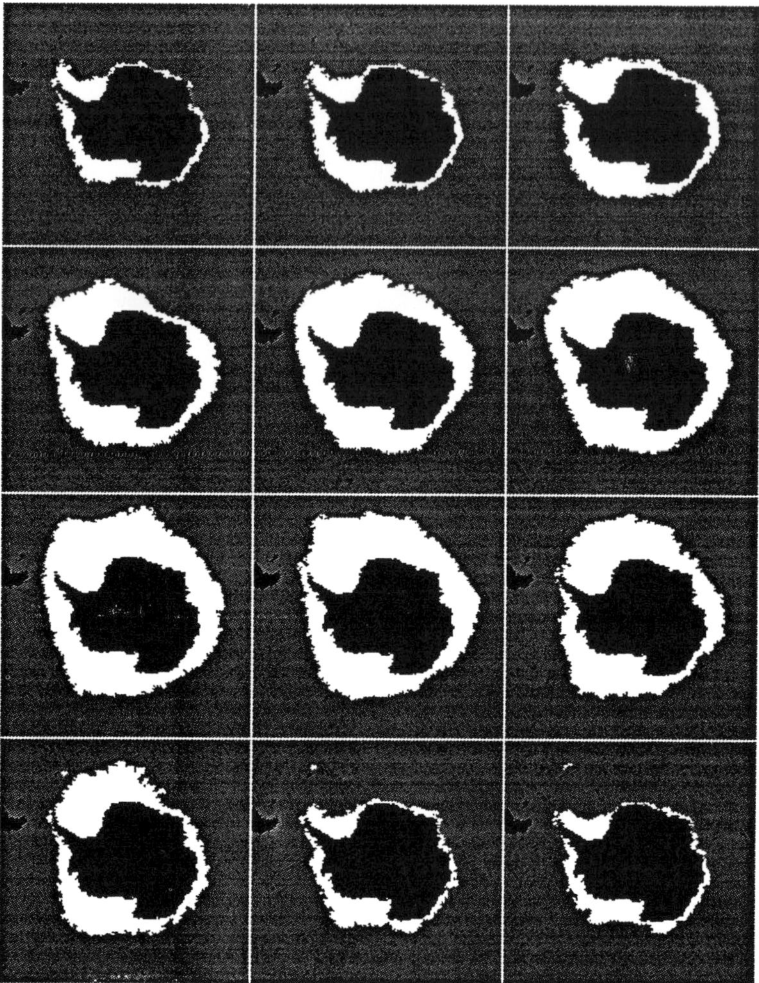

**Figure 23.** *The annual growth/decay cycle of Atlantic sea ice mapped by radar altimetry (courtesy of S W Laxon, MSSL)*

$$\frac{\partial H}{\partial \theta} = \frac{1}{2|\dot\theta|}(\dot H_a + \dot H_d)$$

The east and north components of vertical deflection are related to geoid slope by the following (Heiskanen and Moore 1967):

$$\eta = -\frac{1}{a\cos\theta}\frac{\partial H}{\partial\phi} \xi = -\frac{1}{a}\frac{\partial H}{\partial\theta}$$

Geoid height, gravity anomaly and gravitational potential can all be considered as deviations from some reference earth model such as a spherical harmonic expansion. Using Laplace's equation, the vertical gravity gradient can be calculated from the vertical deflections:

$$\frac{\partial \Delta g}{\partial z} = -g_0 \left( \frac{\partial \eta}{\partial x} + \frac{\partial \xi}{\partial y} \right)$$

Fourier analysis is then required to construct gravity anomalies.

Recent work published by Laxon and McAdoo (1994) takes this approach to derive a map of the gravity field of the Arctic Ocean, an area for which little data has been available before. This was made possible by the orbital inclination of ERS, which takes it to higher latitudes than other altimeter missions. The presence of sea ice at high latitudes confuses the on-board processing, and reprocessing of wave form data is necessary outside of areas of open ocean. A leading edge retracking algorithm is applied, after which outliers are removed as are data points which fail quality checking. Prior to calculating the height profiles, a low pass filter is applied to smooth signals with very short wavelengths.

## 3.4   Inland water applications

Radar altimetry can also be used to study inland water bodies, such as lakes and enclosed seas. Monitoring of closed (*i.e.* with no significant surface or subsurface outflow) and climatically sensitive open lakes can provide useful information for hydrological applications and, over longer time scales, climate change. Since lake volumes respond to changes in precipitation integrated over their catchment basins, they can act as important indirect indicators of changing long term weather patterns. The Mullard Space Science Laboratory has a program for monitoring the relative changes in volume of several hundred lakes worldwide (Birkett, 1994). Lake areas are estimated using imaging radiometer data, and altimetry provides the level data. The footprint size of pulse-limited radar altimeters limits the minimum area of lakes monitored to 100 square km. The data used in studies to date have been taken from Geosat GDR products. These provide height estimates at 10Hz based on Brown model fitting, and hence care must be taken to ensure that quality is good. To determine relative height changes, collinear height differencing was used. This technique resamples data from all passes along a particular ground track at a set equally spaced points. This mean height profile is then used as a reference to estimate height differences for the successive repeat passes. Results have been validated against contemporaneous gauge data available for the Great Lakes. Figure 24 shows a gauge/altimeter comparison across Lake Ontario, where the altimeter data are reference to a University of Texas precision orbit with a quoted global mean accuracy of 17cm RMS. The two datasets match to within 10cm, which demonstrates that the altimeter data can be reliably used for lake level monitoring.

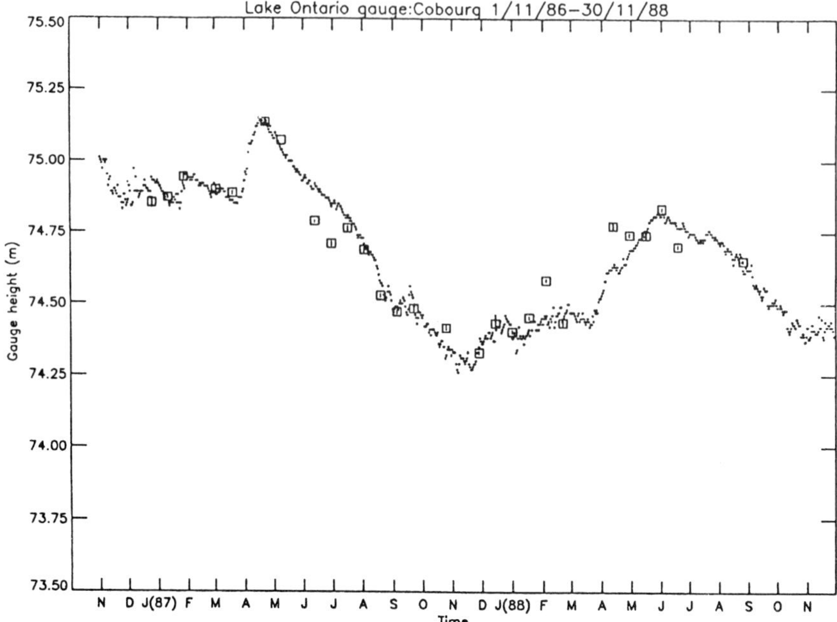


**Figure 24.** *A comparison between ground gauge data and Geostat altimeter data for Lake Ontario. The gauge data (small dots) have been referenced to the International Great Lakes Datum (9155). The Geostat results (squares) are derived from a collinear track analysis. (Birkette, 1994)*

# 4 Applications to topographic mapping and surface characterisation of land and ice

## 4.1 Introduction

Despite the radar altimeter's prime purpose of retrieving geophysical parameters from ocean surfaces, it has been shown that many applications exist over other terrain. Significant results have been achieved in, for example, the areas of polar glaciology, river and wetlands hydrology and digital terrain modelling. Here we consider some of the data processing considerations arising from using radar altimeter data over non-ocean surfaces, and some examples of the results which may be achieved.

## 4.2 Data quality assessment

As we have seen, the radar altimeter experiences difficulties when attempting to track surfaces with significant topography. Also, the effective footprint increases in diameter with increasing surface roughness. Over non-ocean surfaces, we cannot assume that the backscattering properties of facets within the footprint are homogeneous—for example, in a marshy area, echoes from streams and ponds will be much more powerful than echoes from land. It is therefore necessary to take great care in the quality assessment

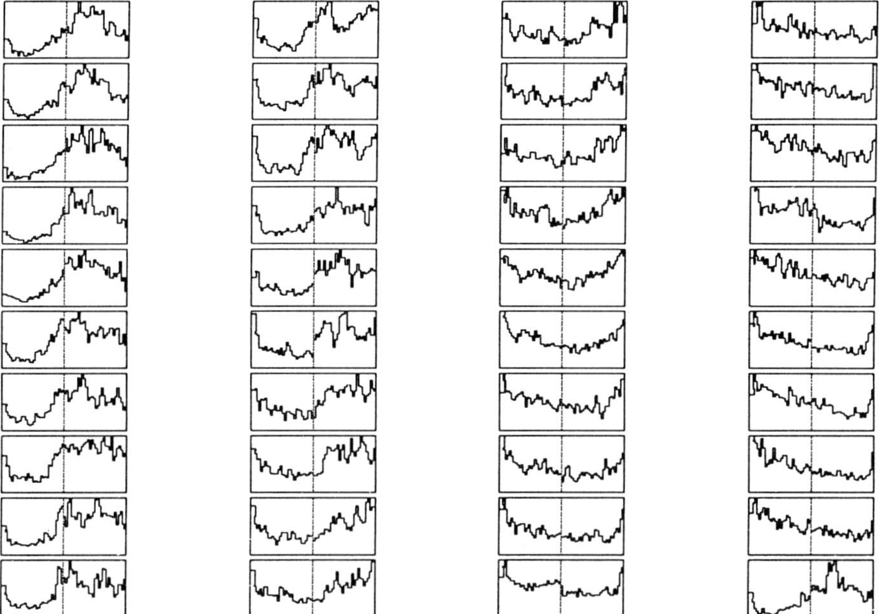

**Figure 25.** *A sequence of echo wave forms recorded by the ERS radar altimeter over Antarctica. The sequence is arranged in columns, from left to right. Note that the leading edge discernible in the first few wave forms is gradually lost as the altimeter loses track of the surface, probably due to a complex surface topography. The altimeter software is designed to carry on recording data for a while, in the hope of regaining a surface without having to drop into acquisition mode. The final wave form in the sequence (bottom right) shows a leading edge re-emerging. However, most of these wave forms do not yield information allowing the determination of a unique surface, and must be discarded.*

of the data, and it is generally the case that applications over non-ocean surfaces require the use of wave form data.

The most foolproof way of quality assessing wave form data is to look at each and every wave form in order to check that is can provide a suitable range estimate. This is a very time consuming process, but it can be accelerated by running automatic checks to remove obviously unsuitable data in the first instance. Most altimeter data products include a 'tracking flag', but this is not necessarily reliable, as criteria for declaring 'loss of lock' tend to be fairly lenient, so as to increase the amount of data recorded in marginal areas (Figure 25). Various techniques have been used for filtering out unsuitable wave forms, and one example is that used by Scott *et al.* (1994) in their comparison of the performance of the ERS radar altimeter's performance in ice and ocean modes over non-ocean surfaces. This method, illustrated in Figure 26, is a heuristic test used to determine if a leading edge is present in the range window. Other

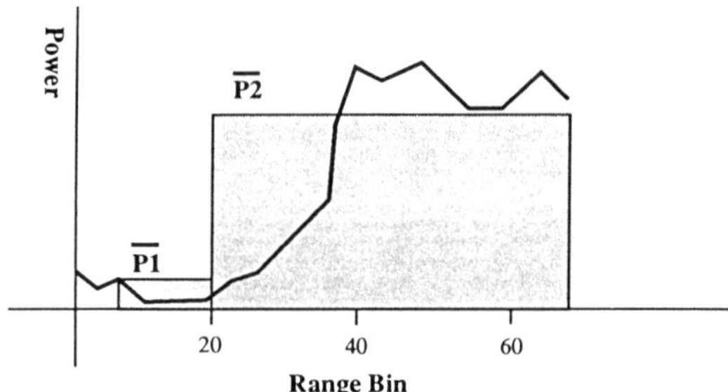


**Figure 26.** *The heuristic quality assessment method. The means of the range window bin values are calculated from 5 to 19 (P1) and 20 to 64 (P2). A leading edge is deemed present if P1 ≤ 0.25 P2.*

methods based on thresholding, on analysing wave form shape, or fitting to models are used. It is important for the user to determine which method is best suited to the application.

## 4.3   Arid and semi-arid deserts

One third of the Earth's land area is arid or semi-arid. Information on the surface topography, characteristics and features of these regions is difficult to obtain due to their general inaccessibility. Satellite radar altimetry provides measurements of surface elevation, surface roughness and surface backscatter characteristics. The elevation profiles reveal many topographic features not recorded on existing maps; surface roughness can be interpreted in terms of dune height; surface backscatter coefficient is related to small-scale roughness, vegetation cover and moisture.

Altimeter data therefore have an application both in mapping hitherto poorly-known regions, and in monitoring change taking place in these regions. Conversely arid regions can provide reference surfaces for radar altimetry, acting as a natural target for external backscatter calibration and orbit error reduction.

It is generally difficult to carry out an absolute radiometric calibration of radar systems, and such calibrations are usually based on pre-launch engineering measurements of the instrument's characteristics. Calibration of backscatter across missions has usually involved the equalisation of histograms of ocean backscatter measurements over time. However, this has drawbacks since for best results overlap of the missions is required, or else it must be assumed that average wind field distributions remain the same from year to year. A more suitable approach would be to identify an invariant, predictable and well understood surface. One area identified as a potential backscatter calibration reference is the Simpson Desert in Australia. A field campaign to gather in-situ measurements to validate the modelling required for this calibration was conducted in 1993 (Ridley *et al.* , in press).

A desert backscatter model was developed (Figure 27) which takes into account all of the factors influencing return power. A microwave pulse incident on a completely smooth surface will be reflected at an angle equal to the angle of incidence (the Fresnel

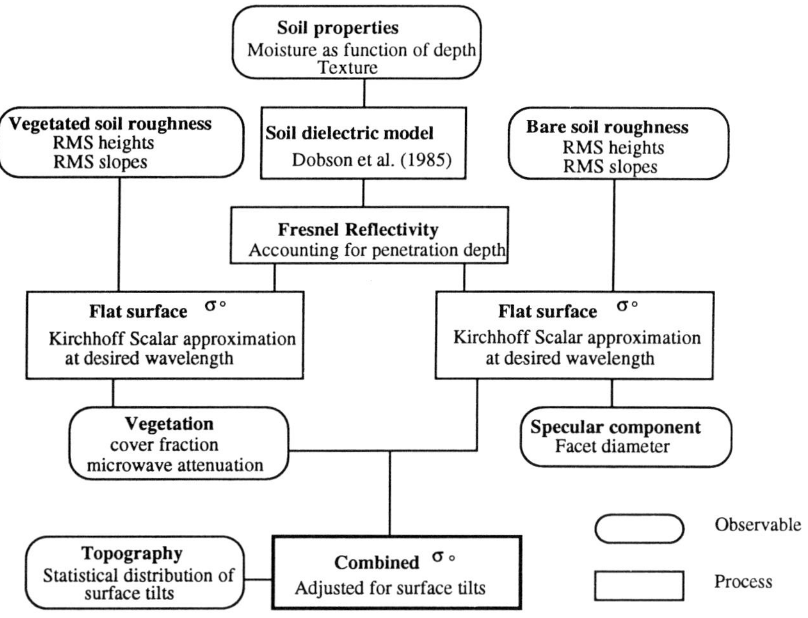

**Figure 27.** *The components of the method used to model backscatter from the Simpson Desert (courtesy of J K Ridley, MSSL)*

direction). If the surface is rough, some of the radiation will be scattered in all directions. The radiation reflected in the Fresnel direction is known as the coherent component, and the scattered radiation as the incoherent component. If the surface is rough, the coherent component becomes very small, and the incoherent component dominates. A surface is considered smooth to incident radiation of wavelength $\lambda$ if $h < \lambda/32 \cos\theta$, where $h$ is the surface height and $\theta$ the angle of incidence (the Frauhofer criterion).

We are interested in the amount of radiation reflected back towards the altimeter, which is described by the backscatter cross-section, $\sigma_i$. The backscatter cross-section is defined as the imaginary area intercepting that amount of power which, when scattered equally in all directions, produces an echo equal in power to that received from the actual target. If this parameter is normalised by dividing by a surface area element $A_i$, then dependencies on radar characteristics are removed, and a parameter $\sigma_0$ (expressed in dB) is defined which is a function of the surface characteristics alone:

$$\sigma_0 = \frac{\sigma_i}{A_i}$$

A well known backscattering coefficient model is the Kirchoff scalar approximation (see Ulaby *et al.* 1982), which splits $\sigma_0$ into coherent ($\sigma_c$) and non-coherent ($\sigma_n$) terms. The coherent component is strongly dependent on the RMS heights of the surface, and decreases rapidly in magnitude as wavelength scale surface roughness increases. The incoherent component depends mainly and inversely on the surface roughness correlation length.

Scattering is modelled as Fresnel reflectivity, which for near normal incidence may be given in terms of the complex dielectric constant of the surface. The dielectric properties of a soil are influenced principally by volumetric water content, but also by soil micro structure and chemistry. Moisture is present both as bound water and free water. The bound water is held tightly to the soil particles, and moves less easily than free water which is located several molecule layers from the soil particles. Hence the free water component has the dominant effect upon the dielectric constant. Coarser grained, sandy soils have a smaller total surface area of particles, hence will have a greater proportion of free water for a given volumetric water content than fine grained soils

The complete equation for the backscatter coefficient is:

$$\sigma_0 = \sum_{\theta=0}^{\theta=\pi/2} (\sigma_c(\theta) + \sigma_n(\theta)) f(\theta)$$

where $\theta$ is the local normal. The magnitude of incoherent scattering decreases as incidence angle increases from local normal, and the coherent component is negligible for non-zero incidence. Therefore as incidence angle increases, the magnitude of the backscatter coefficient drops. A surface within the altimeter footprint with significant topographic relief will have local normals with a range of angles, reducing the backscatter coefficient compared to a flat surface of similar area.

Vegetation cover has an attenuating effect on backscatter coefficient, and individual vegetation elements make a contribution to total scattering which is dependent on their geometry and moisture content. Dealing with vegetation effects analytically is an extremely complex topic. A simplified approach is to make in-situ measurements of the effects on backscatter coefficient, and then derive an empirical attenuation factor from these experiments.

The Simpson Desert lies in the centre of Australia, between the MacDonnell mountains to the north, and Lake Eyre to the south-east. It is a landscape dominated by NNW-SSE tending longitudinal sand dunes, extending up to 100km in length and 25m in height, and spaced at intervals of between 200 to 600m. The surface is covered by spinifex, a spiny evergreen plant. The suitability of this area for backscatter coefficient cross-calibration between different radar altimeters was evaluated by field experiments in August 1993. The area chosen for the experiment lies at a cross-over point for Topex/Poseidon, the joint US/French altimeter satellite. Various data were acquired in order to characterise the surface: GPS surveying was used to measure medium and large scale surface roughness; small scale surface roughness was characterised using a pin profilometer; soil moisture was measured from soil samples acquired at a series of depths; soil dielectric constant was measured directly using a dielectric probe; vegetation cover by species was surveyed over a characteristic area. In addition, a field scatterometer was deployed which allows direct measurement of backscatter at 13.8GHz. In this instance it was found that variability in backscatter due to soil moisture variability makes this surface less than ideally suited for absolute calibration of backscatter. However, the results did show a good, validated correlation with rainfall, demonstrating the use of radar altimetry for monitoring rare rainfall events in such regions (Figure 28).

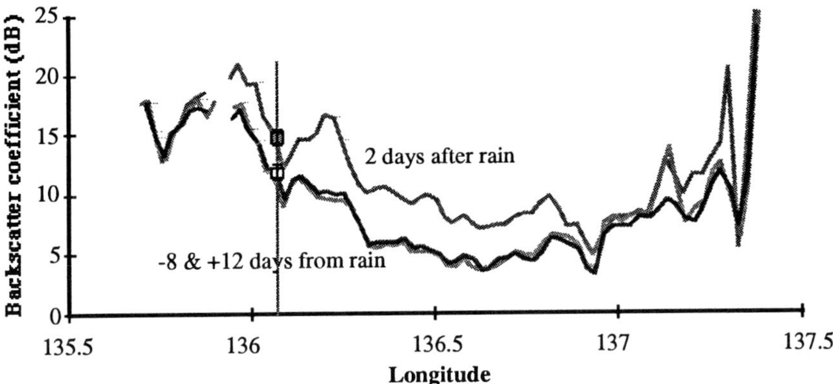


**Figure 28.** NASA *Radar Altimeter measurements compared with model results at 13.6GHz from 3 separate passes across the Simpson Desert, showing the increase in backscatter coefficient caused by increased soil moisture following the rainfall event on 8th September, (Figure courtesy of J K Ridley, MSSL).*

## 4.4 Hydrology

Wetlands, *i.e.* swamps, marshes, flood plains, deltas, etc cover 6% of the world's land surface, and as they are relatively flat they can be observed with pulse-limited radar altimeters. The area and depth of the water in wetlands change continuously in response to inflow, precipitation and evapo-transpiration, and since their hydraulic response tends to be relatively slow, they may be used as proxy indicators fort local climate. Many rivers around the world are monitored routinely for water level and discharge rates. However, although such measurements are usually made to a high degree of accuracy and dense temporal sampling, spatial sampling is usually low, and data availability is poor. In certain conditions, and especially for wider rivers, radar altimeters can be used to determine river levels to a precision of up to 10cm. However, technical problems have restricted such use for much of the data recorded to present. The Seasat and Geosat radar altimeters suffered from two problems: first, when tracking narrow-peaked echo wave forms, such as those received from smooth water, the tracker generated an erroneous height signal causing a jump in the location of the range window, and leading to blurred or double-peaked wave forms from which no accurate elevation can be estimated. Secondly, narrow-peaked wave forms can cause saturation of the on-board analogue-to-digital converters, especially when there is a sudden rise in surface backscatter. This leads to an underestimation of the true surface backscatter coefficient.

## 4.5 Polar glaciology

### Overview

Interest in polar glaciology has grown over recent years in line with growing concerns over global climate change. Radar altimetry has been used to study several aspects of polar glaciology. These include:

- Topographic mapping

- Surface state mapping

- Mass balance assessment

- Ice shelf grounding line position

Interpretation of the results requires consideration of the interaction of electromagnetic waves with snow and ice surfaces, which exhibits a wide range of behaviour over different environmental conditions.

## Special characteristics of ice covered surfaces

Radar altimetry of ice covered surfaces is complicated by several factors. Penetration occurs when the surface reflectivity is low enough to permit significant pulse energy to be transmitted through the surface/air interface. The penetrating pulse undergoes volume scattering from inhomogeneties beneath the surface. The echo wave form is then made up from surface scattering and volume scattering, with contributions from the later delayed by the longer travel time through the sub-surface material, and attenuated by scattering and absorption. We have seen that surface scattering can be modelled as a convolution of the flat surface impulse response, the height distribution of scattering facets and the system point target response. Volume scattering is modelled as Rayleigh backscatter. As the spherically-expanding pulse penetrates the surface, energy is scattered from a volume of material of thickness $dR$ and area $dS$ defined by the pulse width and the instantaneous footprint diameter. Each point of the pulse sphere at a given time will have a depth of penetration determined by its angular distance from nadir. It can be shown that at a specific range beneath the surface, the return power is given by:

$$P(t) = \frac{A'(1 - \exp(2k_e(h - R)))}{R^2}$$

where

$$A' = 2\pi R^4 \left( \frac{P_t G^2 \lambda^2}{(4\pi)^3 R^4} \right)$$

with $P_t$ the transmitted power, $G$ the antenna gain, $\lambda$ the wavelength, $k_e$ the extinction coefficient of the material, $h$ the altimeter height above mean surface, and $R$ the range to any particular depth. An expression for the flat surface response at any time $t$ can be obtained by integrating over the pulse width $PW$:

$$P(t) = A' \int_{(R-PW)}^{R} \frac{(1 - \exp(2k_e(h - R)))}{R^2} dR$$

The existence and importance of penetration depends very much on the surface and subsurface conditions. The maximum depth from which echoes can be received depends first on the width of the range window, and then upon the attenuation of the radar pulse. The penetration depth is defined as the depth to which the echo power has dropped to $1/e$ of that just beneath the surface.

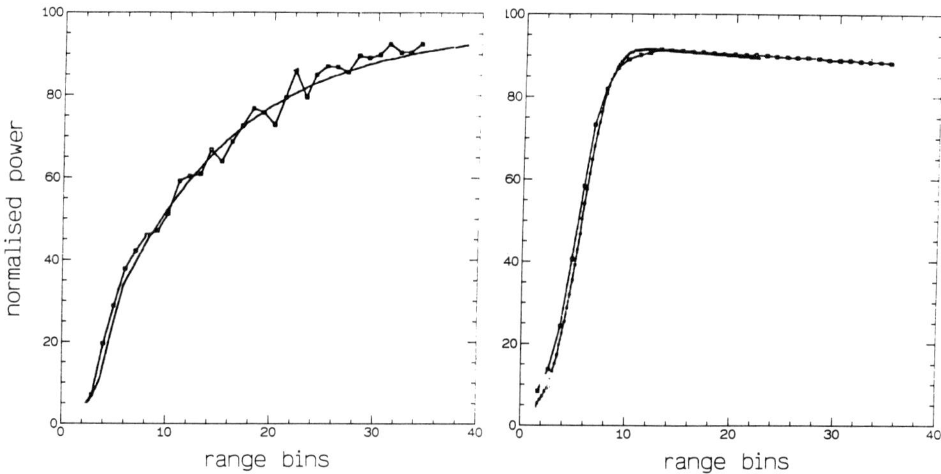

**Figure 29.** *Mean echo wave forms from central Wilkes Land, East Antarctica, fitted by Ridley and Partington model (left) and from Larsen Ice Shelf, West Antarctica, fitted by Brown model with Significant Wave Height of 2m, (adapted from Ridley and Partington 1989)*

When trying to determine surface elevation, it is important to identify areas in which volume scattering dominates. This can be done by analysis of the wave form shape. Ice sheet returns were modelled by Ridley and Partington (1988), demonstrating that where volume scattering dominates, wave form rise time is slower, and where surface scattering dominates, wave forms are more ocean-like (Figure 29). Surface elevation can then be estimated either by applying appropriate retracking methods, which can be automated through intelligent software techniques such as neural networks, or applying an integrated volume and surface scattering model, or by echo migration. The choice of method depends very much on the scientific objectives.

Off-ranging is another factor that must be considered when analysing echo wave forms from ice sheets. Since the first echo received will be from the nearest surface point to the altimeter, if that surface is not flat then this echo will not be from nadir. The offset of the target to nadir is proportional to the surface slope, and lies in the direction of the steepest slope. When uncorrected this offset leads to an error in the estimate of surface elevation, usually known as "slope-induced error". This error can be treated either as an error in range or as an error in location, and is corrected for accordingly (Figure 30). Note that where migration techniques are used in data processing, this problem does not arise.

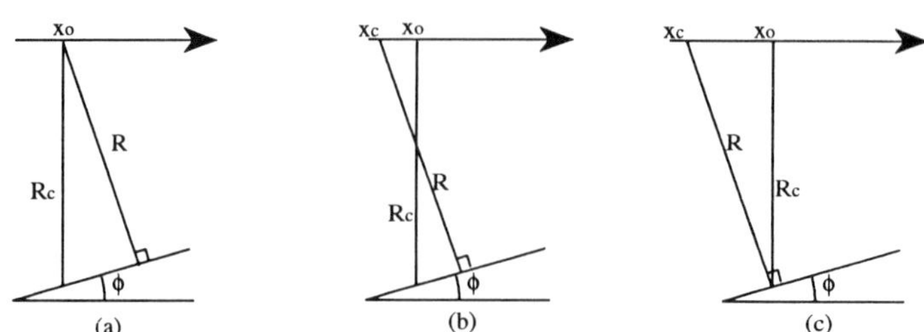



**Figure 30.** *The three different methods of correcting for slope induced error, (a) the 'direct method' where no correction is made to position, and a corrected range to nadir is calculated (b) the intermediate method where no correction is made to range and a position is found for which the range is valid and (c) the relocation method where both range and position are corrected. The range is $R_c$, $X_0$ is the position and $X_c$ the corrected position. All methods require some a-priori knowledge of the terrain shape, which can be derived iteratively from the altimeter data themselves.*

## Applications

Results from radar altimetry related to glaciology can be split into two categories: applications to the field of ice dynamics, usually model inputs, and surface state analysis, usually direct observables. The study of ice sheet and ice shelf dynamics attempts to describe the evolution of these systems, both past and future, and to relate evolution to global climate. Ice shelves are more sensitive to climate warming than ice sheets, as they are generally much thinner, have mean surface elevations of only around 100m above sea level, and are subject to interactions with ocean as well as atmospheric circulation. Furthermore, it is often suggested that the Ross and Filchner-Ronne Ice Shelves (see location map, Figure 31) play a vital role in retaining the West Antarctic Ice Sheet, which is for the most part grounded well below sea level. The smaller ice shelves surrounding the Antarctic continent, whilst not playing any major dynamic role, may act as 'early warning signals' to climatic warming.

Radar altimeter data have been used for ice sheet topographic mapping since the Seasat mission in 1978 (*e.g.* (Zwally, Bindschaler *et al.* 1983)). The profile of the Antarctic ice sheet is only approximately known. The best available conventionally-derived digital terrain model is quoted to be accurate to 30m (Scott Polar Research Institute), but in many parts of the continent is derived from very sparse ground survey data. Although SeaSat and Geosat orbits extended only to 78°S, important advances have been made. In particular, the Geosat Geodetic Mission, during which the satellite was placed in an orbit giving very dense ground track spacing, allowed a detailed map of the periphery of the Antarctic continent to be produced (Figure 31). With the availability of ERS data, mapping can be extended to 81°S, providing the first realistic detailed topography ever produced for large areas of East Antarctica (Figure 31b). Models describing the long term response of the ice sheets to changes in temperature and accumulation require a good description of their current shape in order to determine the extent and shape of drainage basins derived topographies actually provide a finer description than current models actually need.

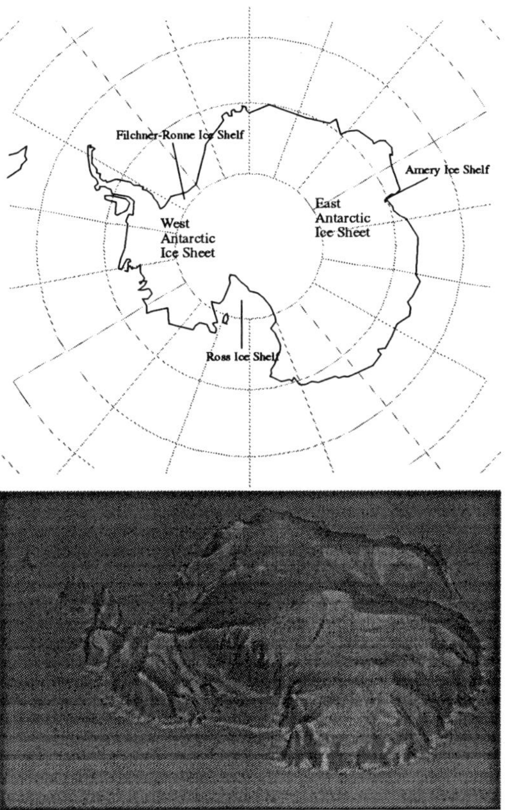


**Figure 31.** *(a) Antartica Location map, and (b) Isometric plot of Antartica derived from ERS radar altimeter data. The central circular area not covered by ERS (south of 81°S) is filled in using data published by Scott Polar Research Institute, derived from ground and airborne surveys. Note the much lower detail present in this area, and the discontinuity at the boundary. The SPRI data underestimate the surface elevation to an order of 100m. (Figure courtesy of J L Bamber, MSSL)*

Much interest has also focused on the determination of the mass balance of the Antarctic and Greenland ice caps. Although these are still responding to the retreat of the last glaciation, it is suspected that they may also react to shorter time scale climatic change, in particular anthropogenically generated change. The interaction of the ice caps (cryosphere) with the oceans and atmosphere is complex, and includes both response and forcing components. It is thus very difficult to determine a single cause for any observed change, say for example the change in surface elevation of Greenland reported from time series of radar altimeter observations from Seasat and Geosat, or to associate regional observations with overall behaviour. Furthermore, although attempts have been made to detect mass balance change, so far quoted Figure for mass balance itself have error bars far greater than any signal we expect to detect (Jacobs, Helmer *et al.* 1992). A new analytical approach being applied to ERS data is to integrate large

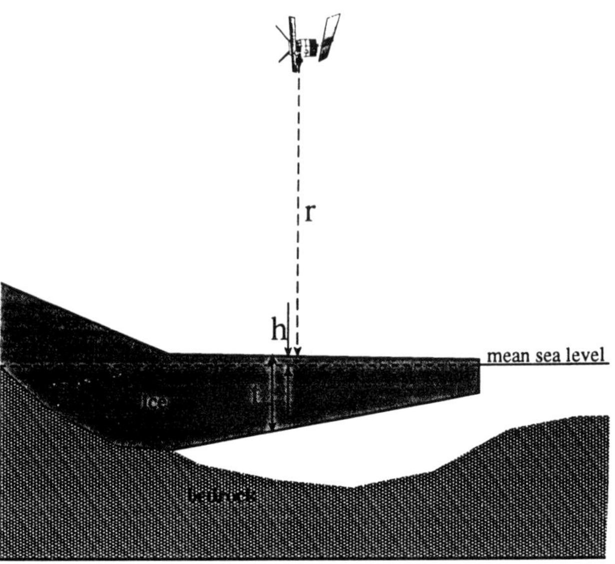


**Figure 32.** *The thickness of a floating ice shelf can be derived from the surface eleva-*
*tion simply by the relationship where $\rho_i$ and $\rho_w$ are the densities of ice and sea water*
*respectively and h is the surface elevation relative to the mean sea level.*

numbers of wave forms over the whole ice cap in order to derive a single average Figure
for elevation change referenced to an arbitrary, defined surface.

Radar altimeter data are also providing important advances in the area of ice shelf
modelling. The contribution here is again the provision of dense accurate height measurements, allowing high quality topographic maps to be produced. Since ice shelves
are floating, it is a simple matter to invert surface elevation to ice thickness, provided
that carefully considered assumptions about ice density profiles are used (Figure 32).

However, this puts much higher constraints on the accuracy requirements than for
ice sheet mapping. To provide a significant improvement over other methods of determining ice thickness, we need to estimate surface elevation to around 50cm vertically,
and although this is made somewhat easier by the very low surface slopes under consideration, new complications are introduced by the ocean tides, which cause the ice
shelves to move up and down. Ice thickness is important, as it governs the dynamic
behaviour of the ice shelf. Additionally, ice thickness defines grounding line position,
where the ice is floated off the bedrock by hydrostatic pressure, and the extent and
importance of grounded areas within the body of the ice shelf, such as ice rises or
ice rumples. The recent catastrophic retreat of Wordie Ice Shelf (Doake and Vaughan
1991) has been correlated with the decoupling of the ice shelf from the bed in a number
of areas, thus reducing the backstress. Various results, *e.g.* those of (MacAyeal 1987)
indicate that ice streams within West Antarctica draining into the major ice shelves are
sensitive to changes in ice shelf backpressure. Accurate topography is also important
to our understanding of processes at the ice shelf base. The largest ice shelves produce
important quantities of Antarctic Bottom Water, cold dense water which plays a signifi-

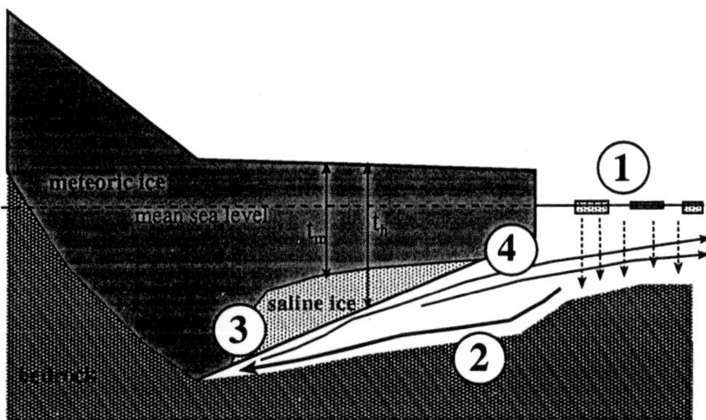


**Figure 33.** *Simplified diagram of the formation of Antarctica bottom water as a result of thermal exchanges at the base of the ice shelf. As the open ocean is exposed to cold air temperatures, sea ice forms (1) and salt is rejected. The salt sinks to the ocean floor, and is absorbed into deeper water resulting in a dense current which sinks under gravity into the sub-ice shelf cavity (2). As this water comes into contact with the ice shelf base near the pressure melting point of ice, the ice starts to melt and lowers the mean density of the water, which then rises as a plume along the base of the ice shelf (3). As the plume rises, the pressure drops, as does the melting temperature. Saline ice platelets start to form along the base of the ice shelf, leading to the build up of the saline ice layer. As the plume continues to rise, it eventually becomes neutrally buoyant, and detaches from the ice shelf, emerging as a deep, low salinity, cold stream. The thickness of the saline layer can be determined from total ice thickness $t_k$ derived from radar altimetry and meteoric ice thickness $t_m$ derived from surface radar.*

cant role in global ocean circulation processes. The production of this water, illustrated in Figure 33 is a result of sea ice formation at the ice front, coupled with the morphology of the ice shelf basin. Part of this process results in the accumulation at the ice shelf base of large amounts of high salinity ice, which contributes to the overall ice thickness. This ice cannot be penetrated by traditional surface and airborne radars, which detect rather the freshwater ('meteoric')/ saline ice interface. Comparing ice thickness maps produced in this way with those derived from radar altimetry (Figure 34) gives us a good idea of the extent and volume of these areas, and provides another potential way in which we can detect change.

All the above results are based on the exploitation of the range measurement. The intensity of the return signal, as measured by the Automatic Gain Control, provides information on the nature and structure of the scattering surface. Remy (1990) used this information to attempt to define areas subject to important, gravity driven katabatic winds. Ridley and Partington (1988) discuss analysing return power together with wave form shape to derive information on the proportion of volume to surface scattering, and hence the sub-surface structure. Antarctica has also been proposed as a site for radiometric calibration. An experiment similar to that carried out on Simpson Desert was conducted in 1992 on Filchner Ice Shelf. Better agreement was observed between in-situ and satellite observations than for the desert, because it is easier to obtain

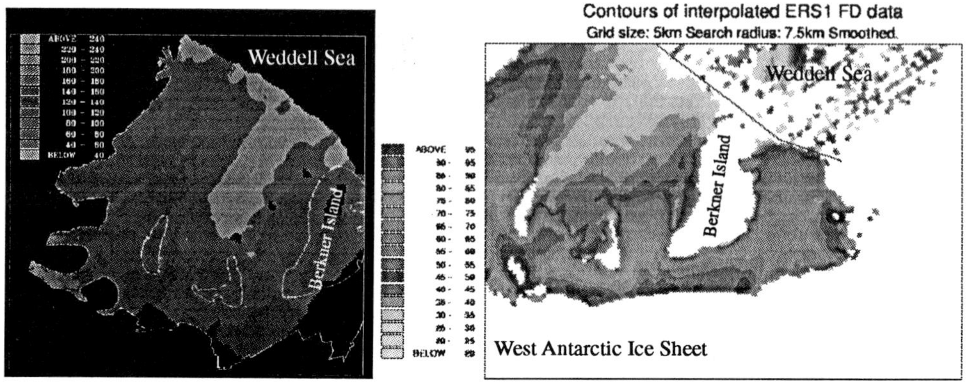


**Figure 34.** *Surface elevation maps of Filchner-Ronne Ice Shelf derived from airborne radar echo sounding (left) and ERS-1 satellite radar altimetry (right). Although the altimeter data used are the lower precision Fast Delivery (FD) data, the detail is still much higher than that achieved from the airborne data. Note in particular the central area of the ice shelf, where the apparent lower elevations derived from airborne surveying cover a much wider area than seen from radar altimetry. The discrepancy is thought to arise from basal freezing processes, creating a layer of saline ice impenetrable by the ice sounding radars.*

representative in-situ measurements due to the relatively featureless nature of the ice shelf. However, sharp variations in backscatter have been observed, thought to be due to transformations in the structure of the snow surface, arising from snowfall events or changes in air temperature. These surfaces are therefore not felt to be suitable for long term calibration purposes, but, as with the Simpson Desert data, results show that it may be possible to detect snowfall and structural changes related to weather.

# 5   References & Further Reading

Birkett, C. M. (1994), Radar Altimetry: A New Concept in Monitoring Lake Level Changes, *EOS* (June 14, 1994)

Brenner A C, Bindschadler R A *et al.* , 1983, Slope-induced errors in radar altimetry over continental ice sheets, *Journal Geophys Res* **88(C3)** 1617

Brooks R L, Campbell W J *et al.* , 1978, Ice Sheet Topography by satellite Altimetry, *Nature* **273** 539

Challenor P G, 1993, Spatial Scales of Wave Height. First ERS-1 Symposium: Space at the Service of our Environment, Cannes, France, ESA SP-359

Doake C S M, 1987, Antarctic ice and rocks. Antarctic Science. (Cambridge, Cambridge University Press)

Doake C S M and Vaughan D J, 1991, Breakup of Wordie Ice Shelf, Antarctica, *Nature* **271** 328

Elachi C, 1987, Introduction to the Physics and Techniques of Remote Sensing, (New York, John Wiley and Sons)

Guzkowska M A J, Rapley C G *et al.* , 1990, Developments in Inland Water and Land Altimetry.

Hawkins J D and Lybanon M, 1989, GEOSAT Altimeter Sea-Ice Mapping, *IEEE Journal of Oceanic Engineering* **14(2)** 139

Jacobs S S, Helmer H H *et al.* , 1992, Melting of ice shelves and the mass balance of Antarctica, *Journal of Glaciology* **38(130)** 375

Laxon S W and McAdoo D, 1994, Arctic Ocean Gravity Field Derived From ERS-1 Satellite Altimetry, *Science* **265** 621

Lingle C S, Schilling D H, Fastook J L, Paterson W S B and Brown T, 1991, A Flow Band Model of the Ross Ice Shelf, Antarctica: Response to CO2-Induced Climatic warming, *Journal of Geophysical Research* **96(B4)** 6849

Lingle C S, Brenner A C, Zwally H J and DiMarzio J P, 1991, Multiyear Elevation Changes near the West Margin of the Greenland Ice Sheet from satellite radar altimetry, Symposium Volume, International Conference on the Role of Polar Regions in Global Change.

MacAyeal D R, 1987, Ice Shelf Backpressure: Form drag vs dynamic drag. The Dynamics of the West Antarctic Ice Sheet. Hingham, Mass., D. Reidel. 141-160.

Massom R A, 1991, Satellite Remote Sensing of Polar Regions. (Belhaven Press, London).

Partington K C, Cudlip W *et al.* , 1987, Mapping of Amery Ice Shelf, Antartica, surface features by satellite altimety, *Annals of Glaciology* **9** 183

Rapley C G, 1990, Satellite Radar Altimeters. Microwave Remote Sensing for Oceanographic and Marine Weather-Forecast Models. (Kluwer Academic Publishers)

Rapley C G, 1990, Satellite Radar Altimeters. Microwave Remote Sensing for Oceanographic and Marine Weather-Forecast Models. (Kluwer Academic Publishers) 45-63

Remy F, Mazzega P, Houry G, Brossier C and Minster J F, 1989, Mapping of the topography of continental ice by inversion of satellite altimeter data, *Journal of Glaciology,* **35(119)** 98

Remy F, Brossier C and Minster J F, 1990, Intensity of satellite radar-altimeter return power over continental ice: a potential measurement of katabtic wind intensity. *Journal of Glaciology,* **36(123)** 133

Ridley J K, Cudlip W, McIntyre N M and Rapley C G, 1989, The topography and surface characteristics of the Larsen Ice Shelf, Antarctica, using satellite altimetry, *Journal of Glaciology* **35(121)** 299

Ridley J K, Laxon S, Rapley C G and Mantripp D R, 1992, Topography of Antarctic Ice Sheet mapped with the ERS-1 Radar Altimeter, *Earth Observation Quarterly* **(37-38)** 14

Ridley J K and Partington K C, 1988, A Model of Satellite Altimeter Return from Ice Sheets, *Int J Remote Sensing,* **9(4)** 601

Robinson I S, 1985, Satellite Oceanography: an introduction for oceanographers and remote sensing scientists. Ellis Horwood Series in Marine Science, (Ellis Horwood Ltd, Chichester)

Sandwell D T, 1992, Antarctic marine gravity field from high-density satellite altimetry, *Geophys J Int,* **109** 437

Srokosz M, 1986, Surface Elevation and Slopes for a Nonlinear Random Sea, *Journal of Geophysical Research,* **91(C1)** 995

Wakker K F, Zandbergen R C A *et al.* , 1988, From satellite altimetry to ocean topography. A survey of data processing techniques, *Int J Remote Sensing* **10/11**

Wakker K F, Wisse E *et al.* , 1993, ERS-1 Radar Altimetry over the North Atlantic, *Proc First ERS-1 Symposium: Space at the Service of our Environment,* Cannes, France. ESA SP-359 439

D.L. Witter and D.B. Chelton (1991), A Geosat Altimeter Wind Speed Algorithm and a Method for Altimeter Wind Speed Algorithm Development. *J Geophys Res,* **96** 8853

Zwally H J, Bindschaler R A *et al.* , 1983, Surface Elevation Contours of Greenland and Antarctic Ice Sheets, *Jnl Geophysical Research,* **88(C3)** 1589

Zwally H J, Stephenson S *et al.* , 1987, Antarctic Ice Shelf Boundaries and Elevations from Satelitte Radar Altimetry, *Annals of Glaciology,* **9** 229

# Remote Sensing of Water Quality: Shallow Water Bathymetry

A C Roberts, R Atkins & J P MacDonald

Simon Fraser University
Canada

## 1 Introduction

Most water quality remote sensing research in the past decade has concentrated upon satellite imaging systems for turbidity studies and airborne active systems (microwave and laser) for the detection and monitoring of oil spills. Although this research has been productive, the relatively low spatial resolution, problems associated with atmospheric scattering and related distortions of the reflected electromagnetic radiation for these orbital sensors has resulted in reduced accuracy levels and often complex calibration procedures.

"In view of the fact that remote quantitative measurement of turbidity is considered operational by some investigators (the same cannot yet be said for types of sediment and/or phytoplankton species), it is surprising that remote sensing of turbidity is not in wider use" (Johnson and Munday 1983). Problems with applications have been part of remote sensing over the past two decades. Although there exists a substantial literature dealing with system performance characteristics, there has been very limited routine application of multispectral remote sensing technology. This can easily be explained by the high costs of data acquisition considered in conjunction with the high probability of failure as most procedures are still experimental rather than operational. In order to make remote sensing operational, in the same sense that aerial photography and photogrammetry are operational, it is necessary to:

- bring the costs down to an affordable level,

- develop suitable airborne systems for installation in conventional aerial survey aircraft,

- develop conventional procedures for resource management applications, and

- develop cost effective image analysis systems that will be accessible to resource managers.

The microcomputer revolution has made a significant impact on image processing and a variety of image processing systems are at present available for 80486 microprocessors. These image processing systems contain most of the conventional software used in remote sensing applications. As a result, the problems of access to image analysis systems and access to conventional processing procedures may be solved. The problems of costs and airborne systems availability have not as yet been solved from a consumer viewpoint. However, some research has been directed into these areas and the use of a suitable airborne system that is capable of meeting these requirements in an operational test is described in this paper.

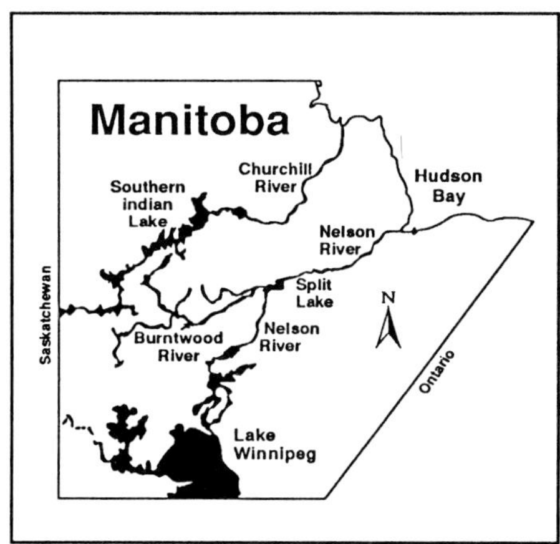


**Figure 1.** *Province of Manitoba showing the Churchill and Nelson river systems. The diversion took place from Southern Indian Lake through the Burntwood River system to join the Nelson River system at Split Lake*

# 2  Water quality in Split Lake, Manitoba

## 2.1  Operational test: Churchill river diversion

In 1976, the diversion of approximately 75% of the flow of the Churchill River into the Nelson River system was undertaken by Manitoba Hydro (see Figure 1).

Water quality parameters, such as alkalinity, turbidity, suspended solids, nutrients and metals, have increased significantly since this diversion became fully operational in 1977 (Guilbault *et al.* 1979; Playle and Williamson 1987; Duncan and Williamson 1988). The lack of a complete water quality dataset from the pre-development period

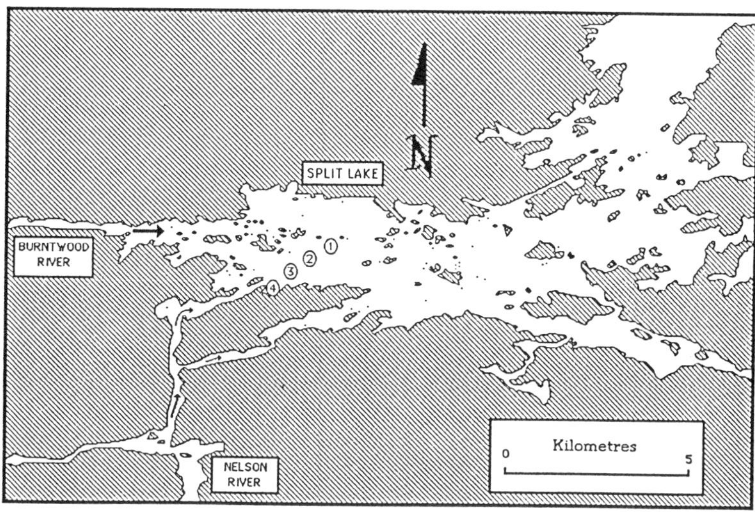


**Figure 2.** *A portion of Split Lake, Manitoba showing the inlets of the Burntwood and Nelson rivers and the four surface sampling stations used for this study*

of this diversion has prompted an investigation into the application of remote sensing as a water quality management tool. This remote sensing study was undertaken to evaluate the possibility of detecting pre- and post-development water quality changes (principally turbidity, non filterable residue (NFR) and chlorophyll-a).

The Split Lake area (Figure 2) was selected as the primary study area for several reasons: (i) it has been affected by the Churchill River diversion through the Burntwood River; (ii) in-flow to Split Lake from the Nelson River has been unaffected by the diversion; (iii) the mixing of these waters permits a direct comparison of the altered system with the unaltered system, and; (iv) Split Lake is a sufficiently large body of water to permit satisfactory spatial resolution of these water quality targets from orbital altitudes.

This project was initially addressed in three phases.

1. Acquisition of initial multispectral airborne data with surface observations of water quality and a coincident or near-coincident Landsat 5 overpass under cloud free conditions.

2. Analytical comparison of the airborne multispectral imagery, the coincident Landsat imagery and the surface observation water quality data.

3. Comparisons between the results from phase 2 and previous Landsat imagery for the target area acquired both before and after development.

Portions of the phase 1 and 2 projects have been completed and work has commenced on phase 3; however, it was not possible to obtain coincident airborne and orbital data in 1987. In addition, further data will need to be acquired to more accurately control for bottom reflection and aquatic vegetation interference with the water quality parameter estimates. This is not unusual for most remote sensing projects; in addition, weather and data acquisition constraints can often terminate a short term project. Although the amount of acquired data was less than desired it could still make some useful contributions to the study and these data constraints and related problems are an important aspect of an operational test.

## 2.2   Multispectral airborn system

The Simon Fraser University (SFU) airborne system has both photographic and video multispectral capability. The greater sensitivity (minimum illumination: 3 lux) and radiometric resolution (higher contrast) of the video system makes it the more suitable of the two camera systems for medium bandwidth (50nm) filtration and subsequent digital multispectral image analysis. The photographic system provides superior spatial resolution and image geometry and is most suited for photogrammetry and feature identification. This system can be installed in any light aircraft with standard camera and light meter ports.

**Aerial Photography System:**

    2 Nikon F-250 cameras with 24mm f/2.8 Nikkor lenses
    Infrared Aerochrome 2443 (No 15 filter)
    Ektachrome 5036 (L1c filter)

**Video Imagery System:**

    3 Sony XC-37 CCD cameras with 12.5-75mm f/1.8 lenses
    2 Oriel bandpass filters (550 and 650nm)
    1 88A infrared filter
    1 Sony CCD-G5 colour camera with 12-72mm f/1.4 lens
    1 L1c haze filter
    4 Sony 8mm metal VCRs
    1 Toshiba wave form monitor

The utilised optical filtration, in both the video and photographic systems, can be satisfactorily compared with data from both Landsat sensor systems (MSS, the Multispectral Scanner and TM, the Thematic Mapper) and has been proven to be good for this type of water quality remote sensing (the most important spectral region for water quality research is 360–1100nm). Although the selected spectral bands of the video imagery do not exactly duplicate the Landsat bands, they can be used to provide a remote sensing estimate of the surface truth and to permit target identifications and reasonably accurate empirical radiometric corrections for the Landsat MSS and TM sensors.

## 2.3   Phase I: Airborne data acquisition and surface truth

Although turbidity was selected as the principal operational water quality parameter for this resource management test, estimates of chlorophyll-a content were also important in terms of the overall research project. The imagery was flown during the summer because chlorophyll-a can best be determined during the summer period (although the fall and spring periods are best for peak flow), and a more useful measure of turbidity change as a result of development would be made during the non-peak summer period when discharge would not be affected by spring melt and run-off.

In addition to atmospheric, spatial resolution and system problems there are a number of target characteristics that can result in interpretation errors if they are not controlled. When measuring turbidity from remote sensing data, chlorophyll components and bottom reflections can interfere with such estimates of suspended particulate mineral material. There are conventional procedures for providing some control for such parameters (Roberts and Liedtke 1986).

An examination of available Landsat imagery from 1972 to 1986 showed that July to August was the most favourable period for cloud free or minimal cloud cover conditions. Since a principal objective of this project was to compare present water quality values with pre- and post-development values, the surface observation and airborne data were to have been collected as close in time as possible to a usable (*i.e.* cloud free) Landsat overpass. In addition, during the months of July and August, when seasonal requirements are most suitable, the largest number of previous relatively cloud free Landsat images are available for most years. This provided the optimum circumstance for comparisons between pre- and post-development conditions.

The most suitable satisfactory Landsat image for the summer of 1987 was recorded on July 26, 1987 between 09:30 and 10:00 local standard time. For the later August and September Landsat overpasses cloud free conditions were not encountered. Although no coincident aerial imagery were acquired it was felt that it was still possible to make some useful analytical comparisons between this imagery and the multispectral SFU airborne imagery collected on September 1, 1987.

The airborne imagery included colour and colour infrared photography and multispectral (medium band: infrared, red and green) video imagery. This imagery was recorded at six different flying heights (150, 300, 600, 1200, 2400, and 4800m *above ground level*) over Split Lake, Manitoba on September 1, 1987 between 13:00 and 14:00 hours local time. As the imagery was for purposes of water quality analysis, exposure settings were adjusted for water rather than for land and vegetation.

The aerial photography was used for some digital image analysis as well as for photographic interpretations, presentations and as control imagery for the multispectral video data. The video data were recorded using a waveform monitor for correct gain adjustment and separate tape recorders for each camera. They were input into a Intel/Xybion image analysis system for digital analysis. Although the coincident Landsat imagery include both TM and MSS imagery, due to its superior spatial resolution only the TM imagery was acquired for comparisons with the airborne remote sensing data and the surface observations. The multi-altitude video and photographic imagery provided high spatial resolution (approximately 0.25m at 150m), precise ground target identifications and accurate empirical estimations for inferring Landsat image degradation as a result of atmospheric scatter, point spread functions, edge effects and system problems. These correction data should prove to be very useful when estimating water quality conditions from the earlier MSS and TM Landsat imagery (without supporting ground observations) for the pre- and post-development conditions.

The surface observation data were collected and analysed by the Water Quality

Branch, Inland Waters Directorate, as part of their water quality testing programme. These 'surface truth' data included: major ions, turbidity, NFR, chlorophyll-a and water depth for shallow areas.

Depth integrated samples (secchi depth) were collected at each sampling site using sampling irons and laboratory prepared bottles (Environment Canada 1983). Laboratory procedures for major ions, turbidity, NFR and chlorophyll-a are outlined in Environment Canada (1986).

## 2.4  Phase II: Analyses

A comparison of the three principal datasets (surface observations, airborne multispectral data and Landsat orbital imagery) is central to developing satisfactory confidence in airborne and Landsat water quality estimates. Both the orbital and airborne remote sensing data required a certain amount of preprocessing and image enhancement prior to conducting analytical examinations for water quality parameters.

This principally involved a redistribution of the digital image values across the full 8 bit (0–255) range from the 6 to 7 bit range (0–64 or 128) of the original orbital imagery. The MSV (multispectral video) imagery was digitised and stretched to 8 bit resolution using a modified linear procedure (original values were equally stretched to fit the full 0–255 range and the video pedestal value was set to zero) and used in the analyses. The colour and colour infrared aerial photography was video digitised onto an International Imaging System (IIS) image analysis computer using a wave form monitor for precise gain adjustment. The colour separations were performed using Wratten No 25, 47A and 58 filters in combination with a 301A infrared cut-off filter to eliminate near-IR input to the video camera. It was digitised to 8 bit resolution and, for some analyses, subsequently stretched using a linear procedure similar to that employed with the video imagery.

The orbital imagery was input into the IIS image analysis computer from 9 track tape at 8 bit resolution level, as received from the Canada Centre for Remote Sensing (CCRS). These data were stretched to 8 bit resolution at the receiving station using standard CCRS procedures. The principal spectral bands used in the analyses were: 4, 5 and 7 for the MSS and 2, 3 and 4 for the TM (in both instances green, red and near infrared). These spectral bands approximate those from the airborne MSV and the colour infrared (2443) aerial film. In all instances the single spectral band containing the most water quality information was red (approximately 600–700nm).

Although surface observation data were collected on August 31, 1987 for the four sampling sites and on September 1, 1987 for only sites 1 and 4 there appears to be a strong agreement between sites 1 and 4 for the two days (see Table 1a). Turbidity and NFR are in excellent agreement for the two days and chlorophyll-a comparisons are quite satisfactory. There were clearly defined differences between sites 1 and 4 water quality values for both days. Sites 2 and 3 from August 31 showed intermediate values between sites 1 and 4 (see Table 1b) When transmission densitometer values from the September 1, 1987 colour aerial photography were ratioed and mean values compared for each of the four sites, their correspondence was very strong.

| Date | Station | Temperature | Turbidity | Colour | NFR | Chlorophyll-a |
|---|---|---|---|---|---|---|
| Aug 31 | 1 | 13.3 | 23.0 | 15.0 | 16.1 | 0.005 |
| Sept 1 | 1 | 14.2 | 24.0 | 13.8 | 18.3 | 0.005 |
| Aug 31 | 4 | 13.9 | 9.8 | 10.0 | 10.2 | 0.010 |
| Sept 1 | 4 | 14.0 | 8.0 | 7.5 | 8.8 | 0.008 |

| Station | Turbidity | Chlorophyll-a | Red DN | Green DN | R:G ratio |
|---|---|---|---|---|---|
| 1 | 23.0 | 0.005 | 84.04 | 87.57 | 0.960 |
| 2 | 15.4 | 0.009 | 125.93 | 116.50 | 1.081 |
| 3 | 12.7 | 0.010 | 129.90 | 116.59 | 1.114 |
| 4 | 9.8 | 0.010 | 124.72 | 99.75 | 1.250 |

**Table 1.** *(a) Mean water chemistry data comparisons at sites 1, 4 on August 31st and September 1st, 1987. (b) Mean values at all 4 sites.*

In Figure 3 we gave an illustrative plot of turbidity versus the R:G ratio for the four sites. To produce a red-green ratio value the red band densitometer readings were divided by corresponding green band readings; such a procedure is necessary to control for the optical distortions to the radiometric quality of photographic and video imagery as a result of light fall-off and vignetting (Lillesand and Kiefer 1987 pp 390-393).

An examination of the red image bands from the all three remote sensing data sources confirmed these sampling station differences. Figure 4 and Figure 5 show chlorophyll-a and NFR values for these surface sampling sites on August 31 and September 1, 1987. Figure 6 shows airborne digital red band photographic images from three different altitudes with transect digital values from the site 4 to site 1 vicinities. There is a clear separation between station 1 and 4 values in Figure 4 and Figure 5 and these differences correspond with the Figure 6 data along the same transect.

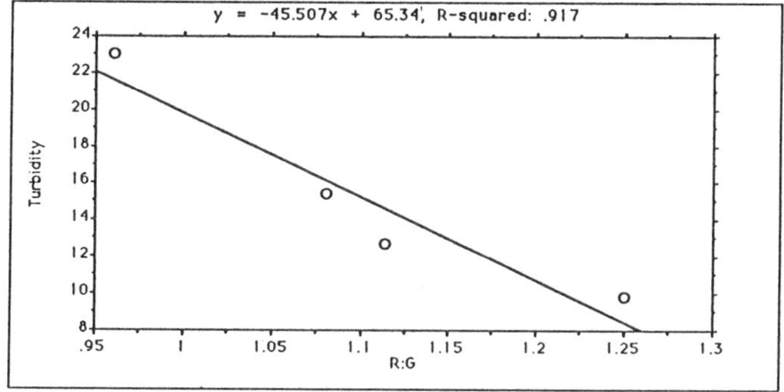


**Figure 3.** *Shows the relationship between turbidity and the ratioed (R/G) densitometric optical count values from the Ektachrome colour transparency film.*

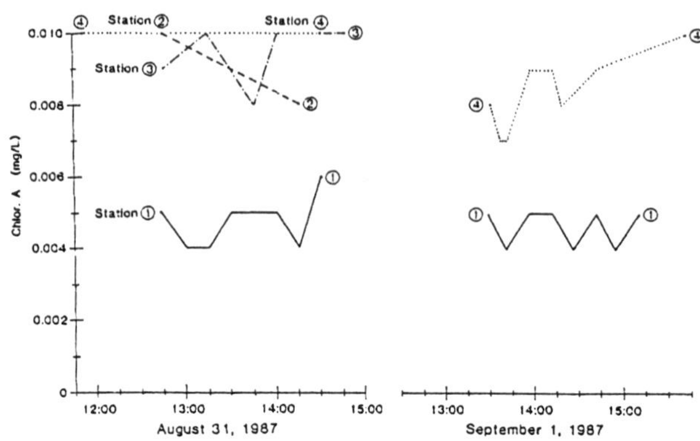



**Figure 4.** *Chlorophyll-a surface sampling data*

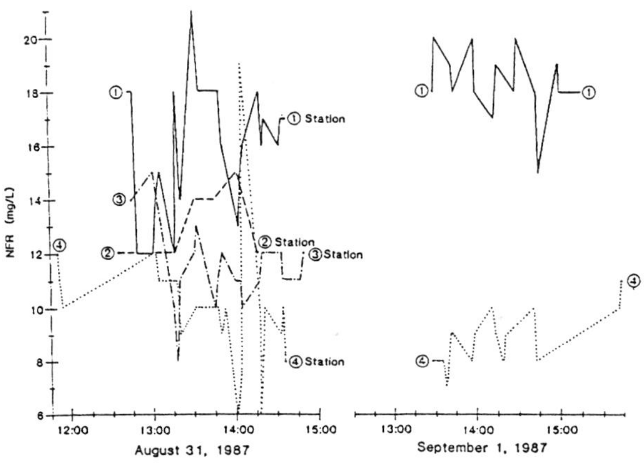



**Figure 5.** NFR *surface sampling data*

Three airborne digital red band photographic images showing water quality changes across the sediment plume boundary from sampling site 1 to 4 are presented in Figure 6 (left): top, centre and bottom images were taken at heights of 4800m, 1200m and 600m respectively. The corresponding digital profiles are shown at the right.

At 4800m the transect crosses from the low site 4 values (40–50 digital number values) to the higher sediment values (65–90) in an area comparable to site 1.

At 1200m a shorter transect across the same area from the site 4 relatively clear water into the edge of the sediment laden area, has distinctly lower absolute values (15–28DN for clearer site 1 water and 45–70DN for the mixing zone and sediment laden water) due to a reduction in atmospheric scatter. It, however, clearly shows the

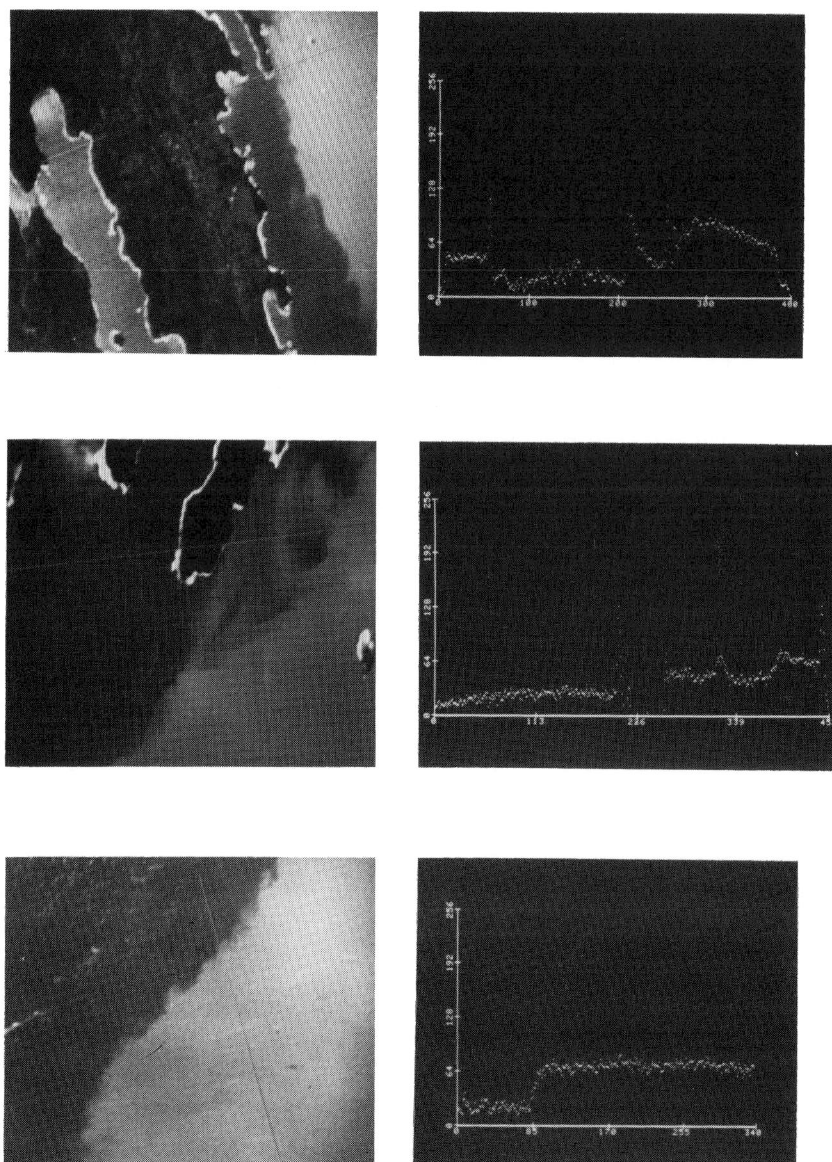

**Figure 6.** *Airborne digital red band photographic images showing water quality changes across the sediment plume boundary from sampling site 1 to 4. The images were taken from heights of 4800m (top), 1200m (centre) and 600m(bottom). At the right are shown the corresponding profile of reflectivity values along the transect shown on the image They show clearer water to the left and more sediment laden water to the right.*

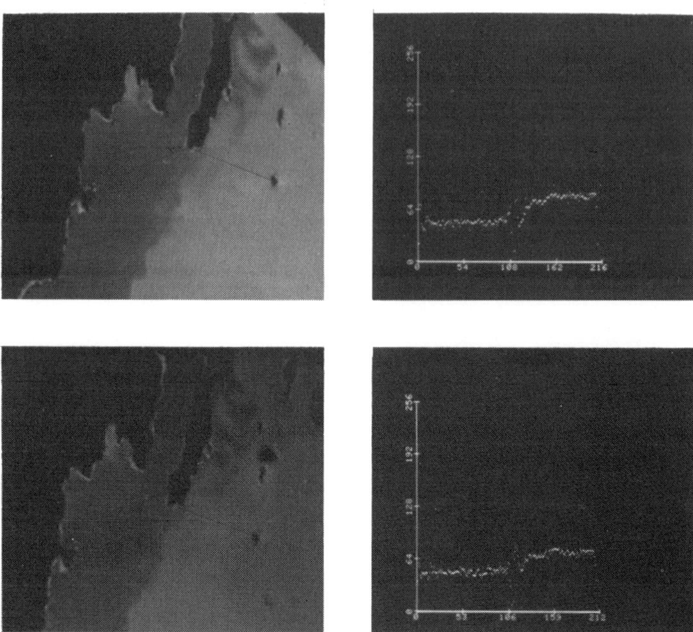

**Figure 7.** *Airborne digital video images from 4800m altitude of the same portion of Split Lake shown in Figure 6; upper is a 50nm bandwidth video image centred on 600nm (red); lower is a 50nm bandwidth image centred on 500nm. Data transect plots are shown at the right. They show clearer water to the left and more sediment laden water to the right.*

transition into the sediment laden water and some of the intermediate values in the mixing zone.

At 600m a transect across the sharp transition from site 1 to site 4 shows the distinct DN value jump (from 15–28 to 60–70DN) from clearer water to sediment laden water. As there was not much atmospheric scatter for the red band between 1200 and 600m there was little change in absolute DN values. The tight grouping (minimum variability) of these low altitude values clearly illustrates how distinctly these water quality differences can be mapped.

Figure 7 shows two 4800m video images over the same area; the upper is centred on 600nm and the lower on 500nm. The corresponding DN plots are shown on the right. Although the sediment laden water is visible on both images the red band image provides superior contrast. The DN plots for the same transect as shown on the 4800m red photographic digital image (Figure 6). These un-stretched digital values cover a greater dynamic range than the photographic red band and provide a distinct separation between the clearer site 4 water (80–90DN) and the sediment laden water in an area in the vicinity of site 1 (125–139DN). The green band, in comparison, shows similar values for clearer water (85–95DN) but a weaker value increase (104–112DN) over the sediment laden water.

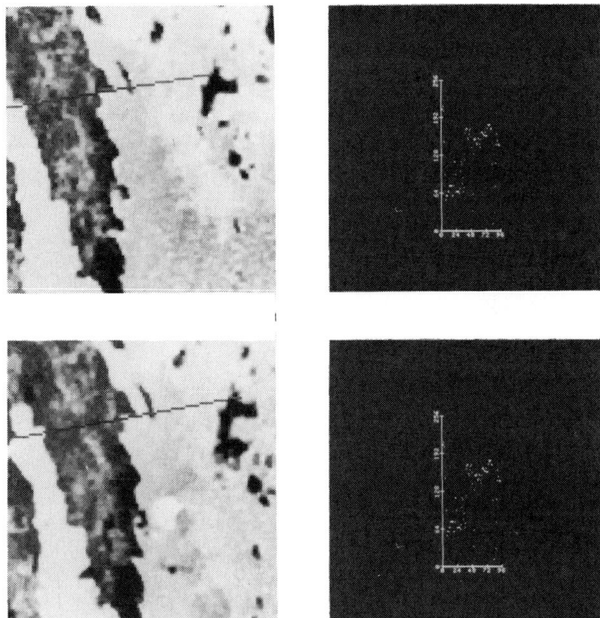

**Figure 8.** *Landsat* MSS *sub-scene images of a portion of Split Lake taken in band 5 (red) in years 1973 (upper) and 1986 (lower). DN values across the data transect are at the right showing clearer water on the left and more heavily sediment laden water on the right.*

## 2.5 Phase III: Comparisons

It is now possible to consider the analytical potential of the orbital Landsat imagery. A total of five Landsat images have been digitally examined for evidence of a pre- to post-development water quality change. The two pre-development Landsat MSS images were recorded on July 27, 1973 and June 16, 1974. The two post-development Landsat MSS images were recorded on July 10, 1978 and June 21, 1986. The fifth Landsat image was the July 26, 1987 TM image.

Figure 8 shows the same area on Split Lake as the previous imagery—at approximately the same scale. Although the MSS images show the poorest spatial resolution they can clearly be used to distinguish water quality between the two islands at the northern end of their transects. Increased sediment concentrations can be interpreted from the steeper DN increase between pixels 75 and 90 on the post development 1986 MSS imagery (bottom) in comparison with the 1973 pre development MSS imagery (top).

Figure 9 compares the 1987 TM (top) and airborne digital photographic imagery (bottom). The striping problem on this red (band 3) TM image is clearly evident and quite severe. This makes it extremely difficult to reliably quantify changes in water quality between the striped areas For example: the clearer water (site 4 area) values are displayed between pixels 110 and 120 on the TM image (top right) and pixels 220 and 245 on the airborne digital image (bottom right); more turbid values (associated with

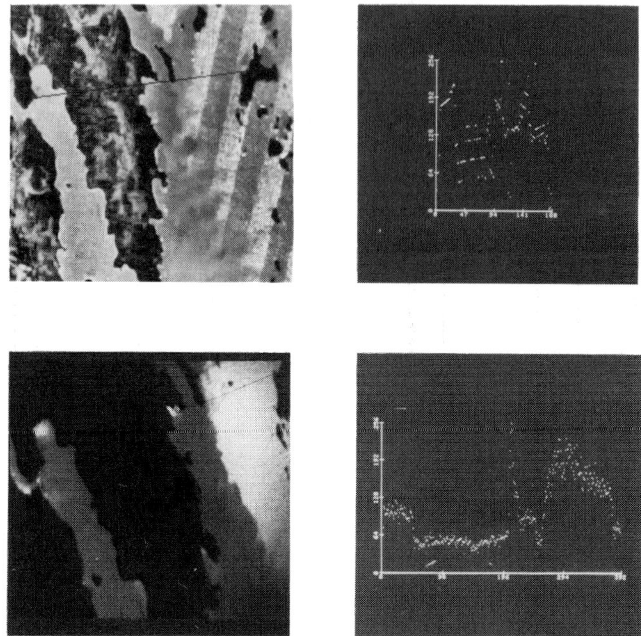

**Figure 9.** *Landsat* TM *band 3 (red) image of the same portion of Split Lake shown in Figure 8, showing data transect (top). At the bottom is an airborne digital [green–red]/[infrared–red] ratioed photographic image from 4800m. DN values are shown at the right.*

the site 1 area) are displayed between pixels 135 and 180 on the TM image and pixels 280 and 380 on the airborne digital image. The airborne imagery provides distinct separation and good quantitative detail regarding these changes in water quality. The defective TM imagery is much more difficult to interpret but it does provide adequate separation between the grosser water quality values.


## 2.6   Summary

An examination of the pre- and post-development Landsat MSS imagery clearly shows changes in both the extent and degree of suspended sediment concentrations in the Burntwood River plume entering Split Lake. The 1987 TM imagery also shows similar post development patterns but, due to sensor problems, the quality of the most important red band was poor.

The airborne multispectral video and photographic imagery was generally of high quality and showed close correspondences with the surface truth water quality data for turbidity, NFR, chlorophyll-a and colour. Atmospheric degradation of the airborne imagery was most evident in the changes in the digital image values between 16000ft and 8000ft flying heights. The video and photographic imagery both showed superior radiometric resolution over the orbital imagery. This was as expected due to the smaller scale and increased atmospheric degradation of the Landsat imagery. The band ratioed digital airborne imagery could display seven different water quality classifications and, although the collected surface observation data could not permit accurate quantitative evaluations of these seven water quality levels, they were clearly related to the water quality changes illustrated by the ground observation data and digital airborne analyses.

## 2.7 Conclusions

Multispectral airborne and orbital imagery were digitally examined for water quality differences in Split Lake, Manitoba. Orbital Landsat MSS imagery recorded before and after the construction of the Churchill River diversion clearly depicted the changes in distribution and concentrations of suspended sediment within Split Lake. Post-development Landsat TM imagery also confirmed these changes. The airborne multispectral video and photographic imagery was flown in conjunction with a surface sampling programme to obtain water quality parameters. These results showed satisfactory correlations between turbidity related parameters and digital reflectivity values from the red spectral region. Both the radiometric and spatial resolution of the airborne imagery was superior to the orbital Landsat imagery.

Although it has been possible to indicate that there have been some substantial changes in the water quality of Split Lake, a number of logistical, financial and data difficulties have made this test an incomplete operational project. Due to local weather conditions it was not possible to obtain coincident airborne and orbital imagery. This is a problem that can be solved only through patience, funding to support further operations and a degree of luck. Financial constraints limited the number of airborne missions, the surface sampling and airborne crew time on site, the access to orbital imagery and the level of assistantship support for analyses. All important considerations for an operational project. The principal data difficulty after weather and logistical problems for data acquisition was the poor image quality of the 1987 TM red band image. This was a problem beyond the project control and, even if it had been possible to obtain coincident surface and airborne data on that day, these data defects would have considerably weakened the results and would have forced a repeat attempt for adequate coincident imagery.

Operational cost effective remote sensing may be considerably closer to reality than it was a decade ago but it is still plagued with difficulties that require experience, expertise, time and funding to solve what has been judged to be a simple operational procedure: the remote sensing of turbidity differences.

# 3    Shallow water bathymetry: theory

## 3.1    Introduction

Shallow water coastal surveys are restricted in their accuracy because they depend on the application of point-sampled data to an areal environment. The determination of water depth for all locations is one of the prime concerns of such hydrographic surveys, along with the identification and location of shipping hazards such as shoals or bars (Theurer 1959). The shifting of many of these underwater hazards, e.g. sand bars, occurs at such a rate that constant revision of nautical charts by traditional depth sounding methods is not feasible (Geary 1968). As a consequence of these inadequacies bathymetric charts are frequently outdated, requiring updating by slow and expensive traditional acoustic surveys on a regular basis to meet the needs of such activities as dredging (McKim *et al.* 1984).

Multispectral remote sensing of such shallow water environments provides an alternative method, allowing quick, repetitive coverage of large areas at a fraction of the cost of shipborne surveys (McKim *et al.* 1984, Tanis *et al.* 1984). Remote sensing techniques may be useful to help maintain navigable water depths in important areas through the calculation of depth values for all the image pixels over the area of interest instead of only at point samples (McKim *et al.* 1984).

The calculation of depth with multispectral data, however, requires the correlation of changes in imaged data with changes in water depth. As long as the water mass is very clear and has little backscatter from suspended material, it is possible to relate upwelling irradiance to water depth in shallow water (Robinson 1985).

## 3.2    Theoretical background

In order to be able to calculate water depth from remotely sensed imagery, the sensors have to be able to record bottom reflectance because the reflectance reaching the sensor is a function of water surface reflectance, volumetric water reflectance as well as bottom reflectance (Lyzenga 1979; Suits 1984).

The total reflectance of the ocean is determined by four factors: 1) nature of the ocean surface or sea state, 2) altitude of the sun, 3) relative amount of daylight present, and 4) optical properties of the water (Williams 1970; Lyzenga 1979). The optical properties of the water can be broken down into two components: 1) light scattered by the oceanic volume and 2) bottom reflectance in shallow water (Lyzenga 1979).

Sea state, altitude of the sun, and relative amounts of daylight contribute to surface reflectance. If the water surface is disturbed, by wind for example, the resultant waves increase the surface area of the water and also cause the apparent angle of incidence of sunlight to fluctuate over time (Williams 1970). The effect of this changing angle of incidence on surface reflectance is small but if the waves begin to form whitecaps, the non- selective scattering of the froth reflects most of the incident light at the water surface (Williams 1970). In the case of such waves, the bottom reflectance component of the total ocean reflectance is minimal.

The sun angle changes throughout the day, with surface reflectance changing as the solar altitude changes. The reflectance of the water surface increases from solar zenith to conditions with the sun on the horizon (Williams 1970)(see Table 2). At high sun

| Solar Altitude (degrees) | <45 | 60 | 70 | 89 | 90 |
|---|---|---|---|---|---|
| Surface Reflectance (%) | 2 | 6 | 13 | 35 | 100 |

**Table 2.** *Changes in surface reflectance with changing solar altitude*

angles the disturbance of the surface waves tends to increase the angle of incidence, although at low sun angles waves tend to decrease the angle of incidence (Jerlov 1968, 1976). This change in the angle of incidence is not a large effect for high sun angles, but surface reflectance for low sun angles is considerably reduced by wave action (Jerlov 1968; 1976).

The relative amount of light on the water can determine depth penetration limits as a function of the sensitivity of the measuring instruments (Jerlov 1968, 1976). If little light is present to penetrate the sea, then the reduction of that light by absorption and scattering may reduce the internal reflectance component of the overall surface reflectance to such a low level that the sensors cannot pick up any bottom signal. On a clear, bright day it should be possible to have better water depth penetration if the sensitivity of the sensor is a relevant factor (Jerlov 1968, 1976).

The internal reflectance component of the water is affected by attenuation which can be broken down into two main processes: 1) absorption of light energy and its conversion to heat energy and 2) scattering of light energy by foreign particulate matter (Williams 1970). These functions of absorption and scattering are affected by both the water mass and the particles suspended in that mass. Water temperature and pressure seem to have little effect on internal reflectance of relatively pure water (Jerlov 1968, 1976). If submerged aquatic vegetation is present, the internal reflectance component is not affected by varying canopy height (Ackleson et al. 1986). Internal reflectance is solely a function of the inherent optical properties of the water mass itself (Jerlov 1968; 1976).

The absorption of light in pure sea water does not differ from that of pure water. Sea salts do not act as absorbing media and do not change the absorption characteristics of the water (Jerlov 1968; Williams 1970; Jerlov 1976). Pure water attenuation is the result of molecular scattering by deflection, refraction, and reflection of the light by the water molecules. Of the three effects, only light deflection is independent of particle type, but all three are relatively unaffected by changing particle shape (Jerlov 1968; 1976). The attenuation characteristics of sea water are largely determined by the various dissolved and particulate substances present. The variations in the attenuation characteristics of sea water, when compared with pure water, are chiefly a function of light scattering by large ($> 2\mu$m) particles and they are virtually independent of wavelength (Jerlov 1968; 1976).

## 3.3 Optical models

Various mathematical simulations of ocean reflectance have been proposed as models of the light response of oceans to changing optical conditions. Here we discuss three.

## The model of Williams

Williams (1970) suggested that the surface reflectance of the ocean could be represented as component functions of the initial incident light of the form:

$$E_0 = A_w + B_w \tag{1}$$

where $E_0$ is the incident sunlight at water surface, $A_w = p_s E_0$ is the reflected component of light at the surface, $p_s$ is the surface reflectance value and $B_w = (1 - p_s)E_0$ is the penetrating component of light at the surface

The penetrating component, $B_w$, was assumed to reach a maximum depth where it was either reflected by the bottom or attenuated to such an extent that the light no longer penetrated. The reflected or scattered light was directed back to the surface as $B_b$ such that:

$$B_b = U_p E_0 (1 - p_s) e^{-kd} \tag{2}$$

where $B_b$ is the light component at surface returning from depth, $U_p$ is the portion of light $B_w$ returning from depth, $k$ is the extinction coefficient and $d$ is the depth of water (light limit or sea floor)

When $B_b$ reached the surface, Williams (1970) suggested that it was scattered and reflected at the water-air interface so that the internal reflectance of the water upwelling on the air side of the interface could be expressed as:

$$B_s = U_0 E_0 (1 - p_s)^2 \tag{3}$$

where $B_s$ is the internal light reflected at the surface and $U_0$ is the portion of originally penetrating light which returns to the surface from depth.

By this model, the light received at a sensor, $E_s$, was equal to the addition of components $A_w$ and $B_s$ such that:

$$E_s = A_w + B_s = p_s E_0 + U_0 E_0 (1 - p_s)^2 = [p_s + (1 - p_s)^2 U_0] E_0 \tag{4}$$

The model of Williams (1970) was inherently dependent on the initial light conditions represented by $E_0$ and the change in the water depth affecting $U_0$.

## The model of Austin

In another model, Austin (1974) suggested that the radiance viewed at the surface could be modelled by:

$$L_z = L_0 T_a + L^* \tag{5}$$

where $L_z$ is the radiance viewed at altitude $z$, $L_0$ is the inherent radiance of the sea viewed from above, $T_a$ is the atmospheric transmission of image-forming light and $L^*$ is the addition of scattered atmospheric light. In this formula, the component $L_0$ could be broken into two components, the underwater radiance $L_w$ and the reflected radiance $L_r$, such that:

$$L_0 = L_w + L_r \tag{6}$$

Substituting (6) into (5) yielded:

$$L_z = (L_w + L_r)T_a + L^*  \tag{7}$$

but the radiance at the surface, $L_0$, could also be represented by:

$$L_0 = t(n_w)^{-2}L_v + rL_s  \tag{8}$$

where $t$ is transmittance coefficient of water, $n_w$ is refractive index of water, $L_v$ is inherent radiance of water, $r$ is reflectance coefficient of water surface and $L_s$ is radiance of the sky. Hence, from (5) and (8):

$$L_z = [t(n_w)^{-2}L_v + rL_s]T_a + L^*  \tag{9}$$

and $t$ was a function of depth of the form

$$t = t(d) = e^{-kd}  \tag{10}$$

where $k$ is the attenuation coefficient and $d$ is the water depth

The model of Austin (1974) suggested that $L_z$ was a function dependent on water depth as an input variable.

## The model of Lyzenga

The model of Lyzenga (1979) considers a given depth of water $D$, and bottom type $R_b$, together with the assumption that

$$E_d = E_0 e^{-kd}  \tag{11}$$

where $E_d$ is the irradiance at depth $d$ and $E_0$ is the initial incident irradiance, held for both downwelling and upwelling light. He proposes that the irradiance of the sea could be determined by:

$$R = K R_b e^{-2kd} + R_s  \tag{12}$$

where $R$ is the total reflectance viewed above the surface, $K$ is the transmittance of water surface and $R_s$ is the reflectance of water surface.

The crudeness of this model did not detract from its accuracy and Lyzenga (1979) showed that it worked well for MSS based studies yielding calculated depths of reasonable accuracy for an assumed constant bottom reflectance over a ratio of two wavelengths. In order to counter the approximate values given by (12), Lyzenga (1979) also proposed a more complex reflectance model of the form:

$$R = \frac{(1 - R_s)(1 - R_s^1)}{(1 - R_s^1 R_w^1)}R_w^1 + R_s - R_s^1 R_w^1  \tag{13}$$

where $R$ is the total reflectance viewed above the surface, $R_s$ is the external reflectance at water surface and $R_s^1$ is the internal reflectance of water surface $R_w^1$ is the subsurface reflectance.

Under diffuse light conditions, $R_s^1$ was calculated to be 0.475, and $R_s$ to be 0.067 (Lyzenga 1979) thus reducing (13) to:

$$R = \frac{0.490 R_w^1}{1 - 0.475 R_w^1} + 0.067 \tag{14}$$

where only $R_w^1$ was needed as an input under diffuse light conditions, and $R_w^1$ was some function of depth.

Lyzenga (1979) suggested that (14) would work well so long as $R_w^1$ could be measured or modelled by a robust theoretical model. The model used in this application would have to include the complete set of water parameters: spectral absorption coefficient of water, scattering coefficient of water, volume scattering function, and bottom reflectance (Lyzenga 1979). The changes of the reflectance $R$ observed above the surface by means of $R_w^1$ could then be linked to a function of depth since $R_w^1$ was dependent on depth.

Following from (14) and assuming that both the surface reflectance and the optical properties of the water do not change over the entirety of the images, the changes in reflected energy, sensed over water of changing depths, are due to the changes in the reflectance from the water and from the bottom (Lyzenga 1979). Shallow water often appears redder than the blue-green colour of deeper water due to the bottom reflectance component and this colour difference should be noticeable by means of remote sensing in different wavelength bands (Hollinger *et al.* 1975). In deeper water, incident energy will have further to travel through the water in order to reflect off the bottom, and thus will be more attenuated by the longer water column it has to pass through. In deeper water, also, the reflectance value from the water itself will have changed due to the larger amount of water in the vertical profile that can reflect energy (Suits 1984). Since the reduction of the bottom reflectance component can be related to the depth of the water, changes in reflectance can be related to changing water depth (Lyzenga 1979).

## 3.4   Wavelengths

Since these changes in reflectance are wavelength dependent (Philpot 1989), those wavelengths of the electromagnetic spectrum which allow the deepest water penetration will allow the deepest measurement of water depth. For water depths greater than a few metres, only the wavelength range from 400–600nm can obtain sensible data. Wavelengths shorter than 400nm are rapidly absorbed in all but the clearest waters, and wavelengths longer than 600nm are quickly attenuated (Austin 1974).

In pure water, blue light penetrates best (Jerlov 1968; 1976) providing a reflectance peak in the 450–500nm range, but the presence of suspended particles shifts this reflectance peak to the blue-green (490–550nm) portion of the spectrum (Roberts and Liedtke 1986). In sea water, the optimum single bandwidth for penetration is 470–570nm (Tanis *et al.* 1984). Practical work has shown that, for water penetration, the green band (500–600nm) works best for sea water, and that the use of the red band (600–700nm) for discrimination of bottom reflectance from suspended sediment (combined with the green band water penetration) yields useful results (Helgeson 1970, Leu 1977, Miller *et al.* 1977, Roberts and Liedtke 1986, Liedtke *et al.* 1986, Ackleson *et al.* 1987).

The decline in transmittance of the red band (560–600nm) in pure water (Jerlov 1968; 1976) is reflected in the characteristic absorption length of water. The characteristic absorption length of water is defined to be the depth at which $1/e$ of the downwelling energy from the surface remains. This characteristic length is dependent on the wavelength of light being used to calculate the value; about 0.6m over the red band (600–700nm) and about 6m over the green band (500–600nm): an order of magnitude difference (Williams 1970). Thus high red returns and low green returns may indicate the presence of either very shallow water or large quantities of suspended sediment (Roberts and Liedtke 1986).

Comparison of intensity values are, therefore, subject to alteration by suspended sediments, and reflectance comparisons can only be made when the water is absolutely clear in all cases, or has a constant, but small, suspended sediment component in all imaged areas. Clear water gives the optimal conditions for water penetration (Roberts and Liedtke 1986), but these conditions are rarely ever met outside the laboratory. Maximum depth penetration is about 30m for optical sensors and accuracy decreases rapidly for water greater than 20m deep, but this is not too great a problem since most chart-using ships have drafts of less than 20m (Robinson 1985).

## 3.5  Problems

Penetration limits effectively negate the use of the infrared (IR) band (700–1100nm) over water, although it may be used to discriminate between subaerially exposed regions of the data because it is mainly reflected by vegetation and soil (Roberts and Liedtke 1986). Water saturated plants tend to reflect less IR energy than dry plants (Gausman *et al.* 1983), so even submerged vegetation yields little IR signal beyond certain depths. Clear water, best suited to green and red reflectance, yields low IR returns (Liedtke *et al.* 1986), so IR is only useful in the very shallow water areas where vegetation provides some reflectance. The IR band is useful, however, in helping to distinguish between vegetated and barren bottom types in very shallow environments in case the presence or absence of vegetation affects the returning bottom reflectance.

The presence of chlorophyll in the water may cause some IR reflectance, but chlorophyll also increases green reflectance and decreases blue reflectance by increasing absorption around 475nm and increasing scattering around 550nm (Miller *et al.* 1973, 1977). Due to the effect of chlorophyll, the quantity of chlorophyll should also be constant across the imagery if comparisons of reflectance need to be made (Roberts and Liedtke 1986). The presence of chlorophyll degrades the return signal for bathymetric data (Miller *et al.* 1973), so the use of the IR band may help identify imaged regions where bathometric data could be largely in error due to the presence of chlorophyll.

In addition to the problems of ensuring penetration in the water, and the identification of contaminated regions, other difficulties arise during the calculation of depth. Changing bottom reflectance, caused by changing bottom type, makes the calculation of depth from single band measurements unreliable (Paredes *et al.* 1983). The use of multispectral imagery overcomes this in some ways, but the calculation of depth using band ratioing may still be affected by changes in bottom reflectance. These changes in bottom reflectance are not easily distinguishable from colour changes due to water depth variations. This problem may mean that two different bottom surfaces at the same depth may yield different depth results unless the changes in bottom reflectance are taken into account (Lyzenga 1978).

Specular reflectance on the water surface, caused by the incident solar radiation, can confuse and disrupt calculations of water depth (Lyzenga 1978). Specular reflectance changes the intensity of the sensed light without any influence from water depth and reduces the water penetration (by drowning it out with surface brightness) so that the depth calculated will be in error (Lyzenga 1978; Roberts and Liedtke 1986).

In addition to the above difficulties, increasing suspended sediment concentrations shift the peak reflectance of water towards the red portion of the spectrum and towards the peak reflectance value the sediments would have were they subaerially exposed to the sensor (Roberts and Liedtke 1986). This shift in peak reflectance could alter any band ratio values such that depth values would be calculated in error. Such spectral peak shifts may be corrected by comparing IR reflectance values in a manner similar to the correction for chlorophyll (Roberts and Liedtke 1986). Transmittance through water decreases as the turbidity of the water increases, so penetration is greatly affected by suspended sediment.

Bottom reflectance, and hence depth values, are not directly observable due to attenuation and scattering (Lyzenga 1979). Changes in water attenuation can also reduce bottom reflected energy, making the water seem deeper than it really is and altering the calculation of the water depth (Lyzenga 1979). Increased water attenuation in a marine environment is not a function of the dissolved sea salts but of the contamination in the water due to such pollutants as suspended particulate matter (Jerlov 1968; Williams 1970; Jerlov 1976). The effects of such changes may be calculated if the water depth and the optical parameters of the water are known at each point (Lyzenga 1981). Bottom albedo cannot be used directly as a measure of reflectance because the relationship between zero-depth reflectance and albedo is not a linear function due to complication of the relationship by light scattering effects of water and internal reflections with changing depth (Lyzenga 1977). Models assuming that reflectance is a linear function of albedo are often in error (Lyzenga 1977). Not only do the optical properties of water affect the reflectance signal, but since the transmission of the atmosphere varies with time (Austin 1974), it too must be taken into account for a rigorous theoretical model to be accurate

# 4　Shallow water bathymetry: methodology

## 4.1　Introduction

Bathymetric measurements depend on finding a relationship between measured water depth and the reflected radiances in one or more wavelengths (Tanis *et al.* 1984). The accuracy of those depth calculations will depend on how well atmospheric, air-water interface, water, and water-bottom interface optical properties are known, and how they contribute to the modelled radiance values (Weidmark *et al.* 1981). Of considerable importance is the use of a supervised classification approach since many of the above parameters cannot be accurately determined in most operational circumstances. As long as the variance is within acceptable depth limits, the use of ground truth and training areas correlated with the spectral data should be the most effective procedure.

## 4.2 Methods

Several methods of remote sensing of water depths have been proposed in order to bypass or account for changes in bottom reflectance, surface reflectance, attenuation, reflectance shifts, penetration, and other optical characteristics of water.

Pincus (1959) proposed the use of terrestrial horizontal stereo photography to spatially locate graduated rods placed along the coastline, and thus bypass the need for calculating the optical characteristic variables. The rods gave depth at a specific point and the photographs located the rods in two dimensions, thus all three dimensions could be measured. This method suffered from the application of point data to a spatial dimension, but was one of the first methods suggested for accurately locating measured depths using photography.

Musgrove (1969) suggested placing calibrated grey-scale targets at different depths underwater and taking imagery of the targets in addition to taking imagery of the same grey- scale target type in a subaerially exposed position. The change in intensity over each target could then be attributed to the increasing depth of the water column and, hence, yield depth information. The depth information collected by this method could only be relative though because the actual depth of the targets was not known.

Lyzenga (1978) proposed an analytical model whereby depth could be calculated directly from spectral data. The model was based upon the radiance model for water whereby the radiance received at the sensor could be calculated as a function of reflected radiances in the form:

$$L_i = L_{si} + k_i r_{bi} e^{-K(i)fz} \tag{15}$$

where $L_i$ is the radiance received at the sensor, $L_{si}$ the radiance over deep water (external reflection and scattering, $k_i$ the irradiance/transmittance/refraction coefficient, $r_{bi}$ the bottom reflectance, $K(i)$ the attenuation coefficient of water, $f$ the pathlength through water and $z$ the water depth, all for band $i$ and bottom type $b$.

Lyzenga proposed that as long as the ratios of bottom reflectance between bands were constant over the collected imagery, then, in accordance with the above reflectance model (15), the depth of the water at any point could be calculated by:

$$z = \frac{1}{(K_1 - K_2)f} \ln\left(\frac{k_1}{k_2}\right) - \ln\left(\frac{R}{R_b}\right) \tag{16}$$

where $R = (L_1 - L_{s1})/(L_2 - L_{s2})$ and $R_b$ is the bottom reflectance ratio defined as

$$R_b = \frac{r_1 b_1}{r_2 b_1} = \frac{r_1 b_2}{r_2 b_2}$$

In the above, $r_i b_j$ is the reflectance value for band $i$ over bottom type $j$. The equation (16) only worked if the bottom reflectance assumption was valid.

Paredes *et al.* (1983) suggested that a different depth model could be calculated from a similar radiance model. They proposed that, as long as the same bottom reflectance ratio assumption put forth by Lyzenga (1978) held true, then water depth could be calculated by:

$$z = \frac{1}{f(C_1 K_1 + C_2 K_2)}(1 - C_1 X_1 - C_2 X_2) \tag{17}$$

where $f$ is the pathlength through water and the $C_i$ are the ratio coefficients chosen such that:

$$(r_1 b_j)C(r_2 b_j)C = \alpha \tag{18}$$

holds true for all bottom types $j$ , $K_i$ is a function of solar radiance for band $i$ and $X_i$ is defined as $\ln[(L_i - L_{si})/K_i]$ where $L_i$ is the radiance in band $i$ over site and $L_{si}$, is the radiance over deep water in band $i$.

For the equation presented above (17), the pathlength $f$ was commonly set at 2 to account for the light energy having to pass through twice the water depth in order to pass from the surface to the bottom and back to the surface again. The accuracy of the model relied on the exponential adjustment of the bottom reflectance ratios so that the band specific reflectances raised to the power of the ratio coefficient $C_i$ and multiplied by one another equalled some pre-determined value for all bottom types. This value was defined as the $\alpha$ variable and was set equal to the value of $e$ for the sake of convenience. The equality presented in (18) was assumed to hold true for all bottom types (Parades *et al.* 1983).

Philpot (1989) returned to the radiance model that Lyzenga (1978) made use of and proposed that this be adjusted to create a shallow water depth model whereby

$$L_d = L_b e^{-gz} + L_w \tag{19}$$

where $L_d$ is the radiance at detector, $L_b$ the radiance term associated with bottom reflectance, $g$ the attenuation coefficient of water, $z$ the depth of water and $L_w$ the remotely observed radiance from deep water.

In order to back calculate depth from this model, $L_b$, $L_w$, and $g$ were needed in addition to the radiance term $L_d$. If $L_d$ was known at two or more depths then the equation (19) could be used to back calculate $L_b$ and $g$. Philpot (1989) showed that in most cases for coastal green-blue water, the value of $g$ could be calculated as $g = 2K_i$ where $K_i$ had the values of 0.095 for the red band and 0.432 for the green band. Additional values of $K_i$ for green coloured water were calculated to be 0.103 for the red band and 0.392 for the green band. Given that these values could be used anywhere that the appropriately coloured water occurred, the knowledge of $L_d$ at two or more depths along with measurements of $L_w$ meant that (19) could be used to calculate values for $L_b$ by residual.

All three analytical methods (Lyzenga 1978, Parades *et al.* 1983, Philpot 1989) for calculating water depth work well in theory but are operationally difficult to implement particularly as a result of their dependence on the measurement of many hard to measure variables such as albedo (Lyzenga 1977). The values of attenuation and reflectance have not been precisely determined in shallow waters (Philpot 1989), yet these are integral to the viability of the models.

## 4.3  Case studies

Modelling of the reflectance characteristics and determining methods for remote sensing of water data required links between theory and application. Various studies were undertaken using a variety of sensing systems including Landsat TM, SPOT, airborne MSS, and airborne SAR. The case studies revealed different strengths and weaknesses in the different technologies.

Landsat applications to bathymetry were investigated by several test studies. The NASA/Cousteau Ocean Bathymetry Experiment proved that Landsat imagery could be as accurate as 10% for depth estimations in water up to 22m deep (Hammack 1977).

Warne (1978) investigated the feasibility of Landsat bathymetric mapping using the relationship:

$$L = a + be^{-cz} \tag{20}$$

where $L$ is the radiance received at the detector, $a$, $b$, $c$ were coefficients dependent on the optical nature of water and were determined empirically and $z$ is the water depth.

The relationship was dependent on wavelength and the coefficients varied from place to place requiring local calibration of the data. Endeavour Strait off the northwest coast of Australia was selected as the test site where the maximum single-band penetration was determined to be 38.8m. The single-band measurement of water depth yielded accuracies of +0.6 m at 10m and +1.8m at 20m depth. Most errors over the entire depth range averaged to +1.0 m or 10%, whichever was greater (Warne 1978).

Chong *et al.* (1979) proposed a single channel model for Landsat bathymetry of the form:

$$L = L_s + kRe^{-kfz} \tag{21}$$

where $L$ is the radiance observed above water, $L_s$ is the deep water radiance, $K$, $f$ are constants, $R$ is the bottom reflectance, $k$ is the effective attenuation coefficient of water and $z$ is the depth of water. This formula could then be rearranged to yield:

$$z = Ax + B \tag{22}$$

where $x = \ln(L - L_s)$ and $A$ and $B$ are constants determined by radiance levels for pixels of known depth. Applying the model to a ratio formula one arrives at a ratio method for Landsat bathymetry of the form:

$$Z = A\ln R + B \tag{23}$$

where $Z$ is the depth, $A$, $B$ are correlation coefficients and

$$R = \frac{(L_4 - L_{s4})}{(L_5 - L_{s5})} = \frac{\text{band4}}{\text{band5}} = \frac{\text{green}}{\text{red}}$$

No measurements were given to determine the accuracy of this ratio method, but the technique was used for Landsat imagery over the sea around Singapore.

Polcyn *et al.* (1979) compared the single-band characteristics for Landsat MSS4, blue-green, and Landsat MSS5, red. The accuracy of the water depth estimation was increased to as small an error as 2.5% when variables of water attenuation and bottom reflectance were known. Away from the control points used in the study, accuracy decreased to a mean 10% error at 22m depth. Maximum penetration depths for the sensors were 21.9m and 24.8m for MSS4 in low-gain and high-gain mode respectively, and 5.6m and 6.1m for MSS5 (low-gain and high-gain mode respectively). The test was carried out near the Bahamas under clear-water conditions.

Davis *et al.* (1982) tested the model proposed by Lyzenga (1978), (Equation (16)), with Landsat data imaged off the coast of Tunisia. Depth measurements were accurate to 3m depth using a ratio of MSS4/MSS5. Measurements over deep water yielded returns for the two wavelength regions of 148, 142 respectively on an 8-bit 0–255 digital number scale.

Difficulties with the Landsat techniques arose from the fact that the sensor system had no detector optimally placed in the spectrum for bathymetric measurements (Warne 1978). The best response was returned by the MSS4 and MSS5 bands. The IR channel, MSS7 showed usefulness in distinguishing land from water. Landsat, like other optical sensing platforms, was hindered by suspended sediment and often mistook water turbidity for bottom reflectance (Robinson 1985). The penetration limit of Landsat was determined to average around 20 m for the MSS4 sensor (Warne 1978, Bunn *et al.* 1983), but this penetration was reduced in turbid waters. Davis *et al.* (1982) found that Landsat predicted depths of about 3 m in turbid waters known to be 25m deep. The exponential form of the relationship between depth and penetration caused the depth resolution to decrease as the water depth approached the penetration limit (Warne 1978).

Not only did water characteristics affect Landsat applications to bathymetry, but, having to sense through the entire atmosphere, it was hampered by atmospheric conditions and worked only when periods of cloud-free, fine weather coincided with an overpass (Bunn *et al.* 1983). Calibration differences between detectors could not be entirely eliminated and the variations in calibrated data suggested that those variations were not independent of the scene viewed, possibly due to a memory effect of the photomultiplier detectors (Polcyn *et al.* 1979). The frequency of Landsat overpasses allowed for multi-image comparisons to correct for tidal current and meteorological disturbances in the imagery (Warne 1978).

Landsat TM was investigated by Tanis *et al.* (1984). Their study, over the Great Bahama Bank, yielded a correlation coefficient of 0.825 and a standard error of +1.32m for a two-band measurement technique of water up to 18m deep. A single-band measurement technique run on the same imagery yielded a correlation coefficient of 0.762 and a standard error of +3.53m. Depth soundings for the study were accurate to +0.01m. The TM bands used were TM1 and TM2 for the two-band technique, and TM1 for the single-band technique. The errors for TM2, which approximates the low-gain Landsat MSS4 response, were half those of MSS4 (Tanis *et al.* 1984). TM3 was also used but returned minimal results due to its penetration limit of less than 3m.

A SPOT study undertaken by Ferguson *et al.* (1987) used bands SPOT 1 (500–590nm) and SPOT 2 (610–680nm) to extract water depth information from clear oceanic waters around an atoll in the Pacific Ocean. They used a single-band method using SPOT 2 for water less than 2m deep, SPOT 1, for water greater than 16m deep, and a band ratio technique based on an algorithm similar to that of Lyzenga (1978), (2), which required $V_s$ the deep water signal, $V_0$ the shallow water signal, $k$ the attenuation coefficient of water, and $f$ the path length as input factors. The data revealed that, for water less than 2m deep, sea surface noise and changing bottom reflectance hampered measurements. For water greater than 16m deep, noise from the sensor and surface waves also influenced the received signal. Ferguson *et al.* (1987) noted that although their study appeared to yield accurate results, the estimated depth measurements needed to be checked by calibration against depth soundings made in the waters around the atoll.

A study by Hollinger *et al.* (1975) over Lake Ontario, Canada, using an airborne CCD-pushbroom scanner, yielded depth measurement accuracies of +0.5m or less. The study involved measuring upwelling radiance at the sensor altitude, including atmospherically scattered light estimated by dark object subtraction of the red return as measured over deep water. The water signal was then divided by the atmospheric attenuation coefficient to obtain the total reflected signal at the sea surface, from which was subtracted the reflected surface component, leaving the internal and bottom reflectance signal. This signal was compared against bottom albedo, measured at zero-depth, to determine the water depth at that point.

Wiedmark *et al.* (1981) and Jain *et al.* (1982) tested an MSS pushbroom scanner over the Bruce Peninsula region of Ontario, Canada, yielding a correlation coefficient of 0.89 and a standard error of +0.56m between modelled depth, based on reflectances, and measured depth, measured to an accuracy of +0.01m. The correlation equation linked measured depth to modelled depth by:

$$\text{Measured depth} = 0.98 \times \text{Modelled depth} + 0.32 \tag{24}$$

with a mean error of estimation of ± 0.15m.

Lyzenga (1981) compared Landsat and airborne MSS ERIM-8 data using the model proposed by Lyzenga (1978), (2), to calculate the attenuation coefficients of water for both techniques. Lyzenga (1981) compared an airborne green (520–570nm) to red (620–700nm) ratio to the Landsat MSS4/MSS5 ratio. Attenuation coefficients were calculated to be 0.33 and 0.24 for the ratios respectively. The airborne data proved more suitable to bathymetric measurements than the Landsat data because the accuracy of the airborne imagery was acceptable down to 15 m depth compared with only 5m for the Landsat imagery (Lyzenga 1981).

## 4.4 Reliability

The case studies suggest that research still needs to be pursued to prove the reliability of remotely sensed water-depth data. A fundamental problem for such a proof arises from the reliance on electronic navigation systems for accurately positioning airborne data, although (in the case of airborne sensors) horizontal control may be provided by supplemental aerial photography showing shorelines and/or reference buoys (McKim *et al.* 1984). Concern on the marine side for the reliability of the measured depths stems from the assessment of passive surveys as being unable to produce the accuracy required by hydrographic charts (Tanis *et al.* 1984). In infrequently travelled areas, a ±0.15m accuracy airborne survey would prove better than no survey at all, and if the quality of surveys need not be better than an *rms* error of $\pm 10\%$, then remotely sensed bathymetry can significantly decrease the time in production and the time and money spent surveying by reducing necessary acoustic soundings by a factor of 50 (Warne 1978).

Satellites rarely provide the spatial resolution needed for bathymetric mapping in association with such activities as dredging and, although Landsat MSS and Landsat TM may cover large areas, they need further investigation to determine the reliability of their calculated depths (Bunn *et al.* 1983, McKim *et al.* 1984). Bunn *et al.* (1983) suggested that satellite remote sensing of bathymetry could:

- provide excellent images of shallow seas that change slowly;

- allow precise location of shoals and reefs

- validate old charts, and

- reduce the risks and costs of mapping shallow water.

Both Landsat and airborne MSS results indicate that simple regression equations will perform acceptably with band ratio water depth prediction procedures (Tanis *et al.* 1984). Airborne laser sounders, while hampered by poor penetration in turbid waters, may prove useful for waters up to about 20m depth (McKim *et al.* 1984).

Part of the problem, with the use of optical procedures for proxy bathymetric data, lies with the fact that conventional maximum likelihood classifiers do not accurately map bottom features under changing water depths because of water induced attenuation of the spectral signals (Lyzenga *et al.* 1979). Some success has been had in classifying bottom types by means of optical data. Lyzenga *et al.* (1979) used an ERIM M-8 scanner to classify two categories of bottom types to an accuracy of 82.7% using three bands, and 4 categories of bottom types to an accuracy of 65.1%. Davis *et al.* (1979) used an airborne MSS sensing platform to classify 5 types of beach mineralogy to an accuracy of greater than 99% sensing only for moisture content and grain size. These two studies indicate that optical data may be used to classify bottom types such that optical data, used in proxy for bathymetric data, may be collected as if the bottom type was invariant over the scene because bottom type, in shallow water, can be mapped as well. The model proposed by Lyzenga (1978), (equation (16)), used invariant bottom type as a central assumption.

McKim *et al.* (1984) assessed the applicability of remote sensing technology to bathymetric surveys and arrived at a three-fold classification for the value of the data retrieved: Class 1 for reliable data, Class 2 whenever additional testing is needed, and Class 3 when the data is of limited value. Shipborne sonar and airborne laser-profiler data were placed in the first class because the  data collected was either a result of a tried and true method or showed great promise respectively. Airborne multispectral scanners were placed in the second class partly as a result of limited testing at present. Airborne panchromatic, airborne colour film, and satellite multispectral scanners were placed into the third class due to the poor reliability of the data obtained.

This evaluation suggests that airborne multispectral video (MSV) imagery probably belongs in the first class for shallow water but requires additional testing thus warranting placement in the second class.

# 5  Shallow water bathymetry: a field study

## 5.1  Introduction

A field study was undertaken to evaluate the performance of multispectral video (MSV) imagery for performance in mapping shallow water bathymetry. Shallow water imagery was collected at various altitudes on two separate days, and analysed for the green (500–600nm), red (600–700nm), and infrared (700–1100nm) bands using an airborne MSV remote sensing system. A simple regression procedure was used for the calculation of water depth in a shallow marine environment. The results indicate acceptable performance in shallow water and suggest that with further refinement the system could be used as a quick estimate of near shore bathymetry.

## 5.2  Methods

The illumination, atmospheric transmission, water attenuation per unit water depth, sea state, surface reflectance, bottom reflectance, and water body radiance per unit volume were all assumed to be constant over the imagery. Any changes present in recorded radiance in each band were assumed to be due to changes in water depth. The relation between the depth $z$ and the recorded radiance was assumed to have the form:

$$f(z) = a + b\frac{L_1}{L_2} \tag{25}$$

where $a$ and $b$ are linear regression coefficients and $L_i$ is the radiance measured in band $i$. The equality in (25) was reliant upon a linear form similar to that of Warne (1978) and Chong *et al.* (1979), assumed to hold true for all bottom types in the study area and considered to be independent of sea state at the time of imagery.

The above method was derived from a similar reflectance model to that of Williams (1970), based on the partitioning of the incident light at the water surface. The reflectance model had the form:

$$E_s = a_s(R_0 + R_w) + E_a \tag{26}$$

where $E_s$ is the light received at the sensor, $a_s$ the attenuation coefficient of the atmosphere, $R_0$ the surface reflectance component, $R_w$ the internal reflectance component and $E_a$ the scattered atmospheric light reaching the sensor.

According to the reflectance model used, the incident light, $E_0$, split into two components when incident at the water surface from above:

$$E_0 = R_0 + P_0 \tag{27}$$

where the reflected component is

$$R_0 = r_a E_0 \tag{28}$$

the penetrating component is

$$P_0 = (1 - r_a)E_0 \tag{29}$$

and $r_a$ is the reflectance coefficient of the water surface from above.

The penetrating component, $P_0$, reached the penetration limit or the sea bottom after having been attenuated by the water. The light reaching the light limit, or bottom, could be expressed as:

$$P_i = a_w P_0 \tag{30}$$

where $P_i$ is the penetrating component of light to reach depth and $a_w$ is the attenuation coefficient of water.

This component $P_i$ was reflected off the bottom, if the bottom was visible (the useful case of this model for the proposed study), and was converted to upwelling light calculated by:

$$R_b = l_b P_i \tag{31}$$

where $R_b$ is the reflected light from the bottom and $l_b$ is the bottom albedo.

The light reflected from the bottom would upwell at the surface at an intensity determined by the transmission characteristics of the water. This upwelling light would have the form:

$$R_s = t_w R_b \tag{32}$$

where $R_s$ is the upwelling light incident at the surface and $t_w$ is the transmission coefficient of water.

The upwelling light at the surface would be split into two components due to the reflectance of some of the light arriving from below the water-air interface. The upwelling light would then be split into:

$$R_s = R_w + R_r \tag{33}$$

where $R_w$ is the internally reflected component of light viewed from above the surface and $R_r$ is the portion of light re-reflected back into the water.

$$R_w = (1 - r_b)R_s \tag{34}$$

$$R_r = r_b R_s \tag{35}$$

where $r_b$ is the reflectance coefficient of the water surface for water arriving from below.

Substituting equations (28) through (35) into equation (27) yielded the equation:

$$E_s = [r_a E_0 + (1 - r_b)t_w l_b a_w(1 - r_a)E_0]a_s + E_a \tag{36}$$

Manipulation of (36) resulted in:

$$\frac{E_s - E_A}{E_0 a_s} = r_a + t_w l_b a_w(1 - r_a)(1 - r_b)$$

$$\frac{E_s - E_A - r_a(E_0 a_s)}{E_0 a_s l_b(1 - r_b)(1 - r_a)} = t_w a_w = e^{-z} \tag{37}$$

where the last equality follows by virtue of the attenuation definition: $a_w = e^{-kz}$ and $t_w = e^{-(1-k)z}$.

Simplifying the left side of the equation to factor out $E_s$ and $a_s$, since they are dependent on depth and altitude respectively, resulted in the equation:

$$\frac{1}{l_b(1 - r_b)(1 - r_a)}\frac{E_s}{a_s}\frac{1}{E_0} - \frac{1}{a_s}\frac{E_a}{E_0} - r_w = e^{-z} \tag{38}$$

By taking logarithms and making some simplifying assumptions a relationship of the following form can be obtained:

$$z = A \ln E_s + B \qquad (39)$$

where $A$ is a correction factor for theoretical inaccuracies. The equation given by (39) suggests a linear model for depth versus the natural logarithm of the reflected energy received at the sensor. This linear equation was applied to the data collected for this study.[1]

## 5.3 MSV system

Data were acquired using an airborne MSV remote sensing system installed in a Cessna 185 turbo photo conversion light aircraft. The video system consisted of 3 Black and White Sony XC-37 CCD video cameras with 12mm f1.8 Sony lenses. The cameras were mounted in a transverse mode, and were filtered for red, green, and infrared bands using 500nm and 600nm Oriel bandpass filters, and an 88A 700+nm sharpcut filter respectively. The video cameras were linked to 3 Sony EV-C8u 8mm metal VCRs, and incoming signals were monitored with a waveform monitor, a video switching unit and a Sony KX 4200 100mm Trinitron colour monitor (after Roberts and Evans 1986). The automatic gain control on the cameras was disabled to allow for subsequent band ratioing in analysis.

## 5.4 Data collection

Imagery was collected on two separate occasions. The first data set was collected on April 3rd 1990, during high slack water, at an altitude of 1200ft over a NE-SW trending transect line. The transect was located between the Tsawwassen ferry terminal causeway and the Roberts Bank Superport causeway, in the Strait of Georgia, British Columbia. Video imagery at this altitude gave a nominal pixel resolution of slightly more than 1m with the 12mm lens (Meisner 1986). The transect consisted of eight targets—orange plastic garbage bags placed at regular intervals and anchored to the bottom by rope and stone filled plastic bottles. Depths at each target location were measured using a 3m boat hook and an electronic depth sounder where the water was deeper than 3m.

The second data set was collected on September 21st, 1990 during low slack water, at altitudes of 8000ft, 4000ft, 2000ft and 1000ft over two N-S trending transects in the middle of Boundary Bay, British Columbia. Video imagery taken at these altitudes gave nominal pixel resolutions of 8m, 4m, 2m and 1m respectively with the 12mm lens. A total of 15 targets were placed in the two transects, although not all were imaged on each overpass. Depths were measured at each target location using a plumb-line and an electronic depth sounder. Depths were taken upon placement and retrieval and were averaged for each set of target measurements. The eighth target, on the April 3rd transect, was only measured during placement as a result of a malfunction in the sounder during target retrieval.

---

[1] Editor's note: the derivation of Equation 39 appears highly dubious. The author later points out (see section 5.6 on Linear regression) that the data do not confirm Equation 39 at all.

## 5.5   Analysis

Initial analysis of the imagery was completed using a Xybion image processing system. Each target image was manipulated using a low pass filter to reduce image noise and the effects of spectral reflectance. Intensity measurements, for the April, 1990 data, were taken for 24 samples in an 8 pixel by 8 pixel spaced grid centred at each target location. Intensity measurements for the September, 1990 data were taken for 30 random samples in the area of the target. The average intensity values for the red, green, and IR bands were calculated for each target and were plotted versus depth for each overflight to identify the general relationship between depth and the imaged intensity returns (intensity was based on an 8-bit digital number scale from 0–255 for each pixel).

Average intensity values for the red and green spectral bands were also plotted versus depth separately so that altitudinal effects could be identified. The average intensity values for the IR band were not plotted as the returns for this band were mostly zero with the exception of the 8000ft and 1200ft data. These non-zero IR returns were attributed to optical port glass glare, visible on the imaged data taken at 8000ft, and to extremely shallow water conditions allowing submerged plant IR response for the data less than 1.5m deep on the 1200ft imagery.

Previous preliminary analysis of the April, 1990 data suggested that a red/green ratio would be useful for calculating water depths, this ratio having yielded a regression equation for $X = 1/\sqrt{z}$ of the form:

$$X = \frac{1}{\sqrt{z}} = 0.218 + 0.323(R/G) \tag{40}$$

The correlation value, measure of fit and rejection level for the fit were given by $r = 0.957$, $r^2 = 0.916$, and $p = 0.02$ respectively.

The success of this ratio technique over eight data points suggested that assumptions of uniformity held true in relatively clear homogeneous shallow water, and that the red/green ratio would be a good predictor of water depth. Consequently this ratio was calculated for the all the data sets and was plotted versus depth.

Linear regressions were run on both the original and averaged value data sets using Statview 512+ (Version 1.1 Brain Power Inc.: California) in order to determine the best predictor variables for water depth from the imaged spectral data.


## 5.6   Results

Observation of the plotted intensities versus depth, for each overpass, revealed the general trend in changes in intensity with respect to changes in measured water depth.


**Data variations by altitude**

During the 8000ft and 4000ft overpasses the red and green band intensity values did decrease with increasing depth, but the intensity change per change in unit depth was small when compared to other overpassess. Altitudinal effects, such as a large scattered atmospheric light component due to the distance from the target and such as the large attenuance of the water signal at this altitude compared to lower altitudes, probably overwhelmed what changes in the water signal there were to the point that the red and green returns appeared almost constant over the change in depths.

At 2000ft and 1000ft, it was possible to discriminate between vegetated and non-vegetated bottom types. Consequently, intensities were sampled for both bottom types about the targets, where such sampling was possible. A plot of these intensities showed that both the red and green reflectances were greater over a non-vegetated bottom than they were over a vegetated bottom at the same depth. The intensity plots for each bottom type tracked one another with respect to increasing depth, as did the plot of the red/green ratioed intensities over the two bottom types.

Analysis of the intensity data by bottom type for both altitudes using a paired *t*-test resulted in a 99.999% certainty that the intensity measurements over vegetated and non-vegetated bottom types were indistinct. This result suggested that intensity measurements for the red and green bands were independent of bottom type for the same depths. The 2000ft and 1000ft data revealed a greater change in intensity over the depth range than the higher altitude data: a result attributed to a smaller atmospheric component at this lower altitude.

The 1200ft data were collected at a different time of year and clearly showed a decreasing red intensity value with decreasing depth in a shallower water range than the other data sets. Green intensity stayed fairly constant suggesting that a red/green ratio would reasonably predict depth. The IR return decreased rapidly as the attenuation of the submerged vegetation signal reduced the IR component of the bottom reflectance to zero. The overall shallowness of the water may have been a contributing factor to the marked decrease observed in red intensities by showing the low attenuation in the shallow water of the red component of the bottom reflected light compared with the high red attenuation in the deeper water.

## Data variations by wavelength

Plots of both the red and green intensities over depth revealed that altitude had an affect on the received intensities. Red intensities increased with increasing altitude most likely as a result of a large scattered atmospheric component adding red wavelength light to the bottom signal return reaching the sensor. Green intensities also increased with increasing altitude but not on as large a scale as the red intensities. The green intensities recorded on the 1200ft data seemed unusually high in comparison with the other data, and this difference was attributed to the different light conditions and camera settings under which the Spring data were collected. The low red intensities on the 1200ft data between 1.5m and 5m were also attributed to differences in lighting conditions and camera settings.

## Data variations by ratio

A plot of the red/green ratios versus depth revealed altitudinally adjusted ratios on the September data which, upon inspection, tended to follow the similar trend of decreasing ratioed values with increasing depth.

| Altitude ft | r | $r^2$ | p | Std Error | Error m | Regression Equation |
|---|---|---|---|---|---|---|
| 8000 | 0.808 | 0.653 | 0.0005 | 0.206 | ±1.23 | $Y = +5.307 - 3.789X$ |
| 4000 | 0.801 | 0.642 | 0.0001 | 0.148 | ±1.16 | $Y = +3.852 - 0.025X$ |
| $2000^N$ | 0.853 | 0.728 | 0.0008 | 0.131 | ±1.14 | $Y = +2.364 - 0.011X$ |
| $2000^V$ | 0.749 | 0.561 | 0.008 | 0.167 | ±1.18 | $Y = +2.356 - 0.013X$ |
| 1200 | 0.961 | 0.923 | 0.0001 | 0.171 | ±1.19 | $Y = -0.372 + 0.453X$ |
| $1000^N$ | 0.684 | 0.468 | 0.0142 | 0.197 | ±1.22 | $Y = +2.062 - 0.011X$ |
| $1000^V$ | 0.427 | 0.182 | 0.1665 | 0.244 | ±1.28 | $Y = +1.846 - 0.007X$ |

**Table 3.** *Best single variable predictor equations by altitude for the simple linear regressions. Y is the ln(Depth) and X is the regression variable. The superscripts denote non-vegetated bottom (N) and vegetated bottom (V).*

The ratio plot also revealed a large difference between the April and September data in the red/green ratios as a result of the low red intensities recorded in the Spring. The difference between the two sets was assumed to be a result of different light, sea state, tidal, solar altitude, and locational properties inherent in the difference in data collection times in addition to different camera settings between the two flight times.

| Altitude ft | r | $r^2$ | p | Std Error | Error m | Regression Equation |
|---|---|---|---|---|---|---|
| 8000 | 0.929 | 0.864 | 0.001 | 0.135 | ±1.14 | $Y = 8.015 - 0.044R$ <br> $-0.011(G - IR)$ |
| 4000 | 0.836 | 0.699 | 0.001 | 0.137 | ±1.15 | $Y = -4.930 + 0.067R$ <br> $+8.971(G/R) - 0.094G$ |
| $2000^N$ | 0.930 | 0.864 | 0.001 | 0.099 | ±1.10 | $Y = 3.026 - 0.008R$ <br> $-0.012G$ |
| $2000^V$ | 0.749 | 0.561 | 0.008 | 0.167 | ±1.18 | $Y = 2.356 - 0.013R$ |
| $2000^A$ | 0.947 | 0.897 | 0.004 | 0.099 | ±1.10 | $Y = 6.254 + 0.029R - 0.046G$ <br> $-3.142(R/G) - 0.269(G/R)$ |
| 1200 | 0.983 | 0.967 | 0.001 | 0.137 | ±1.15 | $Y = -3.905 + 0.254(G/R)$ <br> $+0.031R - 0.004(IR)$ |
| $1000^N$ | 0.684 | 0.468 | 0.014 | 0.197 | ±1.22 | $Y = 2.062 - 0.011G$ |
| $1000^A$ | 0.684 | 0.468 | 0.014 | 0.197 | ±1.22 | $Y = 2.062 - 0.011G$ |

**Table 4.** *Multiple stepwise regressions by altitude. The superscripts denote non-vegetated bottom (N), vegetated bottom (V) and all types(A). R is the red band, G the green band, IR the IR band, (G/R) the green/red ratio, (R/G) the red/green ratio, (G − IR) the green band minus IR band.*

## Linear regressions

All linear regressions were run using the natural logarithm of depth as the dependent variable as this manipulation of the depth data yielded the best-fit regression equations.

Regression equations for the natural log of intensity versus the measured depth, as initially proposed, returned statistically insignificant correlation coefficients, so this method of investigation was replaced by a model dependent on regressions between the natural log of the depth and the recorded intensities or intensity ratioes. The failure of the proposed model to accurately predict water depth was ascribed to an over simplification of the variables involved and the possible neglect of some important controlling factor in the derivation of the simplified approach.

Simple linear regressions run on all the data identified single predictors for the different data collection runs. Tables 3 and 4 show the result of linear regression analysis by altitude and by spectral bands. The red band intensities proved to be the best single predictor for the 4000ft and 2000ft altitude data, the green band for the 1000ft altitude data, the red/green ratio for the 8000ft altitude data, and the green/red ratio for the 1200ft altitude data (all predictor variables denoted X: Table 4).

Re-manipulation of the 1200ft data yielded a multiple linear regression equation predicting water depth in metres from the green/red ratio, IR intensity, and the green intensity. The linear regression for this equation (Table 4) gave a better fit than the equation for intensity versus the reciprocal of the square root of the depth (40).

The general success of the simple linear regressions, the 1000ft data collected over a vegetated bottom excluded (for which no statistically significant best predictor could be identified), suggested a model of a different form than the one initially proposed, so the previously derived model was replaced by an alternate model, which fit the collected data, of the form:

$$\ln z = a_i X_i + b \tag{41}$$

where $z$ is the water depth and $a_i$ the regression slope coefficient for $X_i$, the regression variable $i$ , and $b = b_i$, the regression $y$-intercept coefficient.

The above model (41) was run as a multiple regression for the best predictor values identified in the simple linear regressions and the overall stepwise multiple regressions. The most commonly occuring predictor values (Red, Green, Red/Green, Green/Red) were input into a stepwise regression along with altitude to account for altitudinal variation. In order to ascertain the compatibility of the separate sets of data, the above regression (41) was test run over all data sets combined: the largest sample size possible with the collected data. The best regression, over all the data achieved by this method was:

$$\ln(\text{depth}) = 0.812(R/G) + 0.702\ln(\text{altitude}) - 0.019(R) - 3.244 \tag{42}$$

with $r = 0.869$, $r^2 = 0.755$, $p = 0.0001$ and standard error of 0.198. The standard error corresponded to an accuracy of depth prediction of 1.22m over the 8m depth range for the entire data set. The regression line was plotted with 95% confidence intervals for the slope of the graph. The small variation in the slope indicated a definite relationship between returned intensties and water depth, thus the overall size of the depth estimation error was attributed to scatter in the data points.

This formula (42) was applied to each of the data sets in turn, with each equation corrected by the altitude variable. The overall relationship between the independent data sets and the regression equation yielded the correlation coefficients listed in Table 5.

These coefficient values were compared with the correlation coefficients from the

| Altitude (ft) | 8000 | 4000 | 2000 | 1200 | 1000 |
|---|---|---|---|---|---|
| $r^2$ | 0.292 | 0.464 | 0.544 | 0.734 | 0.562 |

**Table 5.** *Multiple regression coefficients for test equation (42).*

stepwise regression (Table 4). The reduction in the correlation coefficients between the separate equations and the application of the universal equation (42) showed that separate equations for each set of conditions tended to work better. The regression equation (used for Table 5) fitted better on the lower altitude data probably as a result of less interference in the water signal by scattered atmospheric light. On all separate regressions, the linear equation tended to over-predict water depth in waters shallower than 4m deep, and under-predict water depth in waters deeper than 4m. Around 4m depth, the equations predicted water depth well, but this was attributed to the variation of sampled water depths about a mean of approximately 4m.

## 5.7   Conclusions

Although, for the data collected at all altitudes above 1000ft, single variable predictors of water depth could be reasonably well determined, and separate sets of variable could be used reasonably well to predict depth, the creation of a universal regression equation, as a test of the uniformity of the data, did not fit the entire set of data well. Although this sort of fit was not expected to be statistically significant, the test regression revealed that separate equations for differing conditions would work better, but that similar, if not identical, variables could be used in these equations. Such intensity measurements for predicting water depth were assessed to be time and site dependent, and the initial success of predicting depths on the 1200ft altitude data failed to be repeated with the September data. Correlation coefficients as high as 0.923 for simple regressions and 0.967 for multiple stepwise regressions were found , but the standard depth estimation error of the regression equations of roughly 1.20m, over a test depth range of 0 to 8m, indicated little applicable use for this technique in the prediction of shallow water depths unless an algorithm could be found to analyse the data without requiring exponential functions which seemed to enlarge the error estimate: the minimum error using the exponential function will always be around +1.00m (since $e^0 = 1.00$). A more accurate method of measuring depth would also be required in order to reduce variability in that component of the equation.

The correlation coefficients generated by regressions between intensity measurements and water depth indicated that reflectances do decrease with increasing water depth. Altitudinal variations between identical target points indicated that altitude played a role in the sensed reflectance value by adding a scattered atmospheric light component to the water signal. Of the Boundary Bay, B.C., data, intensity measurements collected at about 2000ft altitude gave the best overall correlation and standard error combination. Below this altitude variations in the sensed intensities swamped any trends occurring over changing depth, and above this altitude the addition of an atmospheric scattered component to the sensed data tended to drown out any changes in the returning water signal as a proportion of the received intensity.

Based on the collected data, airborne multispectral video imagery applied to remotely sensed bathymetric projects should be rated in the Class 3 limited value category outlined by McKim *et al.* (1984) due to a standard error of depth estimation of 1.22m ($r^2 = 0.755$, $p = 0.0001$) for an overall linear regression given by equation(42).

At best, the technique can be rated as Class 2 requiring additional testing. The failure of this simplified model to accurately predict changing water depth suggests that airborne multispectral video not be used for bathymetric surveys in conjunction with relatively simple algorithms. More complex algorithms, which take many more optical variables into account may prove more robust and accurate, but the results of this study seem to point to the futility of such investigations, a similar conclusion to that of Jain *et al.* (1982) for airborne multispectral scanners.

# References


Ackleson S G and Klemas V, 1986, Two-flow simulation of the natural light field within a canopy of submerged aquatic plants, *Applied Optics*, **25(7)** 1129

Ackleson S G and Klemas V, 1987, Remote sensing of submerged aquatic vegetation in lower Cheasapeake Bay. *Remote Sensing of the Environment*, **22** 235

Austin R W, 1974, The remote sensing of spectral radiance from below the ocean surface. *Optical Aspects of Oceanography*, N.G. Jerlov, E. Steelman Nielson Eds. (Academic Press Ltd., London), 317

Bunn F E, Domb U, Huntley D, Mills H, Silverstein H, 1983, *Oceans from Space: towards the management of our coastal zone.* The Institute for Research on Public Policy: Montreal, 82

Chong Y J, Liang T Y, Yeo A C and Vong V K, 1979, Remote sensing of the sea around Singapore. *Proceedings, 13th International Symposium on Remote Sensing of the Environment*, Ann Arbor, Michigan, **3** 1807

Davis C F, Shuchman R A and Suits G H, 1979, The use of remote sensing in the determination of beach sand parameters. *Proceedings, 13th International Symposium on Remote Sensing of the Environment*, Ann Arbor, Michigan, **2** 775

Davis P A, Grolier M J, Eliason P T and Schultejann P A, 1982, Analysis of bathymetry and submarine topography off the coast of east-central Tunisia with Landsat multispectral data. *Proceedings, International Symposium on Remote Sensing of the Environment, 1st Thematic Conference*, Ann Arbor, Michigan, **2** 859

Duncan D and Williamson D, 1988, Linear regression analysis of water quality in the Northern Flood Agreement study area, Federal Ecological Monitoring Project Report, Ottawa, Ontario.

Environment Canada, 1983, Sampling for water quality, Water Quality Branch, Inland Waters Directorate, Ottawa, Ontario, pp. 1-55

Environment Canada, 1986, Naquadat: Dictionary of parameter codes, Water Quality Branch, Inland Waters Directorate, Ottawa, pp. 1-319

Ferguson K P, Tanis F J and Tyler W A, 1987, SPOT bathymetric image for archeological investigations. *Proceedings, 21st International Symposium on Remote Sensing of the Environment*, Ann Arbor, Michigan, **2** 863

Gausman H W, Escobar D E, Bowen R L, 1983, A video system to demonstrate interaction of near infrared radiation with plant leaves, *Remote Sensing of the Environment*, **13** 363

Geary E L, 1968, Coastal hydrography, *Photogrammetric Engineering,* **34(1)** 44

Guilbault R A, Gummer Wm D and Chacko V T, 1979, The Churchill Diversion: water quality changes in the lower Churchill and Burntwood rivers, Water Quality Interpretive Report, No. 2, Ottawa, pp. 1-10

Hammack J C, 1977, Landsat goes to sea. *Proceedings, 37th Meeting American Congress on Surveys and Mapping,* Washington, D.C., 187

Helgeson G A, 1970, Water depth and distance penetration, *Photogrammetric Engineering,* **36(2)** 164

Hollinger A B, O'Niell N T, Dunlop J T, Cooper M T, Edel H, Gower J F R, 1975, Water depth measurement and bottom type analysis using a two-dimensional array imager, *Proceedings, 19th International Symposium on Remote Sensing of the Environment,* **2:** Ann Arbor, Michigan, 553

Jain S C, Zwick H H, Weidmark W C and Neville R A, 1982, Passive bathymetric measurements of inland waters with an airborne multi-spectral scanner. *Proceedings, International symposium on Remote Sensing of the Environment,* 1st Thematic Conference, Ann Arbor, Michigan, **2** 947

Jerlov N G, 1968, Optical Aspects of Oceanography, *Elsevier Oceanography Series* No 5: Elsevier, Amsterdam, 194

Jerlov N G, 1976, Marine Optics, *Elsevier Oceanography Series* No 16: Elsevier, 231

Johnson R W and Munday J C, 1983, The Marine Environment, Manual of Remote Sensing, Second Edition, Volume II, R.N. Colwell (ed.), American Society of Photogrammetry, Falls Church, pp. 1371-1496

Leu D J, 1977, Visible and near infrared reflectance of beach sands: a study on the spectral reflectance/grain size relationship, *Remote Sensing of the Environment,* **6** 169

Liedtke J, Roberts A, Evans D, 1986, Discrimination of suspended sediment and littoral features using multispectral video imagery. *Proceedings, 10th Canadian Symposium on Remote Sensing, Edmonton,* 739

Lillesand T M and Kiefer R W, 1987, Remote Sensing and Image Interpretation, Second Edition, John Wiley and Sons, Toronto

Lyzenga D R, 1977, Reflectance of a flat ocean, *Applied Optics,* **16(2)** 282

Lyzenga D R, 1978, Passive remote sensing techniques for mapping water depth and bottom features. *Applied Optics,* **17(3)** 379

Lyzenga D R, 1979, Shallow-water reflectance modelling with applications to remote sensing of the ocean floor. *Proceedings, 13th International Symposium on Remote Sensing of the Environment, 1: Ann Arbor, Michigan,* 583

Lyzenga D R, 1981. Remote sensing of bottom reflectance and water attenuation parameters in shallow water using aircraft and Landsat data. *International Journal of Remote Sensing,* **2(1)** 71

Lyzenga D R, Shuchman R A and Arnone R A, 1979, Evaluation of an algorithm for mapping bottom features under a variable depth of water. *Proceedings, 13th International Symposium on Remote Sensing of the Environment,* Ann Arbor, Michigan, **3** 1767

McKim H L, Klemas V, Gatto L W and Merry C J, 1984, The use of remote sensing for the U.S. Army Corps of Engineers dredging program. *Proceedings, 18th International Symposium on Remote Sensing of the Environment,* Ann Arbor, Michigan, **2** 1141

Meisner D E, 1986, Fundamentals of airborne video remote sensing, *Remote Sensing of the Environment,* **19** 63

Miller J R, Shepherd G G, Koehler R A, 1973, Airborne Spectroscopic measurement over water, *Canadian Aeronautics and Space Journal,* **19(10)** 521

Miller J R, Jain S C, O'Niell N T, McNiell W R, Thomson K P B, 1977. Interpretation of airborne spectral reflectance measurements over Georgian Bay. *Remote Sensing of the Environment,* **6** 183

Musgrove R G, 1969, Photometry for interpretation. *Photogrammetric Eng,* **35(10)** 1015

Paredes J M and Spero R E, 1983, Water depth mapping from passive remote sensing under a generalised ratio assumption, *Applied Optics* **22(8)** 1134

Philpot W D, 1989, Bathymetric mapping with passive multispectral imagery. *Applied Optics* **28(8)** 1569

Pincus H J, 1959, Some applications of terrestrial photogrammetry to the study of shorelines. *Photogrammetric Engineering,* **25(1)** 75

Playle, R.C. and D. Williamson. 1987. Water chemistry changes associated with hydroelectric development in northern Manitoba; the Churchill, Rat, Burntwood and Nelson rivers, Manitoba Environment and Workplace Safety and Health, Water Standards and Studies Report, No. 86-8, pp. 1-50

Polcyn F C and Lyzenga D R, 1979, Landsat Bathymetric mapping by multitemporal processing. *Proceedings, 13th International Symposium on Remote Sensing of the Environment,* Ann Arbor, Michigan, **3** 1269

Roberts A and Liedtke J, 1986, Airborne definition of suspended surface sediment and intertidal environments in the Fraser River plume, British Columbia, Geological Survey of Canada, Paper 86-1A, Current Research, Part A, pp. 571-582

Roberts A and Evans D, 1986, Multispectral video system for airborne remote sensing, Proceedings, 10th Canadian Symposium on Remote Sensing, Edmonton, pp. 729-737

Robinson I S, 1985, Satellite Oceanography. Ellis Horwood Limited, Chichester, 455

Suits G H, 1984, A versatile directional reflectance model for natural water bodies, submerged objects, and moist beach sands. *Remote Sensing of the Environment* **16** 143

Tanis F J and Hallada W A, 1984, Evaluation of Landsat Thematic Mapper data for shallow water bathymetry. *Proceedings, 18th International Symposium on Remote Sensing of the Environment,* Ann Arbor, Michigan, **2** 629

Theurer C, 1959, Colour and infrared experimental photography for coastal mapping. Photogrammetric Engineering, **25(4)** 565

Warne D K, 1978, Low cost hydrographic mapping. *Proceedings, 12th International Symposium on Remote Sensing of the Environment,* Ann Arbor, Michigan, **2** 1063

Weidmark W C, Jain S C, Zwick H H and J R, 1981, Passive bathymetric measurements in the Bruce Peninsula region of Ontario. *Proceedings, 15th International Symposium on Remote Sensing of the Environment,* Ann Arbor, Michigan, **2** 811

Williams J, 1970, Optical Properties of the Sea. United States Naval Institute Series in Oceanography: Annapolis, Maryland, 123

# The Earth's Surface Temperature: Derivation from Satellite Data and Prediction from ATI Models

Y Xue and A P Cracknell

Department of Applied Physics and Electronic & Mechanical Engineering, University of Dundee

## 1 Introduction

An important function of the land surface component of a general circulation model (GCM), from the atmosphere's point of view, is to partition the available radiative energy at the land surface into latent and sensible heat fluxes. With the development of dynamical mesoscale models of the atmosphere and of global climate models there arises the demand for data-sets which describe the transformation of energy at the surface of the Earth. To obtain surface (*i.e.* soil and vegetation) parameters such as albedo and temperature etc., which vary on the land surfaces over very small distances, remote sensing by both aircraft and satellites has proved to be a valuable tool in environmental monitoring of the Earth's surface. Their advantage is the coverage of great areas, their disadvantage is the restriction to only one measured quantity, the radiance, which carries different information in different spectral regions and is masked by the atmosphere if airborne or spaceborne measurements (satellite images) are used.

Ground surface temperature is an important hydrological and meterological parameter which affects the exchange of sensible and latent heat between the atmosphere and the Earth's surface. The knowledge of ground surface temperature over large spatial and temporal scales is a necessity for many applications, such as agrometeorology, climatic and environmental studies. The satellite over-pass time ground surface temperature can be determined using split-window methods. Multi-angle and multi-channel methods have been very successful for satellite over-pass time sea surface temperature determination from satellite thermal infrared data. However, for the determination of land surface temperature the emissivity presents a serious problem. The adaptation of these methods for the derivation of land surface temperatures from AVHRR and the

Along Track Scanning Radiometer, ATSR, thermal band data is considered. Thermal inertia is a property of volume that shows the resistance of a material to a change in temperature. The thermal inertia model was originally used in geological mapping as a complement to surface reflectance data as provided by satellite data. The diurnal ground temperature is derived from the advanced thermal inertia model as a solution of the heat diffusion equation with a constant diffusivity under periodic forcing on the ground surface in terms of temperature. The second-order approximation thermal inertia model can be used to calculate the real thermal inertia value of bare soil ground surface no matter the variation of soil moisture. Also any two-pass satellite or airborne data can be used to calculate the real thermal inertia values. The technique of thermal inertia mapping by aircraft and satellite systems has opened a new field of quantitative thermal property interpretation from thermal infrared remote sensing surveys. The advanced thermal inertia model can be used to predict the diurnal ground surface temperature with any two over-pass satellite data quasi-operationally.

In the present article, we intend to dwell on those results of the interpretation of radiation data obtained by satellites that may be used for immediate practical applications. Over-pass time surface temperature derivation from remotely-sensed data and surface temperature prediction from the thermal inertial model are considered. The dynamic aspects study of surface temperature will be useful for Earth surface energy balance study and it will be important input for global climate models and for medium-term and long-term weather forecasting. The thermal models and surface temperature prediction methods that we have developed can be used for bare soil ground areas, sparse vegetated areas such as desert, arid and semi-arid areas and Arctic and Antarctic ice sheets or sea ice. The further development should be detailed thermal modelling of the vegetated Earth's surface and operational thermal modelling of open sea upper mixed layers. With the thermal models of the Earth's surface and over-pass-time surface temperature, 24 hour period surface temperatures can be predicted. The computer scheme of the dynamic aspects study of global surface temperature from remotely-sensed data is given.

## 2  Basic concepts

### 2.1  Temperature, emissivity and thermal sensing windows

We first of all propose to clarify some basic concepts which will be used later.

Matter whose internal temperature is above absolute zero emits thermal radiant energy. Under appropriate circumstances, the use of spaceborne remote sensors to determine the spectral distribution and intensity of that energy may serve to diagnose the identity and condition of individual or combinations of materials on the Earth's surface.

Temperature (or kinetic temperature) is a measure of the internal kinetic energy of a material. The radiant flux of a body is determined by the radiant temperature or brightness temperature. Radiant temperature may be measured remotely by devices that detect electromagnetic radiation in the visible, thermal infrared and microwave wavelength regions. The radiant temperature of materials is always less than or equal to the kinetic temperature because of a thermal property called the emissivity, which will be defined later.

Remote sensing of temperature is performed by sensing radiation emitted from solids, liquids and gases in the thermal infrared region of the spectrum, in which thermal emission is dominant over reflected solar energy. For Earth resources applications, thermal sensing of solids and liquids is performed in two 'windows' of the atmosphere where atmospheric absorption and emission are at a minimum. The windows normally used are in the 3–4$\mu$m and 10.5–12.5$\mu$m wavelength regions, although other windows between 4.2 and 5.1$\mu$m and between 8 and 14$\mu$m are also available. Neither of the windows which are commonly used is perfect because weak absorption by water vapour and carbon dioxide occurs in both windows and absorption by ozone occurs in the 10.5–12.5$\mu$m window (Figure 1). In addition, the 3–4$\mu$m window is contaminated during daylight hours by a mixture of reflected solar energy over Earth surface targets. For this reason, the 3–4$\mu$m window is normally used for Earth resources purposes only when measurements are made at night.

Within the thermal sensing windows, radiation is emitted in an amount that is determined by two properties of material, the temperature and the emissivity. The temperature of a material controls the rate of radiant emission of the material. The Planck blackbody radiation law gives the rate at which radiation is emitted as follows

$$W_\lambda = \frac{C_1}{\lambda^5} \frac{1}{e^{C_2/\lambda T} - 1} \tag{1}$$

where:
  $W_\lambda$ = Spectral radiance emittance in $\mathrm{W\,m^2\,\mu m^{-1}}$
  $C_1 = 3.7427 \times 10^8 \ \mathrm{W\,m^{-2}\,\mu m^4}$
  $C_2 = 1.4388 \times 10^4 \ \mathrm{\mu m\,K}$
  $\lambda$ = wavelength in $\mu$m
  $T$ = temperature in degrees Kelvin

Figure 1 shows the shape of the spectrum of the emitted radiance. The Planck distribution function is only for a blackbody. A blackbody is an ideal material that emits the maximum amount of energy at each wavelength at a given temperature and in all directions. It also absorbs all incidental radiation at each wavelength and in all directions. It can been shown that the relation between the wavelength of peak emission and the temperature of the radiating body is

$$\lambda_{\mathrm{max}} T = 2897.8 \tag{2}$$

which is known as the Wien Displacement Law and $\lambda_{\mathrm{max}}$ is the wavelength of maximum radiant emittance.

A blackbody is a physical abstraction, for no material has an absorptivity of 1 and no material radiates the full amount of energy given in Equation 1. For real materials, in addition to the temperature, the quantity of radiant emission is also controlled by the emissivity ($\varepsilon$) of the object in the spectral region of interest. Emissivity, $\varepsilon$, is a dimensionless number that represents the ratio of the radiant flux from a surface to the radiant flux from a perfect emitter, or blackbody, at the same temperature (for a blackbody $\varepsilon = 1.00$). The emissivities of real materials vary from greater than 0 to less

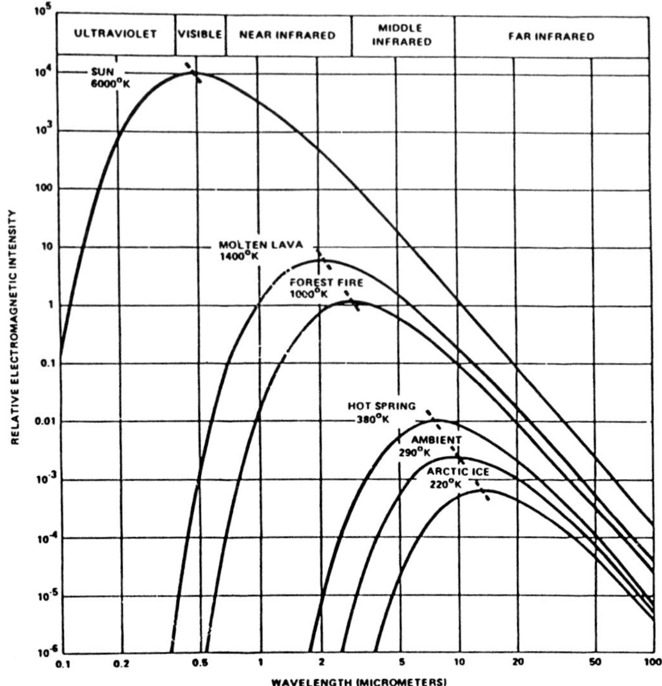



**Figure 1.** *Electromagnetic radiation emitted by familiar objects at various temperatures (ambient = land surface of Earth). (Short and Stuart 1982)*

than 1 and are spectrally dependent, which means that the emissivity of a real material will be different when measured at different wavelengths of radiant energy. Table 1 lists the emissivities of various materials in the 8–12$\mu$m wavelength region, which is widely used in remote sensing.

| Material | Emissivity | Material | Emissivity |
|---|---|---|---|
| metals | 0.01–0.6 | soil (moist) | 0.94–0.95 |
| snow (compressed) | 0.7–0.85 | sand | 0.95–0.96 |
| ice (glacier) | 0.85 | leaves (dry) | 0.96 |
| soil (dry sandy) | 0.88–0.94 | road (tarred) | ≈0.97 |
| wood | 0.90 | snow | 0.97–1.00 |
| granite (rough) | ≈0.90 | ice | 0.98 |
| plaster | 0.91 | skin (human) | 0.98 |
| concrete | 0.92 | peat | 0.98 |
| soil (dry loamy) | 0.92 | grass (green) | 0.98–0.99 |
| brick | 0.93 | leaves (damp) | 0.99 |
| glass | 0.94 | water | 0.99 |

**Table 1.** *Typical values of infrared emissivity (10$\mu$m) (Rees 1990).*

Data from a thermal infrared scanner can be processed to yield the value of the radiance leaving the surface of the Earth. The basic procedures involved in converting the raw thermal infrared scanner data into brightness temperatures are well-established. The four important steps are as follows.

1. The identification and removal of cloud-covered areas in a scene;

2. The digital data have to be converted into radiances in absolute units;

3. Derivation of the brightness temperature from the Planck radiation formula;

4. Geometrical rectification.

Having calculated the brightness temperature for the whole of the area of Earth's surface in the scene, the problem of calculating the atmospheric correction then has to be tackled since the objective of using thermal infrared scanner data is to obtain information about the temperature or the emissivity of the surface of the land or sea. It is probably fair to say that the problem of atmospheric corrections has been studied fairly extensively in relation to the sea but has received little attention for data obtained from land surface areas (Cracknell and Hayes 1991).

The Advanced Very High-Resolution Radiometers (AVHRR) on the National Oceanic and Atmospheric Administration (NOAA) operational meteorological satellites provide radiometric data at wavelengths of $3.7\mu m$ (channel 3), $10.8\mu m$ (channel 4) and $11.9\mu m$ (channel 5) that enable one to make an estimate of sea surface temperature (SST) using a differential absorption technique to allow for atmospheric effects. The operational multi-channel SST (MCSST) algorithms used by NOAA/NESDIS (National Environmental Satellite Data and Information Service) are derived from a regression analysis of coincident satellite and buoy data. The ERS-1 programme is recognised as a major element of an overall European Earth Observation Programme covering a number of disciplines including meteorology, climatology, oceanography, land resource inventory and monitoring, geodesy and geodynamics. The main objective of this programme is to give the participating states the ability to take in both the management of the Earth's resources and the monitoring of its environment. The main part of the playload is a set of active microwave instruments which are similar to those that were flown on Seasat. There is, however, an additional sensor, the Along-Track Scanning Radiometer (ATSR)—a four-channel infrared radiometer used for measuring sea-surface temperature (SST) and cloud-top temperature. The ATSR was designed to provide the following types of data and observations (Vass and Handoll 1991).

- Sea surface temperature with an absolute accuracy of better than 0.5K with a spatial resolution of 50km and in conditions of up to 80% cloud cover.

- Images of surface temperature with 1km resolution and 500km swath, relative accuracy around 0.1K.

The novel feature of the ATSR, as its name implies, is the viewing of the same area both through a near-vertical atmospheric path and through an inclined path of different length some way along the satellite track. The ATSR uses infrared spectral channels

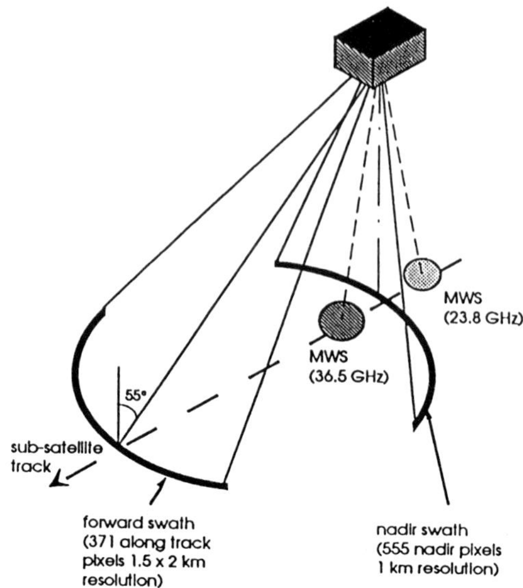


**Figure 2.** *ATSR viewing geometry (Vass and Handoll 1991).*

which are very similar to those of the AVHRR on the NOAA series of meteorological
satellites, with many improvements in accuracy. As an imaging radiometer its four
channels are fully co-registered by a common fieldstop and they operate at wavelengths
of 1.6, 3.7, 10.8 and 12$\mu$m, defined by beam splitters and multi-layer interference filters.

The instantaneous field of view (IFOV) at the nadir on the Earth's surface is a 1km
by 1km square which is imaged on to the detector element via an off-axis parabolic
mirror. The IFOV is scanned over the Earth's surface by a rotating plane mirror in such
a way that it gives two Earth views (0° and 47° to the nadir on the sub-satellite track).
A scan cone angle of 47° gives a zenith angle in the along track view of about 55° (Vass
and Handoll 1991). The two on-board black bodies are viewed between the Earth views
in each scan. The mirror rotation rate is chosen so that each scan is contiguous with the
previous one. The ATSR involves a rotating mirror giving a conically scanned field of
view of the Earth's surface in two curved swaths, 500km wide and separated by about
900km. Figure 2 gives the ATSR viewing geometry. ATSR technical characteristics are
listed in Table 2.

## 2.2	Some simple ideas about thermal inertia

The origins of studies of thermal inertia for the surface of the Earth are derived form a
one-dimensional model of heat conduction with periodic boundary conditions originally
developed by Jaeger (1953) to study the surface temperature of the Moon. Let us try,
to understand in qualitative terms what is meant by thermal inertia.

We all know, from our very early experience what is meant by temperature and it has
been established (Becker 1982, McClain *et al.* 1983, McMillin 1975, Sobrino *et al.* 1993)
that the temperature—at least of the sea—is something that we can now determine

| | |
|---|---|
| spectral channels : | 4 co-registered thermal channels and 3 visible channels; ranges (in $\mu$m) are 11.040–13.104, 10.080–11.712, 3.226–4.138, 1.516–1.748, 0.835–0.895, 0.629–0.685, 0.525–0.585 |
| orbit: | sun-synchronous, near-polar, elliptical orbit, mean altitude 785km, inclination 98.5°, period 100.5 minutes, local mean solar time 10:30am |
| repeat cycle: | 35 days and/or 3 days |
| ground resolution (IFOV): | 1×1 km at nadir; 1.5×2 km for forward view at $\approx$ 53°. |
| scan width : | 500 km |
| scanning method : | mechanical rotating plane mirror; field of view conically scanned. |
| radiometric precision : | < 0.01 degree Kelvin |
| calibration : | two on-board black bodies which are referenced in the scan pattern |
| predicted SST accuracy : | 0.5 degree Kelvin over 50 × 50 area with 80% cloud cover |

**Table 2.** *ATSR technical characteristics*

quite accurately from (thermal) infrared remotely-sensed data. The Sun rises in the morning and it shines more or less equally on different parts of the surface of the Earth, at least on a regional basis. But the response in terms of surface temperature varies enormously. At the end of the day the Sun sets and, again, the response, in terms of cooling varies enormously; generally those areas which experience large temperature rises during the day will suffer large temperature drops at night. Then it all starts again the following morning. The amplitude of the temperature variations of the surface, over a 24 hour period varies enormously for different types of surface. The most extreme differences are between, say, dry sand and bare rock on the one hand and deep water on the other hand. One sees this very clearly in satellite images in the thermal infrared; at night the sea is warmer than the land, but during the day the land becomes warmer than the sea. The amplitude of the temperature variations of the land surface is likely to be several tens of degrees C (the desert can be surprisingly cold at night) whereas the sea surface variations are only on the order of a few degrees (even this can be reduced by the action of waves or tides, which is a separate question).

Basically, therefore, in qualitative terms, thermal inertia is a quantity which is defined to give a measure of the reluctance of the surface of the Earth to respond to a given supply of heat energy (from the Sun).

The concept of thermal inertia was introduced to try to take into account both the specific heat and the thermal conductivity in identifying a quantity that would indicate the reluctance of the surface of the Earth (or whatever) to respond to a given heat input. That is, large thermal inertia gives small amplitude oscillations/variations in the surface temperature over a 24 hour period. Let us consider initially the effects of specific heat and thermal conductivity separately.

1. For a given heat input, a large specific heat, $c$, implies small temperature changes and large thermal inertia.

2. However, for a given heat input, for a large value of the thermal conductivity

(*i.e.* for a good thermal conductor) more heat is taken away to lower layers and so the surface temperature rise is less. That is, a large thermal conductivity, $K$, causes more heat to be conducted to levels below the surface, resulting in a smaller surface temperature rise and large thermal inertia.

Thus it is reasonable to define thermal inertia as

$$P = \sqrt{K\rho c} \tag{3}$$

where $\rho$ = density, $c$ = specific heat capacity and $K$ = thermal conductivity. The $\rho$ is included simply because what matters here is really the specific heat capacity per unit volume whereas $c$, the specific heat capacity, is usually defined as the heat capacity per unit mass.

# 3   Over-pass time surface temperature determination by satellite: split window methods

In this section, we present the theoretical models of split window methods that relate sea and land surface temperature with the temperatures measured by thermal infrared sensors; these include multi-channel methods and also multi-angle methods. We shall consider the MCSST algorithms and then introduce the split window methods for land surface temperature measurement from AVHRR thermal band data. Finally, a method (Simultaneous Split-window Method) which has been developed to retrieve the Earth surface emissivity and temperature from ATSR thermal band data by Cracknell and Xue (1995a) will be presented. With the present model, not only the over-pass time surface temperature but also the surface emissivity can be obtained from ATSR thermal band data.

## 3.1   General theory

For a cloud-free atmosphere under local thermodynamic equilibrium and a Lambertian Earth surface, the signal recorded in channel $i$ of a satellite that observes the Earth's surface under a zenithal angle $\theta$ can be written as follows:

$$I_{\text{sat}} = I_{\text{surf}} + I_{\text{atm}} + I_{\text{refl}} \tag{4}$$

$I_{\text{sat}}$ is the total radiance received by the sensor on the satellite given by

$$I_{\text{sat}} = \int_{\lambda_{i1}}^{\lambda_{i2}} f_i(\lambda) B_\lambda(T_i) d\lambda \tag{5}$$

where $T_i$ is the channel brightness temperature, $f_i(\lambda)$ is the normalised spectral response of the radiometer in channel $i$ and $\lambda_{i1}$ and $\lambda_{i2}$ are the channel wavelength limits: $\Delta\lambda_i$ $(= \lambda_{i2} - \lambda_{i1})$ is the wavelength width of the channel $i$, $B_\lambda(T_i)$ is the black-body spectral radiance for the wavelength $\lambda$ and the temperature $T_i$.

$I_{\text{surf}}$ represents the contribution to the received radiance resulting from the surface emission given by

$$I_{\text{surf}} = \int_{i1}^{i2} f_i(\lambda)\epsilon_\lambda(\theta)B_\lambda(T_s)\tau_\lambda(\theta)d\lambda \tag{6}$$

where $\epsilon_\lambda(\theta)$ is the Earth's surface spectral emissivity and $\tau_\lambda(\theta)$ is the total spectral transmittance of the atmosphere at zenithal angle $\theta$.

$I_{\text{atm}}$ represents the contribution to the received radiance resulting from the upwelling radiance emitted by the atmosphere towards the sensor, thus

$$I_{\text{atm}} = \int_{i1}^{i2} f_i(\lambda)R_\lambda^\uparrow(\theta)d\lambda \tag{7}$$

where $R_\lambda^\uparrow(\theta)$ represents the upward atmospheric spectral radiance at the zenithal angle $\theta$

$$R_\lambda^\uparrow = \int_0^h B_\lambda(T_z)\frac{\partial\tau_\lambda(\theta, z, h)}{\partial z}dz \tag{8}$$

where $\tau_\lambda(\theta, z, h)$ is the spectral atmospheric transmittance between the altitude of the sensor $(h)$ and the altitude $z$ and $T_z$ is the atmospheric temperature at level $z$.

$I_{\text{refl}}$ is the downward atmospheric radiance emitted by the atmosphere that is reflected upward at the Earth's surface and attenuated in its path to the sensor, thus

$$I_{\text{refl}} = \pi\int_{\lambda_{i1}}^{\lambda_{i2}}\int_0^{\pi/2} f_i(\lambda)\rho_{b\lambda}(\theta, \theta')\sin(2\theta')\tau_\lambda(\theta)R_\lambda^\downarrow(\theta')d\lambda d\theta' \tag{9}$$

where $R_\lambda^\downarrow(\theta')$ represents the downward atmospheric radiance emitted by the atmosphere at zenithal angle $\theta'$,

$$R_\lambda^\downarrow(\theta') = \int_h^0 B_\lambda(T_z)\frac{\partial\tau_\lambda'(\theta', z, 0)}{\partial z}dz \tag{10}$$

where $\tau_\lambda(\theta', z, 0)$ is the atmospheric transmittance between the altitude level 0 (surface) and $z$. We assume that there is no dependence on the azimuthal angle.

The solution of the Equation (4) with Equations (5)–(10) for the surface temperature is as follows:

$$T_0 - T_i - \left(L_i(T_i)\frac{(1-\epsilon_i)}{\epsilon_i} + T_i\frac{f_iW}{\epsilon_i(\cos\theta - f_iW)} + \frac{2(1-\epsilon_i)}{\epsilon_i}(T_i - L_i(T_i))(f_iW)\right) = -T_a f_i W \alpha_i \tag{11}$$

(Becker 1987, Becker and Li 1990, Sobrino *et al.* 1991), where $f_i$ can be considered as the absorption coefficient of the whole atmosphere, $T_0$ is the surface temperature of a surface having a brightness temperature $T_i$ and emissivity $\epsilon_i$, W is the total water vapour amount in the atmosphere and $T_a$ is an effective air temperature that takes into account the atmospheric moisture and is defined by

$$T_a = \frac{1}{W}\int_0^W T_z dW(h, z) \tag{12}$$

where $T_z$ is the atmospheric temperature at level $z$,

$$\alpha_i = \frac{2(1-\epsilon_i)}{\epsilon_i} + \frac{1}{\epsilon_i(\cos\theta - f_iW)} \tag{13}$$

and $L_i(T_i)$ is a parameter with dimension of temperature defined as

$$L_i(T_i) = \frac{B_i(T_i)}{\left(\dfrac{\partial B_i(T)}{\partial T}\right)_{T_i}} \tag{14}$$

Several approximations have been proposed for $L_i(T_i)$. Slater (1980), Price (1984) and Becker (1987) use

$$L_i(T_i) = \frac{T_i}{n_i} \tag{15}$$

where $n_i$ is a constant for a given channel.

## 3.2   Multi-channel split window methods

A split-window equation has been obtained writing Equation (11) for two channels ($i$=11 and $i$=12, corresponding to channel 4 at $\approx 11\mu$m and channel 5 at $\approx 12\mu$m respectively)

$$T_s = A_{0,\theta} + A_{1,\theta}T_{11,\theta} + A_{2,\theta}T_{12,\theta} \tag{16}$$

where $A_{0,\theta}$, $A_{1,\theta}$ and $A_{2,\theta}$ are the local split-window coefficients, which depend on the surface emissivity and the atmospheric state (Becker and Li 1990, Sobrino *et al.* 1991). For SST from AVHRR thermal band data, McMillin and Crosby (1984) gave $A_{0,\theta} = -4.588$, $A_{1,\theta} = -2.637$ and $A_{2,\theta} = 3.651$, independent of $\theta$, atmospheric state and, of course, $\epsilon$. Price (1984) gave

$$T_s = (T_{11,\theta} + 3.33(T_{11,\theta} - T_{12,\theta}))\left(\frac{3.5 + \varepsilon_{11,\theta}}{4.5}\right) + 0.75T_{12,\theta}(\varepsilon_{11,\theta} - \varepsilon_{12,\theta}) \tag{17}$$

with all temperatures in Kelvin.

## 3.3   Multi-angle split window methods

The basis of the double-viewing angle method is to use two different absorption path lengths at one wavelength. The downwelling hemispherical atmospheric radiance in channel $i$ was found to be quantitatively equivalent to the directional upwelling radiance at zenith angle $\phi \approx 53°$ (Kondratyev 1969).

$$R_{\text{ati}}^{\uparrow} \approx (1 - \tau_{i,53})B_i(T_a) \tag{18}$$

Equation (18) is important for the development of the double viewing angle technique because 53° coincides practically with the forward view angle of the ATSR (zenith angle 55° in the along track view). For two view angles at one wavelength, Sobrino *et al.* (1993) gave

$$T_s = T_{i,0}\left(\beta_{i,0} + \beta_{i,1}(1 - \varepsilon_{i,0}) + \beta_{i,2}\Delta\varepsilon_\theta\right) + \left(\alpha_0 + \alpha_1(1 - \varepsilon_{i,0}) + \alpha_2\Delta\varepsilon_\theta\right)(T_{i,0} - T_{i,\theta}) \tag{19}$$

with $\alpha_0 = 2.135$, $\alpha_1 = 4.375$, $\alpha_2 = -5.29$, $\beta_{i,0} = 1 + (\beta_0'/n_i)$, $(i = 11, 12)$, $\beta_0' = 0.532$, $\beta_{i,j} = \beta_j'/n_i$, $(j = 1, 2)$, $\beta_1' = 1.631$ and $\beta_2' = -2.545$. The above equation constitutes the single-channel double-viewing angle model for obtaining the surface temperature $T_s$ from ATSR thermal band data, when the surface is viewed both from nadir and at an angle close to 53° from the normal.

## 3.4 Sea surface temperature (SST)

We discussed the use of thermal infrared satellite data for the determination of sea surface temperatures (SST) before turning our consideration to land surface temperatures (LST). This was because the case of LST is considerably more difficult than that of SST.

Surface temperature retrievals from thermal infrared images are of considerable importance for climate research. Global sea surface temperatures are currently being derived from measurements obtained from the AVHRR sensor on the NOAA polar-orbiting satellites with an accuracy of $\pm 0.7$K. The standard technique for deriving sea surface temperature (SST) from satellite data uses the differential absorption of surface-emitted infrared radiation to measure the magnitude of, or to eliminate, the atmospheric absorption. Saunders (1967) was the first to report a use of multi-angle aircraft measurements to correct for atmospheric effects in sea surface temperatures. Then Anding and Kauth (1970) published a paper in which they proposed a method based on the differential atmospheric effects at different wavelengths. Unfortunately, the results of Anding and Kauth (1970) were based on a transmittance model that was not as accurate as some later models. McMillin (1971) started with the radiative transfer equation and developed a theoretical justification for the method.

Several forms of algorithm exist for deriving SST from data supplied by satellite-flown infrared radiometers (McMillin 1971, 1975, Prabhakara *et al.* 1972, 1974, Deschamps and Phulpin 1980, Becker 1982, Chedin *et al.* 1982, Holyer 1984, McMillin and Crosby 1984, 1985). In almost all cases a linear form of algorithm is used that sometimes includes a dependence on view (satellite zenith) angle. The operational MCSST algorithm used by NOAA/NESDIS takes the form

$$(\text{SST}) = aT_x - bT_y + c \tag{20}$$

where $T_x$ and $T_y$ are the satellite measured brightness temperatures in channels $x$ and $y$, and $a$, $b$ and $c$ are coefficients (Barton 1992). McMillin and Crosby (1984) gave the values of the coefficients $a$, $b$ and $c$ for AVHRR channel 3, 4 and 5. McClain *et al.* (1983) gave the different coefficients values for daytime and night-time satellite data. When all channels of AVHRR are used, McMillin and Crosby (1984) gave

$$(\text{SST}) = 3.175 + 0.429T_3 + 2.698T_4 - 2.139T_5 \tag{21}$$

where $T_3$, $T_4$ and $T_5$ are in Kelvin. The standard error of estimate of the model is 1.04K.

## 3.5 Land surface temperature (LST)

For the estimation of land surface temperatures the situation is quite different; the 1.1km field of view of the AVHRR is not adequate to resolve many surface features. Because the temperature of bare soil may vary by several degrees over distances of tens of metres (Vauclin *et al.* 1982) and temperature variations between cropped and fallow areas can be much greater, it is not feasible to seek accurate ground truth verification for AVHRR and ATSR temperature data in agricultural regions where surface temperature data may prove useful. On the other hand, highly accurate values of surface temperature are not required for some purposes (Rosema *et al.* 1978, Carlson *et al.* 1981, Price 1982). Much less work has been done for land surface temperature (LST) than for sea surface temperature because LST is generally not homogeneous within a pixel and land surface emissivity may be quite different from unity and is also dependent on the channel. The effect of emissivity on the measurement of LST has been studied by Price (1984), Becker (1987) and Becker and Li (1990). The split-window technique (Price, 1984, Becker and Li 1990, Sobrino et al. 1991) and the double viewing angle method (Sobrino *et al.* 1993) has been fully studied to obtain LST. Gorodetskii (1985) combined the multi-angle and multi-channel methods to estimate surface temperature. Cracknell and Xue (1995b) used an iterative self-consistent method to retrieve the Earth surface temperature from AVHRR and ATSR thermal band data.

In many applications of satellite-derived temperature, the accuracies required are quite coarse and a simple assumption of emissivity equal to unity is sufficient to give a useful measurement. However, there are other applications related to soil moisture estimation, energy balances at the surface, geological exploration and land use management where a more accurate measurement is required. Temperature at the terrestrial surface is one of the most important parameters to map and monitor in many environmental studies. Remote sensing of surface temperature for large areas has become a practical reality with the advent of environmental satellites. These measurements have satisfactory accuracy and precision for many mesoscale meteorological, biometeorological and other environmental applications. The two principal sources of uncertainty in the temperature observing systems are atmospheric attenuation and emission of thermal radiation and the thermal emissivity of the terrestrial surface (Taylor 1979). A large amount of work has been done to determine sea surface temperature from satellite radiances, however, in contrast to these activities over the ocean, there has been much less work reported on the production of reasonably accurate surface temperature measurements over land. The determination of land surface temperature is, in fact, more challenging for at least three reasons.

- The first reason applies to bare rock, bare soil or desert sand surfaces as well as to vegetated surfaces and this concerns the emissivity. Land surface emissivities may be quite different from unity and spectrally variable and are likely to vary with moisture content, vegetation cover, and surface material (Barton and Takashima 1986). Whereas the emissivity of the sea surface is that of water (albeit salt water) and its value is constant (and therefore known), the value of the emissivity for rock, soil or sand is much more variable than that of water. Thus for any land surface pixel, one has two unknowns, the temperature and the emissivity, whereas for a sea surface pixel there is only one unknown, the temperature.

- Land surface temperature is generally not homogeneous within one pixel. Most land surface areas are much less homogeneous than the sea surface. Even non-vegetated land surfaces, apart from some desert or salt flats areas, are often very inhomogeneous on the scale of the IFOV (or pixels) of the AVHRR. For a vegetated surface it may even be quite difficult to define what is meant by the temperature of the surface of the land. One could consider the temperature of the underlying soil, or the temperature of some part of the vegetation (trunk, stalk, leaf, flower or fruit), or the temperature of the air that is trapped among the vegetation.

- The difference between land surface temperature and air temperature near the surface is larger for land surfaces than for the sea. The air surface temperatures just above the land surface are usually quite different from the land surface temperatures and this may weaken the assumptions behind the split-window method which is widely used for the sea surface.

The last two of these problems have already been mentioned above. The first error source prompted the determination of soil surface emissivity. Several reports exist on the laboratory measurement of the emissivity of various soils and rock types using broad band (8–13$\mu$m) radiometers. There have also been spectral measurements of the emissivity of specific compounds (Taylor 1979). All these measurements form a guide to the assessment of the emissivity of natural land surfaces, but accurate details of emissivities at the AVHRR wavelengths are not yet available. Given the problems associated with the interpretation of satellite measurements the best method of obtaining information on surface emissivity may be from the satellite measurements themselves. Barton and Takashima (1986) assessed the potential of gaining information about infrared surface emissivity from AVHRR data alone, *i.e.* no ground truth data were used. They showed that it is possible to gain some information on the relative values of the surface emissivity at different wavelengths without the need for ground truth data. However, ground truth data must be combined with the satellite measurements before it will be possible to get accurate satellite measurements of land surface temperatures from the AVHRR instruments.

## 3.6 The split-window method for LST

The split window method is being used successfully to retrieve the temperature over the sea surface from satellite radiances in clear sky and has the great advantage of simplicity. However, such a method does not work over the land surface, mainly because the emissivity is not equal to 1 and depends on the channel. However, the situation for the determination of land surface temperatures is not completely without hope. Since the split-window method has, in practice, proved to be so successful for sea surface temperatures and because it has the great advantage of simplicity, several workers have attempted to extend the split-window approach to the determination of land surface temperatures. This problem has been addressed, for example, by Becker (1987) and Becker and Li (1990). For the sea surface the split window method is based on the result, which can be justified theoretically (Becker 1982, Becker and Li 1990, Sobrino *et al.* 1991, 1993), that $T_s$ the surface temperature can be expressed as a linear combination of the channel-3, channel-4 and channel-5 AVHRR brightness temperatures $T_3, T_4$, and $T_5$, respectively, *i.e.*

$$T_s = a_3 T_3 + a_4 T_4 + a_5 T_5 + a_6 \qquad (22)$$

where $a_3$, $a_4$, $a_5$ and $a_6$ are assumed to be constants, on a global scale; in particular their values do not depend on the local atmospheric conditions. During the day a similar relation, but excluding channel 3, applies. Thus, as we have seen, once the values of these coefficients have been determined by using in situ data from a relatively restricted set of data buoys, then these values of the coefficients are applied to the AVHRR data on a global basis. Becker and Li (1990) showed that for the land surface it was also possible to use an equation of the form of equation (21) in which the coefficients are still independent of the local atmospheric conditions. However, the coefficients $a_3$, $a_4$, $a_5$ and $a_6$ are no longer constants with values that are valid globally. Instead the values of these constants become functions of the emissivity. This leads to the replacement of a universal split-window method that is used so successfully for sea surface temperatures by a local split window method for land surface temperatures. In the formulation of Becker and Li (1990) and considering the daytime situation so that channel 3 is excluded

$$T_s = A_0 + P\frac{T_4 + T_5}{2} + M\frac{T_4 - T_5}{2} \tag{23}$$

where

$$A_0 = 1.274 \tag{24}$$

$$P = 1 + 0.15616\frac{(1 - \varepsilon)}{\varepsilon} - 0.482\frac{\Delta\varepsilon}{\varepsilon^2}$$

$$M = 6.26 + 3.98\frac{(1 - \varepsilon)}{\varepsilon} + 38.33\frac{\Delta\varepsilon}{\varepsilon^2} \tag{25}$$

where $\varepsilon = (\varepsilon_4 + \varepsilon_5)/2$ and $\Delta\varepsilon = \varepsilon_4 - \varepsilon_5$. These numerical values are only valid for NOAA-9 since the spectral response of the radiometer is used to compute these coefficients. Slightly different values of the coefficients will be required for AVHRR data from other satellites in the series. Price (1984) presented an expression for the surface temperature as estimated from AVHRR channel 4 and 5 as follows

$$\text{LST} = (T_4 + 3.33(T_4 - T_5))\left(\frac{3.5 + \varepsilon_4}{4.5}\right) + 0.75T_5(\varepsilon_4 - \varepsilon_5) \tag{26}$$

with all temperatures in degrees Kelvin. After analysis of the data set, he concluded that there was satisfactory agreement between the equation resulting from radiative transfer theory and the atmospheric correction algorithm as obtained by analysis of an area of incipient cloud street formation.

Another way of looking at this adaptation of the split-window method is to consider the error $\delta T$ that would arise in the calculated LST by taking $\varepsilon_4 = \varepsilon_5 = 1$ rather than using true values; Becker showed the value of $\delta T$ is significant and is given by

$$\Delta T = 50\frac{(1 - \varepsilon)}{\varepsilon} - 300\frac{(\varepsilon_4 - \varepsilon_5)}{\varepsilon} \tag{27}$$

where $\varepsilon = (\varepsilon_4 + \varepsilon_5)/2$. Typical values of $\varepsilon_4$ and $\varepsilon_5$ for land surfaces give values of $\Delta T$ as high as 6 or 7K. Therefore, while it is possible to derive a formal expression for $T_s$, the (land) surface temperature, it cannot be used in practice unless one has a reasonably accurate set of values of $\varepsilon_4$ and $\varepsilon_5$ for the land surface area being studied. Such information is generally not available.

The local split window method involves using the AVHRR data with equation (23) in association with a land-use map obtained from the visible and near-infrared channels, when the effective emissivities of various surfaces are known. Errors in the assumed values of $\varepsilon$ will lead to errors in the retrieved land surface temperatures, but these errors will be much smaller than the errors $\Delta T$ mentioned above that would be obtained by using $\varepsilon_4 = \varepsilon_5 = 1$ (see Table 6 of Becker and Li 1990). This method has been used with data from two test sites by Kerr and Lagouarde (1989) and Kerr *et al.* (1992); in the latter work Kerr *et al.* demonstrated an accuracy of better than 1.5K. The method proposed by Kerr and Lagouarde (1989) and Kerr *et al.* (1992) was primarily designed for arid and semi-arid areas, where the vegetation can be very sparse and the surface temperature high. Consequently, the method was somewhat similar to the method of Becker and Li (1990) with the addition of two improvements—canopy fraction and angular (shadowing) effects (Kimes and Kirchner 1983). They used different split-window coefficient sets for bare soil and vegetation separately to obtain two temperatures: $T_{\text{bare}}$ for bare soil and $T_{\text{veg}}$ for vegetation. A fractional cover coefficient $C$ is then derived from NDVI values with the expression

$$C = \frac{(\text{NDVI}) - (\text{NDVI})_{\text{bare}}}{(\text{NDVI})_{\text{veg}} - (\text{NDVI})_{\text{bare}}} \tag{28}$$

where $(\text{NDVI})_{\text{bare}}$ is the minimum value of the NDVI for bare soil over the area of interest (typically end of the dry season in the Sahel) and $(\text{NDVI})_{\text{veg}}$ corresponds to the highest NDVI one could expect for a fully vegetated pixel (typically end of the rain season in the Sahel). Finally a surface temperature $(T_s)$ is estimated from a linear combination of $T_{\text{bare}}$ and $T_{\text{veg}}$

$$T_s = CT_{\text{veg}} + (1 - C)T_{\text{bare}} \tag{29}$$

Kerr *et al.* (1992) used a linear combination of the two temperatures (bare soil and vegetation) since it is simple to implement and since no theoretical background justifies a more elaborate formulation. The results gained with this formula indicated that this expression, even though simple, is adequate for their purpose. Figure 3 shows the results they obtained using the data set in a Niger test site. The continuous curve corresponds to ground measurements (time step of 15min) while the 'asterisk' is the satellite estimate.

As an alternative to using the expressions for $A_0$, $P$ and $M$ and estimating the value of $\varepsilon$, one can regard $A_0$, $P$ and $M$ simply as parameters to be determined empirically by a least squares fit to a set of *in-situ* data. This has been done, for instance, by Vidal (1991). However, this method is of limited usefulness because to be so reliant on *in-situ* data considerably negates the benefits of using remotely-sensed data. Attempts have also been made to use the success in studying water surface temperatures to investigate relatively homogeneous nearby land surfaces (salt flats or sand), *e.g.* in some work on AVHRR data for Lake Eyre in Australia (Barton and Takashima 1986). Further work on the development of the theory of the localised split window method, and on the validation of their previous approach, was described by Li and Becker (1993); this involved extending the ideas to the inclusion of channel-3 AVHRR data in addition to using the channel-4 and channel-5 data.

One situation in which the determination of land surface temperature from thermal infrared data is easier than for most cases is when the surface is snow covered. The surface is then fairly homogeneous and the emissivity is more homogeneous and its value is more well known than for most other land surfaces. Some studies of land surface temperature for snow-covered surfaces using AVHRR thermal infrared data are described by Collier et al. (1989) and McClatchey (1992)

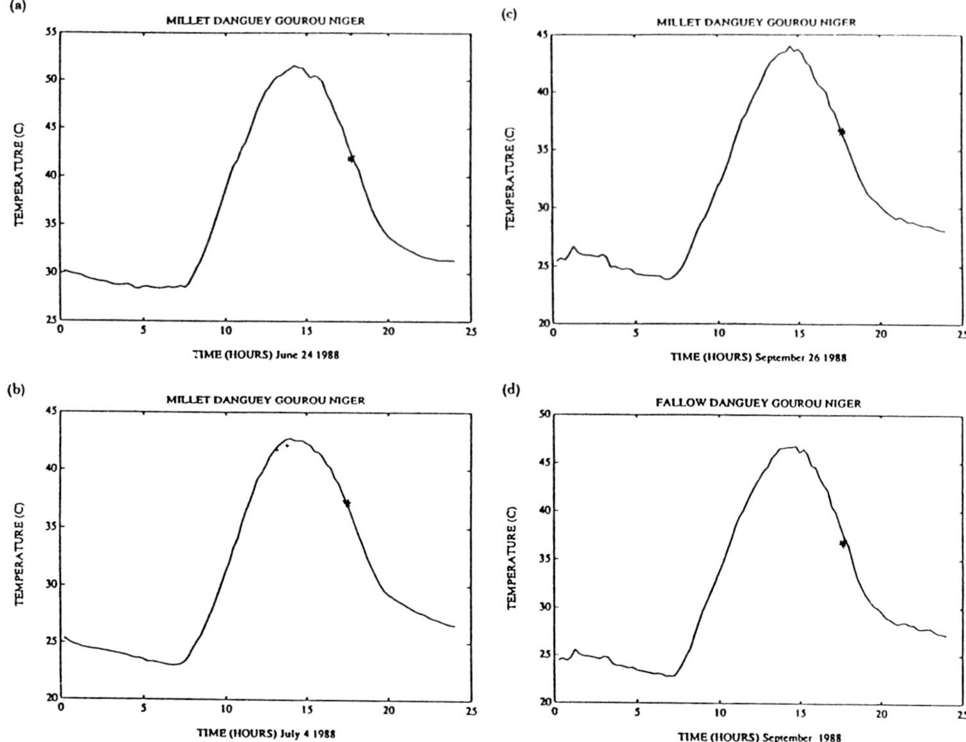

Figure 3. *Niger test site: Estimated and measured surface temperature: (*) satellite-derived surface temperature; (—) ground measurements. (a) 24 June 1988 millet field; (b) 4 July 1988 millet field; (c) 26 September 1988 millet field; (d) 26 September 1988 fallow land (Kerr et al. 1992).*

There have been various attempts to apply thermal infrared data from Earth-orbiting satellites—particularly AVHRR data—to studies of agricultural problems. These include studies of a tall grass prairie by Cooper and Asrar (1989) and to frost damage in orange groves in Spain by Caselles and Sobrino (1989).

Radiative transfer models have been found to be a more reliable technique for removing the effects of atmospheric attenuation on AVHRR-derived thermal radiances than previously used empirical corrections that were a function of scene temperature (Brower *et al.* 1976). Two radiative transfer solutions to atmospheric attenuation are used currently. The first solution, used in the NESS-80 (Wienreb and Hill 1980) and WINDOW (Price 1983), models the attenuation effects of atmospheric gases on thermal radiation; it is computed using direct measurements of atmospheric properties. The second solution takes advantage of the wavelength dependence of atmospheric gas absorption coefficients (split-window models). The radiosonde-driven NESS-80 and Price WINDOW models require atmospheric data as input; however, the location of the input data might not always be ideal. Atmospheric temporal and spatial variability reduces the usefulness of radiosonde input when they are taken at large time differences between radiosonde and AVHRR data acquisition or at large distances from the study site (Price 1983). The split-window technique takes advantage of the frequency-dependent absorption of upwelling radiation in the AVHRR middle and thermal infrared channels. By using AVHRR data for atmospheric correction and surface temperature measurement, spatial and temporal errors associated with radiosonde data are avoided; however, split-window models are dependent upon the accuracy of highly variable water vapour absorption coefficients (Price 1984). Also, as mentioned above, these techniques were developed primary over the sea surface, which has a relatively uniform emissivity and generally small temperature spatial variability. In addition, sea-surface temperatures usually do not exceed 30°C, in contrast to land surfaces which can reach twice as high. Cooper and Asrar (1989) used six atmospheric correction models developed primarily for sea-surface temperature studies to test their utility for the retrieval of radiative temperature over the land surface in the Flint Hills of Kansas. The analysis included an assessment of uncertainties associated with in situ and AVHRR data. They found an uncertainty of ±3.0°C for the AVHRR data and used this limit to evaluate the performance of a given model. When the difference between in situ and AVHRR surface temperatures was smaller than the uncertainty, the model was judged to be adequate.

Details of four of the split-window atmospheric attenuation correction models are given below. $T_c$ is atmosphere corrected surface temperature and Ch.4 and Ch.5 are calibrated surface temperatures for those channels.

D and P (1980) $\quad T_c(4,5) = 2.626(\text{Ch.4}) - 1.626(\text{Ch.5}) - 1.1$
McClain (1983) $\quad T_c(4,5) = 1.035(\text{Ch.4}) + 3.046(\text{Ch.4} - \text{Ch.5}) - 283.934$
Price (1984) $\quad T_c(4,5) = (\text{Ch.4}) + 3.33(\text{Ch.4} - \text{Ch.5}) * 1$
Singh (1984) $\quad T_c(4,5) = 1.699(\text{Ch.4}) - 0.699(\text{Ch.4} - \text{Ch.5}) - 0.240$

Corrected satellite temperatures were compared with surface temperature (ground truth) values in Table 3. Among the six models evaluated, only the NOAA split-window model consistently adjusted the AVHRR surface temperatures within ±3.0°C of the *in-situ* measurements. Of the six atmospheric attenuation models tested in their study, only the McClain split-window model (McClain *et al.* 1983) yielded AVHRR temperatures that were consistently within the defined tolerance interval on the four measurement days. The NESS-80 and Price Window models (Price 1983) worked well when radiosonde data were coincident in space and time with the AVHRR observations. NESS-80 and WINDOW failed the tests on three of the four days when using radiosonde data 77km from the site. This suggests that radiosonde data should be collected as close to the site as possible. However, the Price WINDOW model generated values consistently closer to the ground truth than NESS-80 during the limited evaluation period. The Price split-window model failed on only one of the four study dates. The Deschamps and Phulpin (1980) split-window model exceeded the RSS (Root Sum Square) rejection criterion on two of the four study dates. The Singh (1984) model failed on all four study dates. The Deschamps and Phulpin and Singh algorithms were designed specially to correct AVHRR data over sea surfaces with a limited temperature range; it was not surprising that their performance was not as good as the more generalised models. If evaluated with a larger data base, it might be possible to calculate an offset correction and modify these models for land-surface conditions.

| Day number and model | Temp. actual or AVHRR | Std. error | $t$-SE | $\Delta T$ minus AVHRR | confidence interval | test criteria | |
|---|---|---|---|---|---|---|---|
| | | | | | | RSS | CI |
| **196: actual** | **35.6** | | | | | | |
| D and P | 31.3 | 0.44 | 0.91 | +4.3 | 3.4 to 5.2 | Fail | Pass |
| NESS-80 | 46.9 | 0.46 | 0.96 | -11.3 | -12.3 to -10.4 | Fail | Fail |
| McClain | 34.4 | 0.45 | 0.94 | +1.2 | 0.2 to 2.1 | Pass | Pass |
| Price | 33.0 | 0.45 | 0.94 | +2.6 | 1.7 to 3.5 | Pass | Fail |
| Singh | 30.5 | 0.44 | 0.91 | -5.1 | -6.0 to -4.2 | Fail | Pass |
| WINDOW | 42.4 | 0.48 | 1.01 | -6.8 | -7.8 to -5.8 | Fail | Pass |
| **205 : actual** | **38.4** | | | | | | |
| D and P | 34.1 | 0.50 | 1.03 | +4.3 | 3.3 to 5.3 | Fail | Pass |
| NESS-80/K | 36.3 | 0.51 | 1.04 | -2.1 | -3.1 to -1.1 | Pass | Fail |
| NESS-80/T | 46.6 | 0.52 | 1.06 | -8.2 | -9.3 to -7.2 | Fail | Pass |
| McClain | 38.9 | 0.51 | 1.04 | -0.5 | -1.6 to 0.6 | Pass | Pass |
| Price | 38.5 | 0.51 | 1.04 | -0.1 | -1.1 to 0.9 | Pass | Pass |
| Singh | 31.8 | 0.50 | 1.03 | +6.6 | 5.6 to 7.6 | Fail | Pass |
| WINDOW/K | 40.1 | 0.51 | 1.04 | -1.7 | -2.8 to -0.6 | Pass | Pass |
| WINDOW/T | 43.9 | 0.52 | 1.06 | -5.5 | -6.6 to -4.4 | Fail | Pass |
| **218: actual** | **33.3** | | | | | | |
| D and P | 31.3 | 0.43 | 0.89 | +2.2 | 1.1 to 2.9 | Pass | Pass |
| NESS-80 | 40.3 | 0.45 | 0.94 | -7.0 | -7.9 to 6.1 | Fail | Pass |
| McClain | 35.0 | 0.44 | 0.91 | -1.7 | -2.6 to -0.8 | Pass | Pass |
| Price | 34.4 | 0.45 | 0.94 | -1.1 | -2.0 to -0.2 | Pass | Pass |
| Singh | 29.9 | 0.43 | 0.89 | +3.4 | 2.5 to 4.3 | Fail | Pass |
| WINDOW | 37.6 | 0.50 | 1.04 | -4.3 | 5.3 to -3.3 | Fail | Pass |
| **223: actual** | **29.0** | | | | | | |
| D and P | 26.7 | 0.21 | 0.43 | +2.2 | 1.7 to 2.6 | Pass | Pass |
| NESS-80 | 89.4 | 0.22 | 0.15 | -0.5 | -0.1 to 0.9 | Pass | Pass |
| McClain | 30.8 | 0.22 | 0.45 | -2.0 | -2.4 to -1.5 | Pass | Pass |
| Price | 30.5 | 0.22 | 0.45 | -1.5 | -1.9 to -1.1 | Pass | Pass |
| Singh | 25.0 | 0.21 | 0.43 | +3.9 | 4.3 to 3.5 | Fail | Pass |
| WINDOW | 28.6 | 0.22 | 0.45 | +0.3 | 0.1 to 0.8 7 | Pass | Pass |

**Table 3.** *Statistical analysis of six atmospheric correction models (Cooper and Asrar 1989). The K and T in the NESS-80 and WINDOW models for study date 205 refer to Konza and Topeka operated radiosondes repectively. All temperatures are °C.*

## 3.7 Simultaneous split-window method for earth surface temperature and emissivity from ATSR thermal band data

We now present a method which is a combination of multi-angle and multi-channel methods. The method is based on the theoretical models of Price (1984) and of Sobrino *et al.* (1993). One parameter, the emissivity, is the serious problem in the previous models for the determination of land surface temperature. The present development of the earlier models by Cracknell and Xue (1995a) for ATSR thermal band data, enables not only the surface temperature (LST and SST) but also the surface emissivity to be obtained from ATAR thermal band data using this method. This section is adapted from the paper of Cracknell and Xue (1995a).

For ATSR thermal band data, we have two different channels, each with two viewing angles. We assume that the emissivities for the nadir and forward directions are the same. We select Price's expression, Equation 26, for surface temperature. For nadir two-channel thermal band data we have

$$T_s = (0.992T_{11,0} + 0.001T_{12,0})\,\varepsilon_{11} - 0.773T_{12,0}\varepsilon_{12} - (15.155T_{11,0} - 11.655T_{12,0}) \quad (30)$$

From Equation (19), we have

$$T_s = (7.99T_{11,0} - 6.51T_{11,\theta}) - (4.884T_{11,0} - 4.51T_{11,\theta})\,\varepsilon_{11} - (5.856T_{11,0} - 5.29T_{11,\theta})\,\delta_1 \quad (31)$$

for the 11$\mu$m channel and

$$T_s = (8.013T_{12,0} - 6.51T_{12,\theta}) - (4.901T_{12,0} - 4.505T_{12,\theta})\,\varepsilon_{12} - (5.882T_{12,0} - 5.29T_{12,\theta})\,\delta_2 \quad (32)$$

for the 12$\mu$m channel. $\delta_1$ and $\delta_2$ are constants, the values of which depend on wavelength and viewing angle. The solutions of the simultaneous equations (30), (31) and (32) are surface temperature $(T_s)$, surface emissivities for the 11$\mu$m channel $(\varepsilon_{11})$ and for the 12$\mu$m channel $(\varepsilon_{12})$.

Some ATSR data from the Rutherford Appleton Laboratory (RAL) was used. The date and time of the ATSR data are 20 April 1992 and 21:34 GMT. The area is located at English Channel (0–3°E, 49.5–52°N). The moving window interpolation noise removing method was developed to improve the quality of the ATSR thermal band image (Cracknell and Xue 1995a).

Plate 3 (frontispiece) shows the sea surface temperature (top left). The two pictures below it show the emissivities of the sea surface for the 11$\mu$m channel (middle) and the 12$\mu$m channel (bottom) using the method described above. We can see that the emissivity does depart significantly from unity. The values of emissivities are around 0.988 for the 11$\mu$m channel and 0.975 for the 12$\mu$m channel. This is agreement with the results of Barton and Takashima (1986) and Buettner and Kern (1965).

Plate 3 (top right) shows the the land surface temperature and beneath it the emissivities for the 11$\mu$m channel (middle) and the 12$\mu$m channel (bottom) derived from ATSR thermal band data using the method described above. We can see that the emissivity is not very close to unity. The values of the emissivity are around 0.979 for the 11$\mu$m channel and 0.971 for the 12$\mu$m channel. This is agreement with the result of Taylor (1979) that the surfaces in agricultural areas are nearly black ($\varepsilon = 0.97$). It will be noticed that the emissivity for the 11$\mu$m channel is greater than for the 12$\mu$m channel.

# 4  Surface temperature prediction in inertia models

Typically, the Earth's surface absorbs energy from the Sun during the day, radiating a fraction of it back to space and transferring the balance to the atmosphere as sensible heat and latent heat of evaporation. At night, stored energy is released and the surface cools until dawn, when the heating cycle begins again. A fixed amount of heat arrives at the surface of the Earth; some of it raises the temperature of the surface layer and some of it is conducted down to layers beneath the surface.

The thermal inertia, $P$, as defined in Equation 3, is a lumped parameter or physical variable which can be regarded as describing the impedance to variations of temperature. High thermal inertia values lead to small changes in temperature, for a given transfer of heat, while low thermal inertia values lead to large changes in temperature for the same transfer of heat. Thermal inertia cannot be measured directly, but must be inferred from measurements of the temperature variation during the diurnal cycle, combined with a knowledge of the heating processes which occur during the cycle and of the visible and near-infrared reflective processes during the day.

In the early 1970s the development of high quality visible and near infrared sensors for satellite missions was followed closely by the construction of improved sensors in the thermal infrared region ($10$–$12\mu$m) of the electromagnetic spectrum. The satellite for the Heat Capacity Mapping Mission (HCMM) was launched on 26 April 1978. A mission product 'apparent thermal inertia' was produced for its potential value in determining surface characteristics from the satellite data. The scope of the experimental HCMM has been described by Price (1977). The details of the survey have been published by NASA (1978). The theory underlying the product is quite simplified (Price 1977) and the possibly limited value of the apparent thermal inertia was recognised. But it was a one-off experimental mission. Like Seasat it was useful on a proof of concept basis, but it has all the limitations of such a dataset; in particular it ceased to collect data after a certain period. Cracknell and Xue (1995c) have given a tutorial review which includes a discussion on HCMM and thermal inertia.

The NOAA series of polar-orbiting satellites and the geostationary meteorological satellites, such as Meteosat, provide images of the globe with a high rate of repetition at visible and thermal infrared wavelengths. We had the thought, following the demise of HCMM, that at least the AVHRR has thermal infrared channels and moreover is an operational system which gathers data frequently and regularly, although almost certainly it does not gather data at the times of maximum and minimum temperature. We wondered therefore, whether there was some way in which we might use AVHRR data for thermal inertia studies; this will be discussed later in this chapter.

## 4.1   Thermal model of bare soil ground surface

Here we only discuss the thermal model of a bare soil ground surface. Many thermal models exist for different non-vegetated targets of interest and for planar solid objects. For example, Watson (1971) developed a thermal model for predicting the diurnal surface temperature variation of the ground, and the University of Michigan (1969) developed a model for the prediction of time-dependent temperatures and the radiance of planar targets and backgrounds. A thermal model has been developed at the Jet Propulsion Laboratory to predict the temperature of the Earth's surface as a function of albedo, topography and meteorological conditions (Kahle 1977).

The Energy balance equation applicable to most land surfaces can be written as

$$R_{\mathrm{net}} = G + H + (LE) \tag{33}$$

where $G$ is a constant, $H$ is the sensible heat flux to the atmosphere and LE is the latent heat flux to the atmosphere. Historically, Equation 33 has been used to evaluate the latent heat flux, LE. In the literature, evapotranspiration is frequently referred to as ET. Although LE and ET are used interchangeably, LE is in units of energy, whereas ET is usually used with units of volume per unit area.

Net radiation can be expressed as the sum of the four major components, *i.e.* ,

$$R_{\text{net}} = R_{\text{sd}} - R_{\text{su}} + R_{\text{ld}} - R_{\text{lu}} \tag{34}$$

where $R_{\text{sd}}$ is the downward shortwave radiation from the Sun and atmosphere, $R_{\text{su}}$ is the reflected shortwave radiation by the surface, $R_{\text{ld}}$ is the longwave radiation emitted from the atmosphere toward the surface and $R_{\text{lu}}$ is the longwave radiation emitted from the surface into the atmosphere. As shortwave radiation from the Sun passes through the atmosphere, some of it is absorbed or scattered by aerosols, gases and particulates; therefore, at the Earth's surface, $R_{\text{sd}}$ is composed of both direct solar and diffuse-sky radiation.

## 4.2   Thermal inertia modelling

A mathematical model of the near-surface conductive heat transfer was used to develop a means to interpret thermal infrared images in terms of the relevant physical properties and processes. So far, all thermal models have assumed one-dimensional periodic heating of a uniform half-space (a region bounded by a plane on its upper side and extending downward to infinity) of constant thermal properties. The temperature obeys the diffusion equation:

$$D\frac{\partial^2 T(x,t)}{\partial x^2} = \frac{\partial T(x,t)}{\partial t} \tag{35}$$

where $T(x,t)$ is the temperature at depth $x$ below surface at time $t$ and $D = K/\rho c$ is the thermal diffusivity of the half-space. A detailed glossary of symbols used in this chapter can be found in the Appendix.

Widely known analytical methods permit the solution of the differential Equations governing diffusion of heat in solids (Carslaw and Jaeger 1959). To solve equation (35) one must specify the appropriate boundary conditions. For different forms of the boundary conditions, the methods to solve the equation are different. We use the Fourier series method which is commonly applied to this type of problem. The boundary conditions used here are (Watson 1975, Xue and Cracknell 1992, 1995a)

$$D\frac{\partial^2 T(x,t)}{\partial x^2} = \frac{\partial T(x,t)}{\partial t} \tag{36}$$

$$-K\left.\frac{\partial T(x,t)}{\partial x}\right|_{x=0} = (1-A)S_0 C_t \cos Z' - (A_c + BT(0,t)) \tag{37}$$

and $T(x,t)$ is finite as $x \to \infty$. Here $A_c$ and B are the linearisation coefficients of the linearised boundary condition (36) which is from the dynamic energy balance equation (33) at the ground surface. They are variable for different pixels.

$$-K\left.\frac{\partial T(x,t)}{\partial x}\right|_{x=0} = (1-A)S_0 C_t \cos Z' - R_{\text{earth}} + R_{\text{sky}} - H - (LE) \tag{38}$$

$$\cos Z' = \cos d_{\mathrm{sa}} \cos Z - \sin d_{\mathrm{sa}} \left[\sin\phi\cos\delta\sin\omega t\right. \tag{39}$$
$$\left. - \cos\phi\sin\delta\cos\lambda - \sin\delta\sin\lambda\cos\omega t\right]$$
$$\cos Z = \sin\delta\sin\alpha + \cos\delta\cos\alpha\cos\omega t \tag{40}$$

where $R_{\mathrm{earth}}$ is the Earth emitted radiation, $R_{\mathrm{sky}}$ is the downward longwave sky radiation, $H$ is the sensible heat flux to the atmosphere and LE is the latent heat flux to the atmosphere. The first term of the right-hand side of Equation 36 is the short-wavelength absorbed flux from the Sun (modulated by the atmospheric transmission). The second term is the linearisation expression of the long-wavelength radiation from the sky and the ground.

The solution of the thermal conduction equation (35) subject to the boundary conditions in equation (36) and (37) for the ground surface temperature can be obtained (Carslaw and Jaeger 1959, Xue and Cracknell 1995a). The temperature at depth $x$ is found to be

$$T(x,t) = -\frac{A_c}{B} + (1-A)S_0 C_t \sum_{n=1}^{\infty} A_n \frac{\exp[-k_0\sqrt{nx}]\cos(n\omega t - k_0\sqrt{nx} - \delta_n)}{\left\{\omega n p^2 + \sqrt{2\omega n}\,BP + B^2\right\}^{1/2}} \tag{41}$$

where

$$k_0 = \frac{P}{K}\sqrt{\frac{\omega}{2}} \tag{42}$$

$$\delta_n = \arctan\left(\frac{P\sqrt{n\omega}}{\sqrt{2}B + P\sqrt{n\omega}}\right) \tag{43}$$

$$A_1 = \frac{2}{\pi}\sin\delta\sin\alpha + \frac{1}{2\pi}\cos\delta\cos\alpha\,(\sin 2\psi + 2\psi) \tag{44}$$

$$A_n = \frac{2\sin\delta\sin\alpha}{n\pi}\sin n\psi + \frac{2\cos\delta\cos\alpha}{\pi(n^2-1)}\left(n\sin n\psi\cos\psi - \cos n\psi\sin\psi\right) \tag{45}$$

$$\psi = \arccos(\tan\delta\tan\alpha) \tag{46}$$

Price (1977) compared the surface temperature of all terms with the first term of the series and concluded that the first term in the series is dominant. For areas with variable soil moisture, sparse vegetation cover and ice sheet, the Fourier series solution of surface temperature should take account of second-order terms (Xue and Cracknell 1995a). From Equation (41), we have

$$T(0,t) = -\frac{A_c}{B} + (1-A)S_0 C_t A_1 \frac{\cos(\omega t - \delta_1)}{\left\{\omega P^2 + \sqrt{2\omega}\,BP + B^2\right\}^{1/2}} \tag{47}$$
$$+ (1-A)S_0 C_t A_2 \frac{\cos(2\omega t - \delta_2)}{\left\{2\omega P^2 + 2\sqrt{\omega}\,BP + B^2\right\}^{1/2}}$$

Thus the surface temperature difference between two passes is

$$\Delta T = T(0, t_2) - T(0, t_1) = \frac{(1-A)S_0 C_t}{P\sqrt{\omega}} \frac{A_1[\cos(\omega t_2 - \delta_1) - \cos(\omega t_1 - \delta_1)]}{\left\{1 + \frac{1}{b} + \frac{1}{2b^2}\right\}^{1/2}} \quad (48)$$

$$+ \frac{(1-A)S_0 C_t}{P\sqrt{\omega}} \frac{A_2[\cos(\omega t_2 - \delta_2) - \cos(\omega t_1 - \delta_2)]}{\left\{2 + \frac{\sqrt{2}}{b} + \frac{1}{2b^2}\right\}^{1/2}}$$

where $t_1$ and $t_2$ are the times of two passes. We have also set $b = \sqrt{\omega/2}(P/B)$.

Usually, only the magnitude of the diurnal ground surface temperature wave has been used. If we can use the phase information of the diurnal ground surface temperature wave, we shall be able to obtain some useful information on the ground surface. Now, the second-order approximation thermal inertia model has been obtained as

$$P = \frac{(1-A)}{\Delta T} \frac{S_0 C_t}{\sqrt{\omega}} \frac{A_1(\cos(\omega t_2 - \delta_1) - \cos(\omega t_1 - \delta_1))}{\left\{1 + \frac{1}{b} + \frac{1}{2b^2}\right\}^{1/2}} \quad (49)$$

$$+ \frac{(1-A)}{\Delta T} \frac{S_0 C_t}{\sqrt{\omega}} \frac{A_2(\cos(\omega t_2 - \delta_2) - \cos(\omega t_1 - \delta_2))}{\left\{2 + \frac{\sqrt{2}}{b} + \frac{1}{2b^2}\right\}^{1/2}}$$

(Xue and Cracknell 1995a), where $A_1$ and $A_2$ are the coefficients in the Fourier series. $C_t$ is the atmospheric transmittance in the visible spectrum (typically 0.75), $S_0$ is the solar constant (1367 W $m^{-2}$), $\delta_1$ and $\delta_2$ are the phase differences

$$\omega t_{\max} = \delta_1 = \arctan\left(\frac{b}{1+b}\right) \quad (50)$$

$$\delta_2 = \arctan\left(\frac{b\sqrt{2}}{1+b\sqrt{2}}\right) \quad (51)$$

and $b$ is the function of $t_{\max}$, which is the time of the maximum temperature in the daytime (usually 1:30pm–2:00pm local time),

$$b = \frac{\tan(\omega t_{\max})}{1 - \tan(\omega t_{\max})} \quad (52)$$

This model can be used in the areas with variable soil moisture, sparse vegetation cover, etc with the surface temperature range 280K–310K. From Equation (49), we can see that thermal inertia, $P$, is directly proportional to the apparent thermal inertia.

Figure 4 shows the thermal inertia values obtained using the different models with the times of daytime maximum surface temperature. The value of the albedo is taken as 0.2, the difference of diurnal peak temperature is 18°C, the latitude is 48°N, the solar declination is 7.40° and the times of the two passes are 04:44 GMT and 14:37 GMT. The curve (1) gives real thermal inertia values using our operational first-order

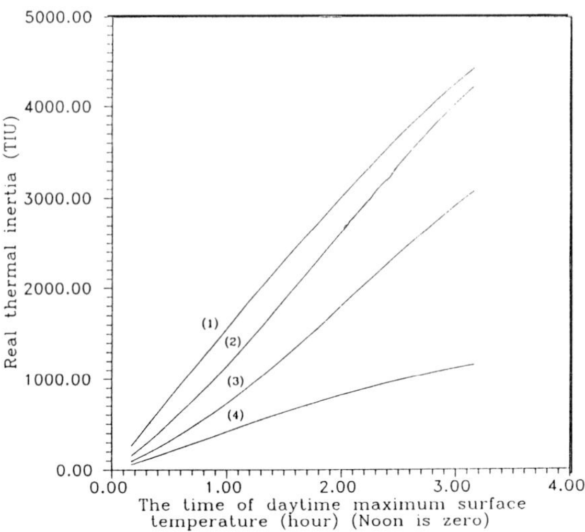

**Figure 4.** *The thermal inertia values obtained by using the different models with the times of maximum surface temperature for the albedo of 0.2, the difference of diurnal surface temperature of 18° C, the latitude of 48° N, the solar declination of 7.40° and the times of two passes of 04:44 and 14:37. Curve 1 is the first-order approximation model with constant times of passes; Curve 2 is the first-order approximation model with real times of passes; Curve 3 is the second-order approximation model with real times of passes (Xue and Cracknell 1995a); Curve 4 is the difference of curves 2 and 3.*

approximation thermal inertial model (Xue and Cracknell 1992). The times of two passes are assumed to be at the times of peak surface temperatures. The curve (2) gives real thermal inertia values using the first-order approximation model but the real times of two passes are used. The curve (3) gives real thermal inertia values using the second-order approximation model (Xue and Cracknell 1995a) and the real times of two passes are used. The curve (4) gives the second-term values of real thermal inertia, that is the difference between the real thermal inertia values using our second-order approximation model (Curve 3) and first-order approximation model (Curve 2). We can see that the relative value of thermal inertia is the same, the difference is only the absolute value of thermal inertia. The difference between curve 1 and curve 2 is from 100TIU to 440TIU. The thermal inertia unit, TIU, is measured in Joules per square meter per °C per second. The difference between curve 2 and curve 3 is from 10TIU to 1000TIU and is in direct proportion to the real thermal inertia value and the time of daytime maximum surface temperature. If we know the real $t_{\mathrm{max}}$ from a meteorological station, not only the apparent thermal inertia but also the real thermal inertia can easily be obtained.

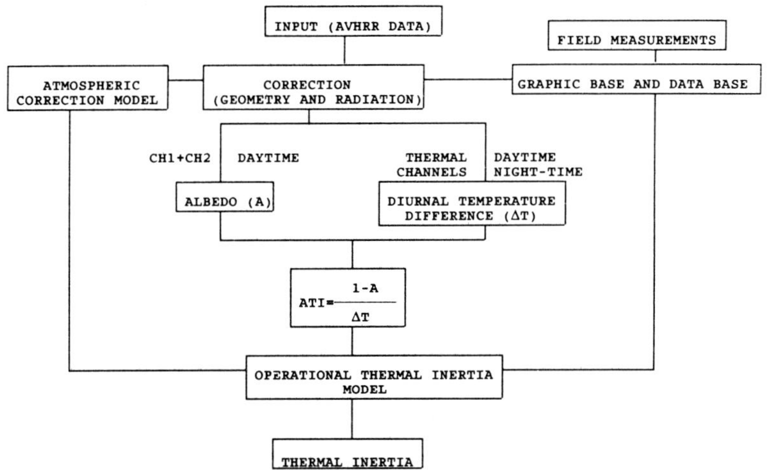



**Figure 5.** *The flowchart of computer mapping (Xue and Cracknell 1992).*

## 4.3   Computer mapping

From NOAA AHVRRdata, we have (Xue 1989),

$$A = W_1 R^1 + W_2 R^2 \tag{53}$$

$$\Delta T = (\text{Ch.5})_D - (\text{Ch.5})_N \tag{54}$$

where $D$ represents noon, $N$ represents midnight. $R^1$ and $R^2$ represent the reflectances of channel 1 and channel 2, which can be obtained using the method developed by Xue and Cracknell (1995b). $(\text{Ch.5})_D$ and $(\text{Ch.5})_N$ represent the surface temperatures retrieved from daytime and night-time channel 5 data. $W_n$ ($n$=1,2) is the weighted coefficient

$$W_n = \frac{S_n}{\sum S_n} \tag{55}$$

where $S_n$ represents the solar spectral irradiation on the ground in the $n$th channel.

The U.S. standard atmospheric model can be used to give $W_1$ and $W_2$ (Xue 1989). We can obtain the time of maximum ground surface temperature from several meteorological stations in the test area. The map of the time of maximum ground surface temperature of the test area can be obtained. Using the above models we can obtain a real thermal inertia map. The flowchart of computer mapping is shown in Figure 5.

One common problem encountered in computing thermal infrared (TI) images is the availability of proper day and night TI data in a diurnal cycle. For mere comparative assessments, the day and night coverages separated by days or weeks can also be used for computing ATI and TI (Kahle and Alley 1985).

## 4.4 Results

Thermal inertia results have already been obtained by aircraft (Kahle *et al.* 1975, 1976). Research was conducted for the development of the necessary image processing techniques for the production of thermal inertia images (Pratt *et al.* 1976, Pratt and Ellyett 1979). Pohn *et al.* (1974) produced thermal inertia contour maps using data from the Nimbus meteorological satellites and showed that it is possible to estimate thermal inertia values for the Earth's surface by using both thermal and panchromatic imagery. Their model was based on diurnal temperature simulation results obtained using a model developed earlier by Watson (1970, 1973, 1975). Watson's model is based on the theory originally introduced by Jaeger (1953) for the lunar surface. The second-order approximation thermal inertia model was tested with data from France (about 47-48°N, 0.8°E–0.4°W) See Xue and Cracknell (1995a) and next section.

## 5  Dynamic aspects study of surface temperature

From Equation (49), we can have

$$
\begin{aligned}
T(0,t) = T(0,t_0) \;&+\; \frac{(1-A)S_0C_t}{P\sqrt{\omega}}\,\frac{A_1[\cos(\omega t - \delta_1) - \cos(\omega t_0 - \delta_1)]}{\left\{1 + \dfrac{1}{b} + \dfrac{1}{2b^2}\right\}^{1/2}} \\[2ex]
&+\; \frac{(1-A)S_0C_t}{P\sqrt{\omega}}\,\frac{A_2[\cos(\omega t - \delta_2) - \cos(\omega t_0 - \delta_2)]}{\left\{2 + \dfrac{\sqrt{2}}{b} + \dfrac{1}{2b^2}\right\}^{1/2}}
\end{aligned}
\tag{56}
$$

where $t_0$ is the time of satellite over-pass during the day. If we know the surface temperature of satellite over-pass time $t_0$, we can calculate the 24-hour period surface temperature from Equation (56).

The model was tested with data from France (about 47°N–48°N, 0.8°E–0.4°W) for 9 April 1992. The thermal inertia map is obtained from the second-order approximation operation thermal inertia model (see Section 4.4 in this chapter). The Earth surface albedos have been obtained using the bi-angle approach developed by Xue and Cracknell (1995b). A split-window method is used to retrieve the surface temperature from NOAA AVHRR channel 4 and 5 data. Here the emissivity is assumed to be 0.96 (Cracknell and Xue 1995a). We used the expression of Price (1984) for daytime over-pass (14:37 GMT) surface temperature determination.

Plate 4 (frontispiece) shows the 24-hour period surface temperature predicted using the method described in this chapter (Cracknell and Xue 1995d). The surface temperatures derived from satellite data are the regional average surface temperatures, at least the average temperature of one pixel size. The values agreed with the data obtained from Meteo-France. We can see that the surface temperature changes very slow during the middle of the night and the middle of the day, but it changes rapidly during near dawn and dusk.

Finally, we note that the comparison of the surface temperature values derived from satellite data with the *in-situ* temperature measurements is very difficult. The ATSR

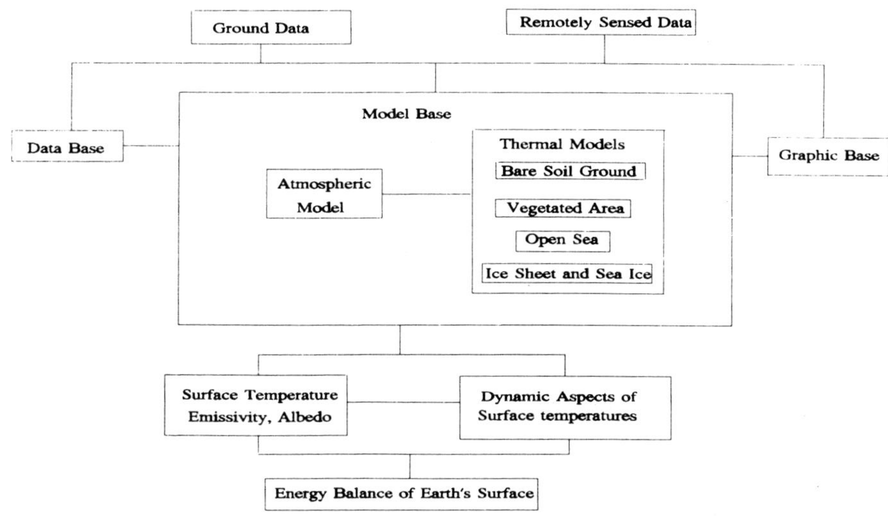


**Figure 6.** *Dynamic aspects study of global surface temperature from remotely-sensed data system (Cracknell and Xue*

and AVHRR have a footprint of $1\mathrm{km}^2$ on the Earth's surface. When the instrument passes over Earth it sees a composite surface temperature. The surface temperature derived from satellite data is the average temperature over the area of the footprint, but the *in-situ* temperature measurement is the point temperature. This is not usually a problem for open sea areas but it is a problem for land areas, coastal regions and marginal sea ice areas. For example, a small percentage of the surface of marginal sea ice areas consists of open or recently refrozen leads with a surface temperature which allows the composite temperature sensed by the instrument to be used to give a measure of the fraction of the icefield occupied by leads. The interpretation of structure within the footprint, or instantaneous field of view, of a remote sensor has to be carried out with great care, taking into account such aspects as the field of view response function and the modulation transfer function of the optical and electronic system; for instance, the response to this structure will depend on its apparent frequency, measured at the instrument output.

# 6    Conclusions

Previous workers used only the magnitude of the diurnal surface temperature change. With the phase information of the diurnal surface temperature change used in remote sensing information modelling (Xue and Cracknell 1992, 1995a), operational thermal inertia mapping is possible. The surface temperature prediction using remotely sensed data with the operational thermal inertia models will be potentially useful for dynamic aspects study of global surface temperature. Figure (6) gives the scheme of dynamic aspects study of global surface temperature from remotely sensed data. The operational thermal inertia model developed by Xue and Cracknell (1995a) can be used for bare soil ground, a sparsely vegetated area, an ice sheet and sea ice. The thermal models of open sea and vegetated areas should be developed. With the launch of the ERS-1 satellite and the ERS-2 satellite, the accurate over-pass time surface temperature can be obtained (Cracknell and Xue 1995a). From NOAA AVHRR data and ERS-1 or ERS-2 data with the thermal models of Earth's surface, it is possible to predict the global surface temperature operationally. Undoubtedly, much time will be required before the above-described global system of meteorological observations will be operative. However, it is important to appreciate that at present various components of this system already exist or are at the stage of development and testing.

# Appendix: list of symbols

| | |
|---|---|
| $A$ | surface albedo |
| $A_1$, $A_2$ | coefficients of Fourier series |
| $A_c$ | linearisation coefficient of boundary condition |
| $ATI$ | apparent thermal inertia $(ATI = (1 - A)/T)$ |
| $B$ | linearisation coefficient of boundary condition $(\mathrm{kg\,s^{-3}K^{-1}})$ |
| $c$ | specific heat |
| $C_t$ | atmospheric transmittance in the visible spectrum (typically 0.75) |
| $dsa$ | surface slope angle measured downward from the horizontal |
| $D$ | thermal diffusivity of the half-space |
| $H$ | sensible heat flux to the atmosphere |
| $K$ | thermal conductivity of half-space $K = Dc$ |
| $LE$ | latent heat flux to the atmosphere |
| $P$ | thermal inertia (for dry ground surface) $P = (Kc)$ |
| $R_{\mathrm{earth}}$ | earth emitted radiation |
| $R_{\mathrm{sky}}$ | downward longwave sky radiation |
| $S_n$ | solar spectral irradiation on the ground in the nth channel. |
| $t$ | time (noon is zero) |
| $T(x,t)$ | temperature at depth x below the surface and time t |
| $TIU$ | thermal inertia unit $(1\,\mathrm{TIU}=1\,\mathrm{J\,m^{-2\circ}C^{-1}s^{-1}})$ |
| $t_{\mathrm{max}}$ | time of the maximum temperature (usually 1:30–2:00pm) |
| $W_n$ | weighted coefficient for the $n$th channel |
| $x$ | depth below the surface |
| $Z'$ | local zenith angle for an inclined surface |
| $Z$ | local zenith angle for a smooth surface |
| $\alpha$ | local latitude |
| $\delta$ | solar declination |
| $\delta_1$, $\delta_2$ | phase differences |
| $\varepsilon$ | mean emissivity (spectral average) of the surface |
| $\rho$ | density |
| $\rho_a$ | density of dry air |
| $\sigma$ | Stefan-Boltzmann constant |
| $\phi$ | azimuth of the slope angle measured counterclockwise from north |
| $\omega$ | angular frequency of earth's rotation |

# References


Anding, D, and Kauth, R, 1970, Estimation of sea surface temperature from space. Remote Sensing Environment, 1, 217-220.

Barton, I J, and Takashima, T, 1986, An AVHRR investigation of surface emissivity near Lake Eyre, Australia. Remote Sensing of Environment, 20, 153-163.

Barton, I J, 1992, Satellite-derived sea surface temperatures  A comparison between operational, theoretical, and experimental algorithms. Journal of Applied Meteorology, 31, 433-442.

Becker, F, 1982, Absolute sea surface temperature measurements by remote sensing and atmospheric corrections using differential radiometry. In Processes in Marine Remote Sensing, edited by F. J. Vernberg and F. P. Diemer (Columbia: University of South Carolina Press) pp. 151-174.

Becker, F, 1987, The Impact of Spectral Emissivity on the Measurement of Land Surface Temperature from a Satellite. International Journal of Remote Sensing, 8, 1509-1522.

Becker, F, and Li, Z. -L., 1990, Towards a local split window method over land surfaces. International Journal of Remote Sensing, 11, 1509-1522.

Brower, A L, Gohrband, H S, Pichel, W, Signore, T L, and Walton, C, 1976, Satellite derived sea-surface temperatures from NOAA spacecraft, NOAA Technical Memorandum NESS 78, U.S. Dept. of Commerce, NOAA, Washington, DC, 74 pp.

Carlson, T N, Dodd, J K, Benjamin, S G, and Cooper, J N, 1981, Satellite estimation of Surface energy balance, moisture availability and thermal inertia. Journal of Applied Meteorology, 20, 60-87.

Carslaw, H S, and Jaeger, J C 1959, Conduction of heat in solids, 2nd edition (Oxford: Oxford University Press).

Caselles, V, and Sobrino, J A, 1989, Determination of frosts in orange groves from NOAA-9 AVHRR data. Remote Sensing of Environment, 29, 135-146.

Caselles, V, and Sobrino, J A, 1991, Shelter and remotely sensed night temperatures in orange groves. Theoretical and Applied Climatology, 44, 113-122.

Chedin, A, Scott, N A, and Berrior, A, 1982, A single-channel, double-viewing angle method for sea surface temperature determination from coincident METEOSAT and TIROS-N radiometric measurements. Journal of Applied Meteorology, 21, 613-618.

Collier, P Punacres, A M E, and McClatchey, J, 1989, Mapping very low surface-temperature in the scottish highlands using NOAAA-AVHRR data. International Journal of Remote Sensing, 10, 1519-1529.

Cooper, D I, and Asrar, G, 1989, Evaluating atmospheric correction models for retrieving surface temperatures from the AVHRR over a tallgrass prairie. Remote Sensing of Environment, 27, 93-102.

Cracknell, A P, and Hayes, L W B, 1991, Introduction to remote sensing. (London: Taylor and Francis).

Cracknell, A P, and Xue, Y, 1995a, Earth surface temperature and emissivity derivation from ATSR thermal band data. (Submitted to International Journal of Remote Sensing)

Cracknell, A P, and Xue, Y, 1995b, Moving window iterative self-consistent approach for Earth surface temperature deriving. (Submitted to International Journal of Remote Sensing)

Cracknell, A P, and Xue, Y, 1996, Thermal Inertia Determination from Space - A tutorial Review. International Journal of Remote Sensing, 17, 431-461

Cracknell, A P, and Xue, Y, 1995d, Dynamic aspects study of land surface temperature from remotely-sensed data using advanced thermal inertia model (SoA-TI Model). (Submitted to IEEE Trasactions on Geoscience and Remote Sensing)

Cracknell, A P, and Xue, Y, 1995e, Estimation of Ground Heat Flux Using AVHRR data from Advanced Thermal Inertia Model. (Submitted to International Journal of Remote Sensing)

Deschamps, P Y and Phulpin, T, 1980, Atmospheric correction of infrared measurements of sea surface temperature using channels at 3.7, 11, and 12m. Boundary Layer Meteorology, 18, 131-143.

Gorodetskii, G J, 1985, Earth surface temperatures determined from angular radiation distribution in atmospheric windows. Soviet Journal of Remote Sensing, 2, 981-996.

Holyer, W B, 1984, A two-satellite method for measurement of sea surface temperature. International Journal of Remote Sensing, 5, 115-132.

Jaeger, J C, 1953, Conduction of heat in a solid with periodic boundary conditions, with an application to the surface temperature of the Moon. Proceedings of the Cambridge Philosophical Society, 49, 355-359.

Kahle, A B, 1977, A simple thermal model of the Earth's surface for geologic mapping by remote sensing. Journal of Geophysical Research, 82, 1673-1680.

Kahle, A B, and Alley, R E, 1985, Calculation of thermal inertia from day-night measurements separated by days or weeks. Photogrammetric Engineering and Remote Sensing, 51, 73-75.

Kahle, A B, Gillespie, A R, and Goetz, A F H, 1976, Thermal inertia imaging: a new geologic mapping tool. Geophysical Research Letters, 3, 26-28.

Kahle, A B, Gillespie, A R, Goetz, A F H, and Addington, J D, 1975, Thermal inertia mapping. Proceedings of the 10th International Symposium on Remote Sensing of Environment held in Environmental Research Institute of Michigan, Ann Arbor, Michigan, on October 1975, (Ann Arbor, Michigan: Environmental Research Institute of Michigan), pp. 985-994.

Kerr, Y H, and Lagourde, J P, 1989, On the derivation of land surface temperature from AVHRR data. EUM-P06, Eumetsat, Rothenburg, FR Germany, pp 157-160

Kerr, Y H, Lagourde, J P, and Imbernon, J, 1992, Accurate land surface- temperature retrieval from AVHRR data with use of an improved split window algorithm. Remote Sensing of Environment, 41, 197-209.

Kimes, D S, and Kirchner, J A, 1983, Directional radiometric measurements of row crop temperature. International Journal of Remote Sensing, 4, 299-311.

Kondratyev, K YA, 1969, Radiation in the atmosphere. (New York: Academic Press) Li, Z. -L., and Becker, F, 1993, Feasibility of land surface-temperature and emissivity determination from AVHRR data. Remote Sensing of Environment, 43, 67-85.

McClain, E P, Pichel, W G, Walton, C C, Ahmad, Z, and Sutton, J, 1983, Multichannel improvements to satellite-derived global sea surface temperatures. Advances in Space Research, 2, 43-47.

McClatchey, J, 1992, The use of climatological observations as ground truth for distributions of minimum temperature derived from AVHRR data. International Journal of Remote Sensing, 13, 155-163.

McMillin, L M, 1971, A method of determining surface temperatures from measurements of spectral radiance at two wavelengths. PhD dissertation, Iowa State University, Ames.

McMillin, L M, 1975, Estimation of sea surface temperature from two infrared window measurements with different absorption. Journal of Geophysical Research, 80, 11587- 11600.

McMillin, L M, and Crosby, D S, 1984, Theory and validation of the multiple indow sea surface temperature. Journal of Geophysical Research, 89, 3655-3661.

McMillin, L M, and Crosby, D S, 1985, Some physical interpretations of statistically derived coefficients for split-window corrections to satellite-derived sea-surface temperatures. Quarterly Journal of the Royal Meteorological Society, 111, 867-871.

NASA, 1978, Heat Capacity Mapping Mission User's Guide. NASA: Goddard Space Flight Centre, Greenbelt, Maryland.

Pohn, H A, Offield, T W, and Watson, K, 1974, Thermal inertia mapping from satellites-discrimination of geologic units in Oman. Journal of Research US. Geological Survey, 2, 147-158.

Prabhakara, C, Dalu, G, and Kunde, V G, 1972, Estimation of sea surface temperature from remote sensing in the 11-13 m window region. X. Doc. 651-72-358, NASA/GSFC, Greenbelt, MD, 15pp.

Prabhakara, C, Dalu, G, and Kunde, V G, 1974, Estimation of sea surface temperature from remote sensing in the 11-to-13 m window region. Journal of Geophysical Research, 79, 11587-11601.

Pratt, D A, and Ellyett, C D, 1979, The thermal inertia approach to mapping of soil moisture and geology. Remote Sensing of Environment, 8, 151-168.

Pratt, D A, Ellyett, C D, McLauchlan, E C, and McNabb, P, 1976, Recent advances in the application of thermal infrared scanning to geological and hydrological studies. Remote Sensing of Environment, 7, 177-184.

Price, J C, 1977, Thermal inertia mapping: a new view of the Earth. Journal of Geophysical Research, 82, 2582-2590.

Price, J C, 1982, Estimation of regional scale evapotranspiration through analysis of satellite thermal infrared data. IEEE Transactions on Geosciences and Remote Sensing, GE-20, 286-292.

Price, J C, 1983, Estimating surface temperatures from satellite thermal infrared data  a simple formulation for the atmospheric effect. Remote Sensing of Environment, 13,353-361.

Price, J C, 1984, Land surface temperature measurements from the split window channels of the NOAA-7 Advanced Very High Resolution Radiometer. Journal of Geophysical Research, 89, 7231-7237.

Price, J C, 1989, Quantitative aspects of remote sensing in the thermal infrared. In Theory and Applications of Optical Remote Sensing, edited by G. Asrar, (New York: John Wiley & Sons), pp. 578-603.

Rees, W G, 1990, Physical principles of remote sensing (Cambridge: Cambridge University Press).

Rosema, A, Blyeveld, J H, Reinger, P, Tassone, G, Blyth, K, and Gurnney, R J, 1978, TELL-US, A combined surface temperature, soil moisture and evaporation mapping approach. Proceedings of 12th International Symposium on Remote Sensing of Environment held in Manila, Philippines, on 20th-26th April, 1978 (Ann Arbor, ERIM), pp 2267-2276.

Sabins, F F, 1987, Remote Sensing: Principles and interpretation (New York: W. H. Freeman)

Saunders, P M, 1967, Aerial measurement of sea surface temperature in the infrared. Journal of Geophysical Research, 72, 4109-4117.

Short, N M, and Stuart, L M Jr., 1982, The heat capacity mapping mission (HCMM) anthology. (Washington, DC: NASA)

Singh, S M, 1984, Removal of atmospheric effects on a pixel by pixel basis from the thermal infrared data from instruments on satellites. The Advanced Very High Resolution Radiometer (AVHRR). International Journal of Remote Sensing, 5, 155- 163.

Slater, P N, 1980, Remote sensing, optics and optical system. (Reading, MA: Addison- Wesley)

Sobrino, J A, Coll, C, and Caselles, V, 1991, Atmospheric correction for land surface temperature using NOAA-11 AVHRR channel 4 and 5. Remote Sensing of Environment, 38, 19-34.

Sobrino, J A, Li, Z-L, Stoll, M P, and Becker, F, 1993, Determination of the surface temperature from ATSR data. Proceedings of 25th International Symposium on Remote Sensing of Environment held in Graz, Austria, on 4th-8th April, 1993 (Ann Arbor, ERIM), pp II-99-II-109.

Taylor, S E, 1979, Measured emissivity of soil in the Southeast United States, Remote Sensing of Environment, 8, 359-364.

University of Michigan, 1969, Target temperature modelling. Evaluation of Contract 30602-68-c-099, Defense Documentation Center.

Vass, P, and Handoll, M, 1991, UK ERS-1 reference manual. EODC Documentation and Information Service, Earth Observation Data Centre, UK.

Vauchlin, M, Vieira, S R, Bernard, R, and Hatfield, J L, 1982, Spatial variability of surface temperature along two transects of a bare soil. Water Resources Research, 18, 1677-1686.

Vidal, A, 1991, Atmospheric and emissivity correction of land surface temperature measured from satellite using ground measurements or satellite data. International Journal of Remote Sensing, 12, 2449-2460.

Watson, K, 1970, Part A, Introduction and Summary, and Part B, Data Analysis Technique, in Remote Sensor Application Studies Progress Report, July 1, 1968 to June 30, 1969, Geological Survey, Washington, D.C., USGS-GD-71-004.

Watson, K, 1971, A computer program of thermal modelling for interpretation of infrared images. U.S. Geological Survey Report PB Washington, DC.

Watson, K, 1973, Periodic heating of a layer over a semi infinite solid. Journal of Geophysical Research, 78, 5904-5910.

Watson, K, 1975, Geological application of thermal infrared images. Proceedings of the IEEE, 63, 128-137.

Watson, K, 1982, Regional thermal-inertia mapping from an experimental satellite. Geophysics, 47, 1681-1687.

Wienreb, M P, and Hill, M L, 1980, Calculation of atmospheric radiances and brightness temperatures in infrared window channels of satellite radiometers, NOAA Technical Report NESS 80, U.S. Department of Commerce, NOAA, Washington, DC, 40 pp.

Xue Y, 1989, Theoretical models of thermal inertia and monitoring soil moisture by automatic recognition of remote sensing data. MSc thesis, Peking University. (In Chinese)

Xue, Y and Cracknell, A P, 1992, Thermal inertia mapping: from research to operation. Proceedings of the 18th Annual Conference of the Remote Sensing Society, UK, held in University of Dundee, Dundee, on 15th-17th September 1992, edited by A P Cracknell and R A Vaughan, (Nottingham: Remote Sensing Society), pp. 471- 480.

Xue, Y, and Cracknell, A P, 1995a, Advanced thermal inertia modelling. International Journal of Remote Sensing, 16,431 - 446

Xue, Y, and Cracknell, A P, 1995b, Operational bi-angle approach to retrieve the Earth surface albedo from AVHRR data in the visible band. International Journal of Remote Sensing, 16, 417 - 429

# Line Detection using Morphological Filters

A I Watson

University of Stirling
Scotland

## 1   Introduction

One of the problems of image processing is the detection of 'lines'. These can arise in
bubble chambers where they record the passage of charged elementary particles. They
can occur in images acquired for geological prospecting where they may represent faults
in the rocks caused by tectonic activity. Fault lines may also occur in the images of
metallic surfaces where they may be associated with stresses. Finally they may occur
in medicine where we are all familiar with the use of x-rays in detecting fractures. In
all these cases it is important that the 'lines' are easily visible to a human analyst,
and hence images are frequently enhanced to render the 'lines' more obvious. In other
cases the task is one of detecting the presence of a line by a computer, and reporting
the 'statistical' properties of the 'lines' within an image. The two aims need not be
achieved in the same way.

In this paper we are going to look a variety of methods of finding lines within an
image. Some are well established and can be found on any system. These are usually
based on the notion that at a discontinuity, differentiation in some form will reveal that
discontinuity. Some are not so well known, and these are based upon morphological
ideas rather than differential ideas. The reasons for introducing these *new* methods are
many, but principally because they always work as well as the 'differential' methods,
and sometimes a great deal better!

# 2   Differential or filter methods

In this section we will outline the common methods of finding lines in an image, using the notion of a filter. A filter is an array of values that is applied to a pixel and its nearest neighbours, and the result written to a new image. The numerical operations and the size of the filter, and consequently the number of nearest neighbours involved, can be varied, and are chosen to suit the purpose for which the filter is being used. For instance, one filter that can be used for smoothing (removing 'errors') in an image consists in finding the average of a pixel and its nearest neighbours. Thus the original values in the input image are converted to their mean values, which are less variable. Whether this removes the 'errors' is a moot point.

In our case we are interested in finding lines in an image. It is assumed that the pixels that form the lines are either lighter or darker (or have a different colour) than the background pixels. The purpose of the filter is therefore to transform the image so that these lines can be clearly seen by an observer. The filter must be able to pick the lines out despite all the other variations in the original image. To illustrate the difficulties that have to be overcome consider the simulated SPOT image of the Ochil Hills in Scotland shown in Plate 2a (frontispiece). The major variations are due to topography and differences in vegetation. Nonetheless, the area does have several 'lines' or 'lineaments' that are of interest to a geologist, as these may due to fractures or intrusions in the underlying rocks. Most of these lineaments are very difficult to see, even if you know where to look for them!

## 2.1   The Sobel filter

This is the simplest of the filters and consists in multiplying every pixel and its nearest neighbours by the corresponding values in the array and putting the sum of the results in the output image. Thus if A is the filter array,

$$
A = \begin{array}{ccc} -1 & -1 & -1 \\ 2 & 2 & 2 \\ -1 & -1 & -1 \end{array}
$$

and B is the original image, the pixels of the output image, C, are given by

$$
C_{j,i} = \frac{1}{3} \left\{ \begin{array}{lll} B_{j-1,i-1} \times A_{1,1} & + B_{j-1,i} \times A_{1,2} & + B_{j-1,i+1} \times A_{1,3} \\ + B_{j,i-1} \times A_{2,1} & + B_{j,i} \times A_{2,2} & + B_{j,i+1} \times A_{2,3} \\ + B_{j+1,i-1} \times A_{3,1} & + B_{j+1,i} \times A_{3,2} & + B_{j+1,i+1} \times A_{3,3} \end{array} \right\}
$$

for every non-boundary pixel. Boundary pixels of the image $A$ can be transformed to zeros in the output image $C$. Clearly, where the image is homogeneous the pixel values in $C$ will be zero, and where there is an horizontal line of pixels with values larger than the values in the adjacent horizontal lines, the output values in C will be positive. In most cases the lines may not be quite horizontal and the 'homogeneous' areas may not be entirely uniform, nonetheless, the filter should emphasise the horizontal component of the line at the expense of all other variations in the image. Equally, relatively negative

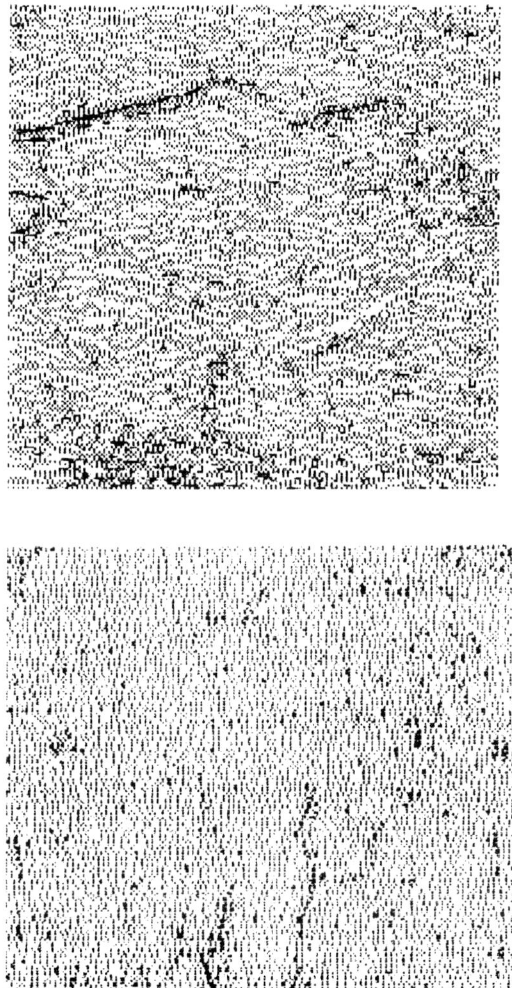

**Figure 1.** *The result of applying Sobel filters to the image of the Ochil Hills shown in Plate 2: (a) Sobel horizontal filter, (b) Sobel vertical filter,*

valued pixels that are aligned horizontally should be emphasised as well. Thus the filter will create an image with any horizontal line in the original line being shown as positive or negative against a background of zero values. Hence these lines should be easily seen, as the original image will have been simplified so that all other variations are suppressed. It is fairly obvious that a filter with values transposed could be used that would pick out the vertical lines in the same way. Figure 1(a) and Figure 1(b) show the results of applying the two filters to the SPOT data.

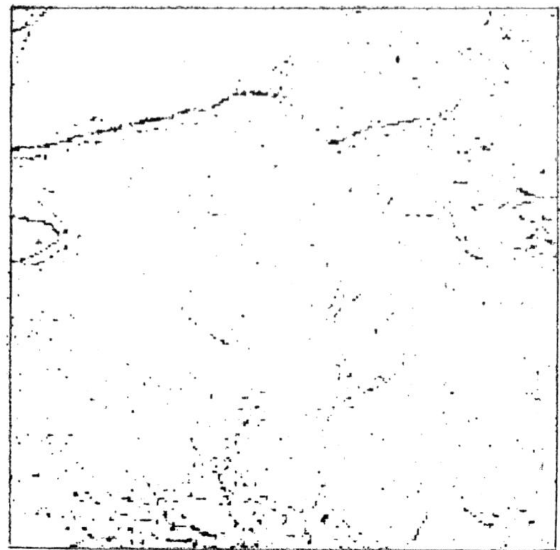

**Figure 2.** *The result of applying a Laplace filter to the image of the Ochil Hills shown in Plate 2(a).*

The results are not very encouraging. The filtered images are certainly simpler than the original, but the lines or lineaments do not in general align themselves horizontally or vertically, and it is difficult to detect any of the lines but the most obvious ones in the filtered SPOT image of Plate 2(a).

Sobel filters can be constructed to enhance lines in any specific direction, and if only lineaments in that direct are of interest, then the Sobel filter will enhance them. However, it is rare that we have any knowledge of which direction to enhance, and hence we require a non-directional filter.

## 2.2  The Laplace Filter

In order to overcome the orientation specific problem associated with the Sobel filters, a generalisation is employed, known as the Laplace filter. This is basically a combined vertical and horizontal Sobel filter. In this case the filter A is

$$A = \begin{array}{ccc} -1 & -1 & -1 \\ -1 & 8 & -1 \\ -1 & -1 & -1 \end{array}$$

It is applied in the same way as the Sobel operators. Again, where the input image is relatively homogeneous, the output image tends to have very small or zero values. Where there is a line of relatively bright (or dark) pixels in the input the filter tends to produce large values (both positive and negative) in the output image. The filter is relatively symmetric and will enhance lines in any direction. Figure 2 shows the result of applying the Laplace filter to the SPOT data.

It is obvious that the Laplace is better than the Sobel and many lines or lineaments can be seen in Figure 2. The question arises, is this the best that can be achieved? It is not easy to answer the question but several points can be made that have some validity. First, images in general do not have many grey levels, and hence all numerical calculations are prone to rounding errors, and any error in a grey level will have a large effect on the results. Further, since the results of a filter involve the grey level values of many pixels, the errors will accumulate to produce a noisy output. Thus the end result of applying the 'differential' type of filters will be a an image with a poor signal to noise ratio. Thus if filters that are not as reliant on errorless continuous data can be found, and they produce a similar enhancement, then they should be better, and have a greater signal to noise ratio.

# 3 Morphological filters

These are based on set theory concept, and originate in the processing of binary images. In the binary image case the pixels within an object have grey level values of one and the background grey levels are zero. Objects are therefore clearly 'sets' of connected 'ones'. The generalisation to grey level images, where there are no obvious objects, is difficult but can be justified quite rigorously. Here we are just going to show how one type of 'morphological' filter can be adapted to enhance lineaments.

Consider the images shown in Plate 2(a). Instead of considering the image as a set of grey levels, it can also be thought of as a relief map with the height at any point being proportional to the grey level. Now the problem of finding lineaments is one of finding narrow ridges or furrows in the image 'topography'. First, let us concentrate on detecting the ridges. Imagine a balloon filled with helium so that it has a tendency to rise. If it is allowed to float beneath the surface of the topography defined by the image grey level values, then depending on its diameter it will define another surface, strictly lower than the image surface, which will be much smoother. As the balloon rolls beneath the surface it will not be able to penetrate the ridges and pimples that are smaller than its diameter. Equally a heavy ball rolled over the surface of the image will not penetrate the cracks and dimples, and hence will define a smoother surface that lies strictly above the original surface. The process of defining these two smooth surfaces, that lie above and below the original surface, is known as 'opening' the surface.

For the purposes of computation, an opening can be decomposed into two simpler processes, erosion and dilation. An erosion $C_{n,m}$ of an image $A_{n,m}$ by a set $B_{i,j}$ is defined as follows

$$C_{n,m} = \min_{(i,j)}\{A_{n-i,m-j} - B_{-i,-j}\}$$

where $i$ and $j$ range from $-r$ to $+r$, and $r$ is the size of the 'ball'. The dilation of $C_{n,m}$ by $B_{i,j}$ completes the opening of A by B. The dilation $D_{n,m}$ is defined by

$$D_{n,m} = \max_{(i,j)}\{A_{n-i,m-j} - B_{i,j}\}$$

                                 *A I Watson*

```
7  7  7  7  7  7  7  7  4  1  1  1  1  1  1
7  7  7  7  7  9  7  7  4  1  3  1  7  1  1
7  7  7  7  9  7  7  7  4  1  1  3  1  1  1
7  7  7  9  7  7  7  7  4  1  1  1  3  1  1
7  7  7  7  7  9  9  9  4  1  1  1  1  1  1
7  7  7  7  7  9  9  9  4  1  1  1  1  1  1
7  7  7  7  7  7  7  7  4  1  1  1  1  1  1
```

**Figure 3(a).** *A simple image with lines*

```
7  7  7  7  7  7  4  1  1  1  1  1  1
7  7  7  7  7  7  4  1  1  1  1  1  1
7  7  7  7  7  7  4  1  1  1  1  1  1
7  7  7  7  7  7  4  1  1  1  1  1  1
7  7  7  7  7  9  4  1  1  1  1  1  1
7  7  7  7  7  7  4  1  1  1  1  1  1
```

**Figure 3(b).** *Figure 3(a) eroded by 3 × 3 square*

```
7  7  7  7  7  7  7  4  1  1  1  1  1
7  7  7  7  7  7  7  4  1  1  1  1  1
7  7  7  7  7  7  7  4  1  1  1  1  1
7  7  7  7  9  9  9  4  1  1  1  1  1
7  7  7  7  9  9  9  4  1  1  1  1  1
7  7  7  7  9  9  9  4  1  1  1  1  1
```

**Figure 3(c).** *Figure 3(b) eroded by 3 × 3 square*

```
 -   -   -   -   2   -   -   -   -   2   -   -   -
 -   -   -   2   -   -   -   -   -   -   2   -   -
 -   -   2   -   -   -   -   -   -   -   -   2   -
 -   -   -   -   -   -   -   -   -   -   -   -   2
 -   -   -   -   -   -   -   -   -   -   -   -   -
 -   -   -   -   -   -   -   -   -   -   -   -   -
```

**Figure 3(d).** *Figure 3(a) – Figure 3(c)*

```
 -    -   -2   -4    14   -2     4    0    -9    14   -4   -2    -
 -   -2   -4    12   -4   -2     4    0    -9    -4   12   -4   -2
 -   -2   14    -6   -6   -6     5   -2    -9    -2   -4   12   -4
 -   -2   -2    -6   -6  -10     9   -2    -9    -    -2   -4   14
 -    -   -6   -10    -   -1    -6   -9     -     -   -2   -2    -
 -    -    -    -4   -6  -10     9   -2     -     -    -    -    -
```

**Figure 3(e).** *Figure 3(a) transformed by a 3×3 Laplace filter*

Let us apply the processes of opening an image, Figure 3(a), with a flat square set B of size $3 \times 3$. The results of the erosion are shown in Figure 3(b), and the dilation of Figure 3(b) by B is shown in Figure3(c). This is the opening of the image by B, and is the smooth surface that a 'rolling ball' would define. When Figure 3(c) is subtracted from Figure 3(a), the remainder, Figure 3(d), consists of all the narrow ridges and 'pimples' that are smaller than the processing element B. Despite the gross variation in grey levels across the image, and a large 'error blob', the process of opening successfully reveals the presence of the 'lines'. As a comparison, Figure 3(e) shows the results of applying the Laplace filter to Figure 3(a). It is fairly obvious that the opening filter gives a much clearer transformation. The example shown in Figure 3(a) contains no 'furrows', but it is easy to open an image from above, and it consists in reversing the operations of erosion and dilation, providing the 'ball' is symmetric.

Plates 2(b) and 2(c) show the result of opening the SPOT image, both from above and below to reveal the lineaments. The 'balls' used were respectively of size $3 \times 3$ and $7 \times 7$ and had the values

$$B = \begin{matrix} 0 & 1 & 0 \\ 1 & 2 & 1 \\ 0 & 1 & 0 \end{matrix}$$

$$B = \begin{matrix} 0 & 0 & 1 & 2 & 1 & 0 & 0 \\ 0 & 1 & 2 & 3 & 2 & 1 & 0 \\ 1 & 2 & 3 & 4 & 3 & 2 & 1 \\ 2 & 3 & 4 & 5 & 4 & 3 & 2 \\ 1 & 2 & 3 & 4 & 3 & 2 & 1 \\ 0 & 1 & 2 & 3 & 2 & 1 & 0 \\ 0 & 0 & 1 & 2 & 1 & 0 & 0 \end{matrix}$$

As can be seen the results are similar, although the larger filter appears to be smoother. It is expected that the larger the 'ball', the more that should remain after subtraction. However, due to the stretching of the image grey level histogram for presentation purposes, the smaller variations are suppressed.

Comparing the images produced using the Sobel and Laplace with those produced using the 'rolling ball', it is clear that the latter are better at revealing subtle lineaments in an image. There is a catch in using them—the processing time is usually 10 to 100 times longer for the 'rolling ball' transforms! In the past this has meant that they could only be implemented on very expensive parallel processors. But now fairly fast PCs and Work Stations exist, and therefore it is feasible to use these morphological filters. The image in Plate 2(c) is 200 by 200 pixels and took a 486 PC about 10 secs to process.

# References

Serra J P, 1981, Mathematical Morphology and Image Analysis, Academic Press, London
Sternberg S R, 1983, Biomedical Image Processing, *Computer* **16** 22

# Retrieval of Atmospheric Parameters from Space-borne Radiometric Measurements

G E Peckham and I H Woodhouse

Department of Physics
Heriot-Watt University

## 1  Introduction

The retrieval of parameters describing the state of the atmosphere from remote measurements of emitted thermal radiation requires the solution of a Fredholm integral equation. There is no unique solution to this equation, so that extra constraints must be imposed to solve it. A number of methods for the solution have been proposed which differ in the nature of these constraints. We compare some of these methods and relate them through a common formalism. The application of the optimal estimation method to the analysis of data from the Microwave Limb Sounder on the Upper Atmosphere Research Satellite is outlined.

## 2  The retrieval problem

### 2.1  The forward model

Measurements of a parameter of a system which is a function of an independant variable such as space or time do not give true values for the parameter, even in the absence of systematic error, for the following reasons.

1. The measuring instrument has a finite resolution so that the measurement does not represent the value of the parameter corresponding to a particular value of the independent variable, but is an average over a range of values.

2. The measurement is contaminated by the presence of *noise*, a random signal originating in the measuring instrument or arising from variations in other system parameters which affect the measurement.

A physical model relates the measured quantity $y_i$ to the state of the system which is described by a parameter $f(x)$ as a function of the independent variable $x$. In many cases this forward model results in a linear relationship described by the integral equation

$$y_i = \int_a^b K_i(x)f(x)dx \tag{1}$$

where the kernel $K_i(x)$ is the resolution function. This equation may be extended readily to cover more complicated situations. For instance, more than one parameter function may be included by summing over a number of terms on the right hand side, introducing a separate kernel for each parameter; non-linear physical models may often be approximated by a linear expansion about some chosen state. The measurements $y_i$ include random noise $\epsilon_i$

$$y_i = \int_a^b K_i(x)f(x)dx + \epsilon_i \tag{2}$$

Equation 2 is a Fredholm integral (since the integral limits are fixed) and of the first kind (since only $f(x)$ appears in the integrand).

## 2.2    Convolution

If the shape of the resolution function is constant over a set of measurements, we may write $K_i(x) = K(x'-x)$ where $y_i = y(x')$ and $x'$ is the value of the independent variable characterising the measurement. The integral then becomes a convolution integral:

$$y(x') = \int_a^b K(x' - x)f(x)dx + \epsilon_i \tag{3}$$

Many measurement systems may be so described. An example is the measurement of spectral transmission by a laboratory spectrometer; the measured spectrum is a convolution of the true spectrum with the resolution function of the spectrometer.

## 2.3    The inverse problem

The act of estimating the function $f(x)$ from a set of measurements $y_i$ is known as *inversion* or in the case of atmospheric science as *retrieval* to avoid confusion with the use of the word *inversion* to describe a class of atmospheric temperature profiles. The measurements $y_i$ (or the measured function $y(x')$) do not define a unique solution $f(x)$. We assume that the resolution function $K(x)$ is peaked so that its width may be defined as the range of $x$ over which the function has a significant value. Thus features of $f(x)$ which are small in width compared to the width of $K(x)$ will make small contributions to the integral which are unmeasurable because of the presence of noise. Even in the absence of noise there is no unique solution, since any multiple of a function which is orthogonal to the $K_i(x)$ over the interval $[a, b]$ may be added to $f(x)$ without making any change in the measured values $y_i$. To obtain a solution, assumptions must be made about the shape of $f(x)$ such as that it is a smooth function. Methods for the solution of the inverse problem are characterised by the particular assumptions made about the form of the solution.

The inverse problem is encountered in many branches of physics and engineering, and much literature has been devoted to its solution (Twomey 1977, Rodgers 1976, Houghton *et al.* 1984, Skilling 1989, Erickson and Smith 1988a, Janssen 1993). Throughout these fields there has been a degree of independent replication as well as a good deal of cross-fertilisation of approaches and methods. The Backus-Gilbert method (Backus and Gilbert 1970), for instance, which was developed for analysing geophysical data has been applied to atmospheric remote sounding; the Chahine method (Twomey *et al.* 1977) used in atmospheric remote sounding problems has been found to be identical to a method used in medical imaging (Depierro 1987). Other examples of inverse problems arise in image processing (Burch 1983, Gull and Skilling 1984), astronomical imaging (Gull and Daniell 1978), determination of particle size distribution (Twomey 1975) and meteorology (Read 1982).

The breadth of literature has inevitably led to an inhomogeneity of symbols, terminology and approach that can lead to confusion. In this paper, we use a generalised approach to describe a number of inversion methods. Included in these is the Maximum Entropy (ME) method, which, although used extensively in some areas of research, has rarely been applied to atmospheric remote sensing.

# 3  The general inverse problem

As has has been emphasised, the inversion of Equation 2 is an ill-posed problem—in the sense that it does not have a direct solution which satisfies the three conditions of *existence, uniqueness and stability.* (Mohammad-Djafari and Demoment 1989). Thus the solution is underconstrained; *i.e.* for a given set of measurements, $y_i$, there exits an infinite number of possible solutions. The fundamental difficulty in inversion problems is therefore not one of finding a solution, but of deciding which of the infinite number of possible solutions is most appropriate to the particular problem at hand. It is convenient to distinguish between these problems by describing the former as the *inverse* problem and the later as the *estimation* problem.

For practical computation, functions are tabulated at a discrete set of values of the independent variable; integrals are replaced by their quadrature approximation and integral equations arc replaced by their matrix equivalent. Equation 2 is then expressed as

$$\mathbf{y} = \mathbf{K}\mathbf{x} + \epsilon \tag{4}$$

Here the elements of the vector $\mathbf{x}$ are the tabulated values of $f(x)$ and elements of the matrix $\mathbf{K}$ are products of the corresponding $K_i(x)$ and the quadrature coefficients.

The main question that estimation theory addresses is that of the uniqueness of the solution. Since there are components of $\mathbf{x}$ which make little or no contribution to the observations, even small changes in $\mathbf{y}$ can lead to large changes in the solution. Practical inversion methods attempt to avoid such instabilities and provide a criterion for choosing a 'best' solution through the application of constraints.

# 4   Constrained linear inversion

Some inverse problems can be solved by a weighted least squares method; the estimate $\hat{\mathbf{x}}$ of $\mathbf{x}$ is that for which the function

$$Q(\hat{\mathbf{x}}) = (\mathbf{K}\hat{\mathbf{x}} - \mathbf{y})^T \mathbf{S}_\epsilon^{-1}(\mathbf{K}\hat{\mathbf{x}} - \mathbf{y}) \tag{5}$$

takes its minimum value. Here the superscript $T$ implies transposition and $\mathbf{S}_\epsilon$ is the error covariance matrix of the measurements. The minimum occurs when

$$\hat{\mathbf{x}} = (\mathbf{K}^T \mathbf{S}_\epsilon \mathbf{K})^{-1} \mathbf{K}^T \mathbf{S}_\epsilon^{-1} \mathbf{y} \tag{6}$$

In many practical applications, the matrix $\mathbf{K}^T \mathbf{S}_\epsilon \mathbf{K}$ is ill-conditioned, so that the solution to Equation 6 is highly sensitive to the exact values of the measurements; the equation may even be insoluble–for instance, if the number of unknown parameters is larger than the number of independent measurements, the matrix $\mathbf{K}^T \mathbf{S}_\epsilon \mathbf{K}$ is singular and cannot be inverted. This problem may be overcome by imposing an additional constraint on the solution which is not a consequence of the measurements, but is derived from prior knowledge of the form of acceptable solutions. (The term *smoothing* to describe this procedure, while in common usage, is too restrictive and is avoided here. While some methods do in fact employ some kind of smoothness constraint, most methods have no correlation between adjacent elements of $\mathbf{x}$. In an attempt to clarify this some authors (*e.g.* Titterington 1985) distinguish between *local smoothness* and *general smoothness*).

Mathematically, the constraint is achieved by minimising a function that is the sum of two positive terms: $\Delta_1(\hat{\mathbf{y}}, \mathbf{y})$, measuring the fit of the estimate $\hat{\mathbf{y}} = \mathbf{K}\hat{\mathbf{x}}$ to the measurement $\mathbf{y}$, and $\Delta_2(\hat{\mathbf{x}})$, being the penalty function which incorporates the constraint. The function $Q(\hat{\mathbf{x}})$ is now

$$Q(\hat{\mathbf{x}}) = \Delta_1(\hat{\mathbf{y}}, \mathbf{y}) + \lambda \Delta_2(\hat{\mathbf{x}}) \tag{7}$$

and is to be minimised with respect to $\hat{\mathbf{x}}$. The positive Lagrange multiplier $\lambda$ determines the trade-off between complying with the measurements and satisfying the constraints on the solution. Some commonly used penalty functions for $\Delta_2(\hat{\mathbf{x}})$ are:

| | |
|---|---|
| variance | $\sum_i (\hat{x}_i - \bar{x})^2$ |
| sum of squares of first differences | $\sum_i (\hat{x}_i - \hat{x}_{i-1})^2$ |
| sum of squares of second differences | $\sum_i (\hat{x}_i - 2\hat{x}_{i-1} + \hat{x}_{i-2})^2$ |
| entropy | $\sum_i \overset{\star}{x}_i \ln \overset{\star}{x}_i,$ |

where $\bar{x}$ and $\overset{\star}{x}_i$ are defined by

$$\bar{x} = \sum_i \hat{x}_i / N, \qquad \overset{\star}{x}_i = \hat{x}_i / \sum_j \hat{x}_j,$$

and $\hat{\mathbf{x}}$ has dimension $N$.

$\Delta_1(\hat{\mathbf{y}}, \mathbf{y})$ is usually assumed to be a weighted quadratic:

$$\Delta_1(\hat{\mathbf{y}}, \mathbf{y}) = (\hat{\mathbf{y}} - \mathbf{y})^T \mathbf{W}(\hat{\mathbf{y}} - \mathbf{y}) \tag{8}$$

where $\mathbf{W}$ is a known positive-definite symmetric matrix, usually the inverse error covariance matrix $\mathbf{S}_\epsilon^{-1}$. This weighted least-squares criterion is used in all the linear methods described here. It is also the most common approach in ME applications (*e.g.* Read 1982).

## 4.1  Twomey-Tikhonov regularisation

The method introduced independently by Twomey (Twomey 1963, Twomey *et al.* 1977) and Tikhonov (Tikhonov 1963) represents the penalty function as a matrix product, $\hat{\mathbf{x}}^T\mathbf{H}\hat{\mathbf{x}}$, where $\mathbf{H}$ is a matrix determined by the form of the penalty function (called the *regularisation function* by Tikhonov).

Equation 7 becomes

$$Q(\hat{\mathbf{x}}) = (\hat{\mathbf{y}} - \mathbf{y})^T\mathbf{S}_\epsilon^{-1}(\hat{\mathbf{y}} - \mathbf{y}) + \lambda(\hat{\mathbf{x}}^T\mathbf{H}\hat{\mathbf{x}}) \tag{9}$$

where $\lambda$ is the Lagrange multiplier and $\mathbf{S}_\epsilon$ is the measurement covariance matrix which, in the usual case of independent measurements, is a diagonal matrix with elements $S_{ii}$ equal to the variance of the measurement $y_i$.

Minimising $Q$ with respect to $\hat{\mathbf{x}}$ gives

$$\hat{\mathbf{x}} = \mathbf{H}^{-1}\mathbf{K}^T(\mathbf{K}\mathbf{H}^{-1}\mathbf{K}^T + \lambda\mathbf{S}_\epsilon)^{-1}\mathbf{y} \tag{10}$$

The matrix $\mathbf{H}$ corresponding to a sum of squares of first differences penalty function is

$$\mathbf{H} = \begin{bmatrix} 1 & -1 & 0 & 0 & \cdots \\ -1 & 2 & -1 & 0 & \cdots \\ 0 & -1 & 2 & -1 & \cdots \\ 0 & 0 & -1 & 2 & \cdots \\ \vdots & \vdots & \vdots & \vdots & \ddots \end{bmatrix}$$

The matrix $\mathbf{H}$ corresponding to a sum of squares of second differences penalty function is

$$\mathbf{H} = \begin{bmatrix} 1 & -2 & 1 & 0 & 0 & 0 & \cdots \\ -2 & 5 & -4 & 1 & 0 & 0 & \cdots \\ 1 & -4 & 6 & -4 & 1 & 0 & \cdots \\ 0 & 1 & -4 & 6 & -4 & 1 & \cdots \\ 0 & 0 & 1 & -4 & 6 & -4 & \cdots \\ \vdots & \vdots & \vdots & \vdots & \vdots & \vdots & \ddots \end{bmatrix}$$

If sufficient information about the likely state of the system is available, a penalty function $\Delta_2(\hat{\mathbf{x}}, \mathbf{x}^{(0)})$ may be used which is a measure of the departure from an *a priori* state or first guess $\mathbf{x}^{(0)}$. Equation 7 becomes

$$Q(\hat{\mathbf{x}}) = (\hat{\mathbf{y}} - \mathbf{y})^T\mathbf{S}_\epsilon^{-1}(\hat{\mathbf{y}} - \mathbf{y}) + \lambda(\hat{\mathbf{x}} - \mathbf{x}^{(0)})^T\mathbf{H}(\hat{\mathbf{x}} - \mathbf{x}^{(0)}) \tag{11}$$

and minimising with respect to $\hat{\mathbf{x}}$ gives

$$\hat{\mathbf{x}} = \mathbf{x}^{(0)} + \mathbf{H}^{-1}\mathbf{K}^T(\mathbf{K}\mathbf{H}^{-1}\mathbf{K}^T + \lambda\mathbf{S}_\epsilon)^{-1}(\mathbf{y} - \mathbf{y}^{(0)}) \tag{12}$$

where $\mathbf{y}^{(0)} = \mathbf{K}\mathbf{x}^{(0)}$

## 4.2   Minimum variance and optimal estimation

If the statistical distribution of $\mathbf{x}$ is known, a suitable form for $\Delta_2$ is the weighted variance

$$\Delta_2(\hat{\mathbf{x}}, \mathbf{x}^{(0)}) = (\hat{\mathbf{x}} - \mathbf{x}^{(0)})^T \mathbf{S}_x^{-1} (\hat{\mathbf{x}} - \mathbf{x}^{(0)}) \tag{13}$$

where $\mathbf{x}^{(0)}$ is the mean of the distribution and $\mathbf{S}_x$ is its covariance. Incorporating $\lambda$ into $\mathbf{S}_x$,

$$Q(\hat{\mathbf{x}}) = (\hat{\mathbf{y}} - \mathbf{y})^T \mathbf{S}_\epsilon^{-1} (\hat{\mathbf{y}} - \mathbf{y}) + (\hat{\mathbf{x}} - \mathbf{x}^{(0)})^T \mathbf{S}_x^{-1} (\hat{\mathbf{x}} - \mathbf{x}^{(0)}) \tag{14}$$

$Q(\hat{\mathbf{x}})$ takes a minimum value when

$$\hat{\mathbf{x}} = \mathbf{x}^{(0)} + \mathbf{S}_x \mathbf{K}^T (\mathbf{K}\mathbf{S}_x\mathbf{K}^T + \mathbf{S}_\epsilon)^{-1}(\mathbf{y} - \mathbf{y}^{(0)}) \tag{15}$$

An alternative approach is to consider the *a priori* information as a *virtual measurement* with mean $\mathbf{x}^{(0)}$ and error covariance $\mathbf{S}_x$. The normal rules for combining measurements (Rodgers 1976, Houghton *et al.* 1984) then give Equation 15 directly. The covariance $\hat{\mathbf{S}}$ of the estimate $\hat{\mathbf{x}}$ is

$$\hat{\mathbf{S}} = \mathbf{S}_x - \mathbf{S}_x \mathbf{K}^T (\mathbf{K}\mathbf{S}_x\mathbf{K}^T + \mathbf{S}_\epsilon)^{-1} \mathbf{K}\mathbf{S}_x \tag{16}$$

If all statistical distributions are Gaussian, Equation 15 describes the maximum likelihood solution. This optimal estimation procedure has been used widely in atmospheric sounding (Westwater and Strand 1968, De Luisi and Mateer 1971, Deepak 1977, Deepak *et al.* 1989) and is used as the routine retrieval algorithm for the Microwave Limb Sounder (Froidevaux 1987).

## 5   Non-linear methods

### 5.1   Background

If $Q$ is a quadratic function of $\hat{\mathbf{x}}$, the minimum value may be found by solving a set of linear equations. However, other forms of $Q$ lead to non-linear equations, and in this case it is usually necessary to iterate the solution. The non-linearity may arise from the forward model or from the constraints imposed on the solution—we consider the latter here (Twomey 1977, Deepak 1977, Janssen 1993).

Non-linear methods have certain advantages over the linear kind. In particular, non-linear iterative methods can be arranged so that at all stages the iterated solution is positive while still satisfying the constraints imposed by the observed data. The other advantages of non-linear methods are that they are, in general, computationally very simple (requiring no matrix inversions, which is a particular advantage when dealing with large kernels) and they have the ability to cope with problems where the dynamic range of the solution is large (ie, ranges over a few orders of magnitude)—the more direct methods often fail in such situations, unless the method is adapted to solve for some transformation of $\mathbf{x}$, rather than $\mathbf{x}$ itself, which does not vary so widely. (This last point is particulary relevant to retrieving water vapour below the tropopause).

The disadvantages of non-linear methods include the increased computing time that is required for the iteration (at least four times larger, according to Twomey, than a constrained linear inversion) and the difficulty in quantifying the uncertainty in the final solution.

## 5.2 Maximum structural entropy

Maximum Entropy (ME methods (Erickson and Smith 1988a, Jaynes 1983) have been applied successfully to all manner of problems involved with the inversion of a non-linear function (Burch 1983, Yee *et al.* 1987, Skilling 1989, Deepak *et al.* 1989). To date it has been noticeably underused in the field of atmospheric remote sounding with the exception of Read (1982) (Read 1982), who applied a ME method to the analysis of ground based umkehr observations of atmospheric ozone, and a recent study by the authors applying a ME method to water vapour retrievals from the Upper Atmosphere Research Satellite Microwave Limb Sounder (UARS-MLS) instrument (Woodhouse and Peckham 1994).

Although some authors have justified the use of the ME method by relating it to Jaynes Principle of Maximum Entropy (Jaynes 1982), it is presented here simply as one of many possible penalty functions that can be used for constraining the solution (Titterington 1985).

The Maximum Entropy method minimises a functional $Q(\mathbf{x})$ that is the sum of two positive terms: $\Delta(\hat{\mathbf{y}}, \mathbf{y})$, measuring the fit to the measurements $\mathbf{y}$, and $-\mathcal{S}(\hat{\mathbf{x}}, \mathbf{x}^{(0)})$, being *minus* the entropy of the solution with respect to some *a priori* solution $\mathbf{x}^{(0)}$. The solution is then found by minimising

$$Q(\hat{\mathbf{x}}) = \lambda\Delta(\hat{\mathbf{y}}, \mathbf{y}) - \mathcal{S}(\hat{\mathbf{x}}, \mathbf{x}^{(0)}) \tag{17}$$

where $\lambda$ is again a positive Lagrange multiplier which determines the trade-off between the two ojectives.

The entropy is defined as

$$\mathcal{S}(\hat{\mathbf{x}}, \mathbf{x}^{(0)}) = -\sum_i \overset{\star}{x}_i \ln\left(\frac{\overset{\star}{x}_i}{\overset{\star}{x}_i^{(0)}}\right) \tag{18}$$

Where $\overset{\star}{x}_i$ is the previously defined normalised $x_i$ and $\overset{\star}{x}_i^{(0)}$ is similarly related to $x_i^{(0)}$.

$Q$ takes a minimum value when $\hat{\mathbf{x}}$ satisfies

$$\hat{x}_i = x_i^{(0)} A(\hat{\mathbf{x}}) \exp\left[-\lambda'(\hat{\mathbf{x}})\frac{\partial}{\partial \hat{x}_i}\Delta(\hat{\mathbf{y}}, \mathbf{y})\right] \tag{19}$$

where $\lambda'(\hat{\mathbf{x}}) = \lambda \sum_j \hat{x}_j$ and

$$A(\hat{\mathbf{x}}) = \exp\left(\frac{\sum_j \hat{x}_j(\ln \hat{x}_j - \ln x_j^{(0)})}{\sum_j \hat{x}_j}\right)$$

With no constraint due to the observations (i.e. $\lambda = 0$), $\hat{\mathbf{x}}$ is proportional to $\mathbf{x}^{(0)}$, the *a priori* profile. The ME solution is thus the solution for which $\mathcal{S}(\hat{\mathbf{x}}, \mathbf{x}^{(0)})$ has its greatest value subject to the constraints imposed by the measurements.

Read (Read 1982) has suggested that Equation 19 be solved by iteration

$$x_i^{(n+1)} = x_i^{(0)} A(\mathbf{x}^{(n)}) \exp\left( -\lambda' \left(\mathbf{x}^{(n)}\right) \left[ \frac{\partial}{\partial \hat{x}_i} \Delta(\hat{\mathbf{y}}, \mathbf{y}) \right]_{\hat{\mathbf{x}} = \mathbf{x}^{(n)}} \right) \tag{20}$$

## 5.3   Chahine-Twomey

The relaxation method introduced by Chahine and later improved by Twomey (Chahine 1970, Chahine 1972, Barcilon 1975, Twomey 1977, Twomey 1977b) consists of a sequence of modifications of $\hat{\mathbf{x}}$ designed to reduce the difference between the values $\hat{\mathbf{y}} = \mathbf{K}\hat{\mathbf{x}}$ calculated from the forward model and the actual observations $\mathbf{y}$. The procedure is started from a first guess $\mathbf{x}^{(0)}$. At the $n$th iteration, a new estimate $\mathbf{x}^{(n+1)}$ is determined from

$$x_i^{(n+1)} = x_i^{(n)} \left( 1 - [\overset{*}{\mathbf{K}}{}^T \xi]_i \right) \tag{21}$$

where $\overset{*}{\mathbf{K}}$ is a scaled version of $\mathbf{K}$ and $\xi_i = (\hat{y}_i - y_i)/\hat{y}_i$. The iteration is continued until a suitable convergence criterion is satisfied—typically when the difference between successive $\hat{y}_i$s is less than the expected error on the measurement of $y_i$. In most applications 5 to 10 iterations are sufficient.

Chahine-Twomey relaxation methods have been used in medical imaging (Daube-Witherspoon and Muehllehner 1986, Depierro 1987), simulated UV ozone retrievals (Twomey *et al.* 1977), limb sounder simulations (Chu 1985, Puliafito 1991, Woodhouse and Peckham 1994), estimating particle size distributions (Twomey 1975), terrestrial temperature sounding (Conrath *et al.* 1970, Shaw *et al.* 1970, Smith 1970), CO soundings of the Venusian atmosphere (Janssen 1993) and ground based microwave measurement of $H_2O$ (Bevilacqua 1987).

The only constraint imposed on the solution arises from the smoothing effect of multiplication by $\overset{*}{\mathbf{K}}{}^T$ in Equation 21. The solution to the Chahine-Twomey method is largely independent of the initial guess $\mathbf{x}^{(0)}$. (For a more detailed discussion on iterative non-linear methods the reader is referred to references (Deepak 1977 and Deepak *et al.* 1989)

The first Chahine-Twomey iteration can be shown to be equivalent to the maximum entropy method in a second order approximation if the variance in the measurement $y_i$ is assumed to be proportional to its estimated value $\hat{y}_i$ (Woodhouse and Peckham, to be submitted).

# 6   Application to the analysis of data from MLS

## 6.1   The Microwave Limb Sounder

The Microwave Limb Sounder (MLS) on the Upper Atmosphere Research Satellite (Barath *et al.* 1993) is designed to determine the concentrations of a number of atmospheric constituents by measuring the spectral power density of thermal radiation emitted by the atmosphere in a number of microwave frequency bands. The atmosphere is viewed along a tangential path (limb view) so that atmospheric emission is seen against the cold space background. Atmospheric constituents are identified by the frequencies of their emission lines and the intensity of the emission is related to the concentration of the constituent. The field of view is scanned in elevation to give a vertical concentration profile whilst the near-polar orbit of the satellite and the rotation of the earth provide global mapping.

## 6.2 Retrievals

The state vector, from which estimates of the measured radiances are found through Equation 4, includes the concentrations of relevant constituents (*e.g.* $O_2$, $O_3$, $H_2O$, and ClO) tabulated at a number of heights in the atmosphere and also atmospheric temperature. The logarithm of atmospheric pressure is used as a vertical coordinate. Since the satellite attitude is not known with sufficient accuracy, the pressure at the lowest point viewed along the antenna axis (*the tangent point*) is also retrieved from the measurements and so is included in the state vector. The retrieval is based on two models:

- **Virtual measurement model** (VMM): the purpose of this model is to calculate an *a priori* estimate of the state vector.

- **Direct measurement model** (DMM): this model relates the measured radiances to the state vector.

These models depend on a number of variables which are classified in table below. A 'x' in the column headed SV indicates that the variable is included in the state vector and will be retrieved from the measurements; entries in the VMM or DMM columns indicate that estimates of the variable are required for the virtual or direct measurement models respectively.

| Description | Examples | SV | VMM | DMM |
|---|---|---|---|---|
| experimental objective tabulated at regular height intervals | mixing ratio of atmospheric constituents and atmospheric temperature | × | × | × |
| independent variables needed by VMM and DMM | tangent pressure | × | × | × |
| Independent variables needed only by VMM | satellite attitude | × | × | |
| other variables required by VMM | antenna elevation angle, climatological data | | × | |

The emitted radiation from an element of the atmosphere is calculated from a knowledge of the absorption line frequencies, strengths and shapes. The emission depends on the concentrations of relevant atmospheric constituents and on the temperature and pressure of the element. The radiation emitted to space is calculated by integration along the atmospheric path, allowing for reabsorption by the air mass between each element and space. Each channel of the MLS accepts a band of frequencies and the emitted radiance must be integrated over the pass band of each channel. Further integration is necessary over the field of view of the antenna. It is not feasible to do this calculation in real time so that a linear approximation is derived by calculating radiance and its derivatives with respect to the elements of an atmospheric state vector for a number of standard atmospheric states. The linear approximation is adequate for microwave measurements where the Rayleigh-Jeans approximation to the Planck function—namely emission proportional to absolute temperature of the source; this is valid as long as the atmosphere is optically thin (transmission through the atmosphere less than say 20%).

The derivatives of the signal strengths in the 90 spectral channels of the MLS with respect to the elements of this state vector form the elements of the matrix $\mathbf{K}$ in Equation 4. The routine MLS retrievals are carried out using an optimal estimation procedure of the form described in Section 4.2.

An advantage of the optimal estimation method is that the retrieval can be applied sequentially. An updated state vector is obtained by applying Equation 15 to each measurement in turn. $\mathbf{y}$ is now a scalar and $\mathbf{K}$ a matrix with a single row. It can be shown (Rodgers 1976) that this proceedure is equivalent to a single application of Equation 15 to the complete vector of measurements. The main advantage of the sequential method is that missing or corrupted data can be omitted easily from the retrieval; also the sequential updates of the state vector do not require matrix inversion (or the solution of simultaneous linear Equations).

## 6.3. UARS data analysis levels

In order to standardise the processing of data from a number of instruments on the UARS spacecraft, four data 'Levels' have been defined:

0: Raw telemetry data for a particular instrument extracted from the total data stream.

1: Calibrated instrument radiances and related (*e.g.*, instrument engineering) data.

2: Geophysical parameters, such as gas concentrations, which are retrieved from Level 1 data. Level 2 data are on a grid selected by the individual experiment team.

3: Level 2 geophysical parameters on standard grids. There are three types of Level 3 data (3AT, 3AL, and 3B) each of which are mapped onto different grids. Level 3AT data are geophysical parameters gridded each 65.536 s time interval onto selected atmospheric pressure 'surfaces' (the $n^{th}$ surface has pressure $p_n = 1000 \times 10^{-[n-1]/6}$ mb, and $1 \leq n \leq 37$). Level 3AL data are the same as Level 3AT, but gridded each 4° of latitude. Level 3B data are a globally mapped representation of Level 3AL data.

4: 'Higher level' products such as zonal means.

# References

Backus G E and Gilbert J F, 1970, Uniqueness in the inversion of inaccurate gross earth data, Philosophical Transactions of the Royal Society of London, Series A, **266**, 123.

Barath F T, Chavez M C, Cofield R E, Flower D A, Frerking M A, Gram M B, Harris W M, Holden J R, Jarnot R F, Kloezeman W G, Klose G J, Lau G K, Loo M S, Maddison B J, Mattaugh R J, McKinney R P, Peckham G, Pickett H M, Siebes G, Soltis F S, Suttie R A, Tarsala J A, Waters J W and Wilson W J, 1993, The Upper Atmosphere Research Satellite Microwave Limb Sounder Instrument, Journal of Geophysical Research, **98:10**, 751.

Barcilon V, 1975, On Chahine's relaxation method for the radiative transfer equation, Journal of Atmospheric Science, **32**, 1626.

Bevilacqua R M, Wilson W J and Schwartz P R, 1987, Measurements of mesospheric water vapor in 1984 and 1985: results and implications for middle atmospheric transport, Journal of Geophysical Research, **92(D6)**, 6679.

Burch S F, Gull S F and Skilling J, 1993, Image restoration by a powerful Maximum Entropy Method, Computer Vision, Graphics and Image Processing, **23**, 113.

Chahine M T, 1870, Inverse problems in radiative transfer: determination of atmospheric parameters, Journal of Atmospheric Science, **27**, 960.

Chahine M T, 1972, A general relaxation method for inverse solution of the full radiative transfer equation, Journal of Atmospheric Science, **29**, 741.

Chu W P, 1985, Convergence of Chahine's nonlinear relaxation inversion method used for limb viewing remote sensing, Applied Optics, **24(4)**, 445.

Conrath B J, Hanel R A, Kunde V G and Parbhakara C, 1970, The infrared interferometer experiment on Nimbus 3, Journal of Geophysical Research, **30**, 5831.

Daube-Witherspoon M E and Muehllehner G, 1986, An iterative image space reconstruction algorithm suitable for volume ECT, IEEE Transactions on Medical Imaging, **MI-5**, 61.

De Pierro A R, 1987, On the convergence of the iterative image space reconstruction algorithm for volume ECT, IEEE Transactions on Medical Imaging , **MI-6(2)**, **174.**

Deepak A, editor, 1977, Inversion methods in atmospheric remote sounding, Academic Press, Inc., London.

Deepak A, Fleming H E and Theon J S, editors, 1989, RSRM '87: Advances in remote sensing retrieval methods, A Deepak Publishing, Hampton, Virginia.

Erickson G J and Smith C R, editors, 1988, Maximum-Entropy and Bayesian Methods in Science and Engineering, Volume 2: Applications, Kluwer Academic Publishers, Dordrecht.

Erickson G J and Smith C R, editors, 1988, Maximum-Entropy and Bayesian Methods in Science and Engineering, Volume 1:Foundations, Kluwer Academic Publishers, Dordrecht.

Froidevaux L, Read W G, Waters J W, Lahoz W A and Drew W A, 1987, Microwave Limb Sounder level 2 data processing: Theoretical basis, Technical Report version 1.4, NASA, Jet Propulsion Laboratory, Pasadena.

Gull S F and Daniell G J, 1978, Image reconstruction from incomplete and noisy data, Nature, **272**, 686.

Gull S F and Skilling J, 1984, Maximum Entropy method in image processing, IEE Proceedings, **131 F(6)**, 646.

Houghton J T, Taylor F W and Rodgers C D, 1984 Retrieval Theory, chapter 7, Cambridge University Press, Cambridge.

Janssen M A, editor, 1993, Atmospheric Remote Sensing by Microwave Radiometry, Wiley and Sons, New York.

Jaynes E T, 1982, On the rationale of Maximum-Entropy methods, IEEE Proceedings, **70(9)**, 939.

Jaynes E T, 1983, Papers on Probability, statistics and statistical physics, Kluwer Academic Publishers, Dordrecht.

De Luisi F J and Mateer C C, 1971, On the application of the optimum statistical inversion technique to the evaluation of Umkehr observation, Journal of Applied Meteorology, **10**, 328.

Mohammad-Djafari A and Demoment G, 1989, Maximum Entropy and Bayesian approach in tomographic image reconstruction and restoration, Maximum-Entropy and Bayesian Methods, Kluwer Academic Publishers, Dordrecht.

Puliafito S E, 1991, Comparison of inversion techniques for limb sounding radiometric measurements, IGARSS '91: Remote Sensing: Global Monitoring for Earth Management, Espoo, Finland, IEEE.

Read P L, 1982, The analysis of Umkehr observations of stratospheric ozone by a Maximum Entropy Method, Quarterly Journal of the Royal Meteorological Society, **108(457)**, 719.

Rodgers C D, 1976, Retrieval of atmospheric temperature and composition from remote measurements of thermal radiation, Reviews of Geophysics and Space Physics, **14(4)**, 609.

Shaw J H, Chahine M T, Farmer C T, Kaplan L D, McClatchey R A and Schaper P W, 1970, Atmospheric and surface properties from spectral radiance observations in the 4.3 micron region, Journal of Atmospheric Science, **27**, 773.

Skilling J, editor, 1989, Maximum-Entropy and Bayesian Methods, Kluwer Academic Publishers, Dordrecht.

Smith W L, 1970, Iterative solution of the radiative transfer equation for the temperature and absorbing gas profile of an atmosphere. Applied Optics, **9**, 1993.

Tikhonov A N, 1963, On the solution of incorrectly stated problems and a method of regularization, Dokl Acad Nauk SSSR, **151**, 501.

Titterington D M, 1985, Common structure of smoothing techniques in statistics, International Statistical Review, **53(2)**, 141.

Twomey S, 1963, On the numerical solution of Fredholm integral equations of the first kind by the inversion of the linear system produced by quadrature, Journal of the Association for Computing Machinery, **10**, 97.

Twomey S, 1975, Comparison of constrained linear inversion and an iterative nonlinear algorithm applied to the indirect estimation of particle size distribution, Journal of Computational Physics, **18**, 188.

Twomey S, 1977, Introduction to the mathematics of inversion in remote sensing and indirect measurements, Developments in Geomathematics 3, Elsevier Scientific Publishing Company, Amsterdam.

Twomey S, Herman B and Rabinoff R, 1977, An extension to the Chahine method of inverting the radiative transfer equation, Journal of Atmospheric Science, **34**, 1085.

Westwater E R and Strand O N, 1968, Statistical information content of radiation measurements used in indirect sensing, Journal of Atmospheric Science, **25**, 750.

Woodhouse I H and Peckham G E, On the relationship between the Chahine-Twomey and Maximum Entropy methods, (to be submitted).

Woodhouse I H and Peckham G E, 1994, A review of atmospheric water vapour retrieval methods using UARS-MLS data, Proceedings of PIERS 1994, Kluwer Academic Publishers, (on CD-ROM).

Yee E, Paulson K V and Shepard G G, 1987, Minimum cross-entropy inversion of satellite photometer data, Applied Optics, **26(11)**, 2106.

# Ozone concentrations: satellite and ground-based measurements

## C Varotsos

University of Athens
Greece

# 1 Introduction

Ozone was first discovered by C F Schenbein in 1839. It is a form of oxygen which has three atoms, compared to the usual two, in each molecule. It is formed in the stratosphere by the action of the solar radiation on normal oxygen in a process called photolysis, by which $O_2$ molecules are broken down to yield atomic oxygen, which in turn combines with molecular oxygen to produce ozone. Ozone is destroyed naturally through a series of catalytic cycles involving oxygen, nitrogen and hydrogen. In 1880, laboratory and sunlight experiments showed that ozone absorbs strongly in the ultra-violet spectrum, which led by 1913 to the proof that most of the atmosphere's ozone was located in the stratosphere. So, looking up through the atmosphere, it has its maximum partial pressure in the lower stratosphere at a level of 19–23km above the Earth (Figure 1)

The total atmospheric ozone at any specific location is variable and is largely determined by large-scale atmospheric dynamics. The stratosphere contains 90% of all the ozone in the atmosphere. If all the ozone molecules in the atmosphere were transferred to the Earth's surface, they would assume a thickness of only 2.5–3.5mm at average surface temperature and pressure. In more detail total ozone is defined as being equal to the amount of ozone contained in a vertical column of base $1cm^2$ at standard pressure and temperature. It can be expressed in units of pressure, a typical value is about 0.3 atmosphere centimetres. More frequently used is the milli-atmospheric centimetre, which is also known as the Dobson Unit. One Dobson Unit corresponds to an average atmospheric concentration of approximately one part per billion by volume (1ppbv) of ozone. Typical amounts range from 230 to 500 Dobson Units, with a world average of about 300, although the ozone is not distributed uniformly through the vertical column. The maintenance of this layer of ozone is essential for the health of humans, animals

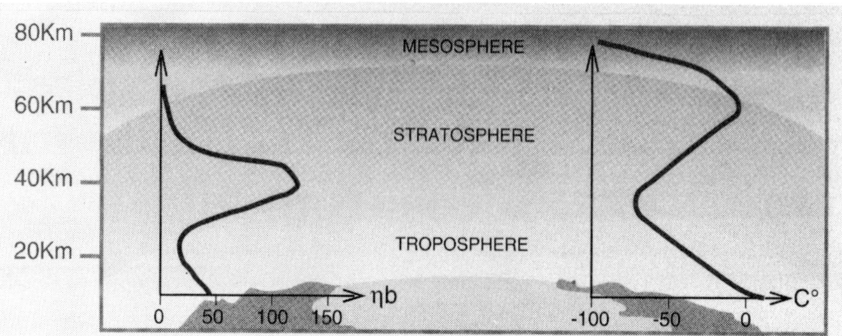



**Figure 1.** *The vertical distribution of the ozone partial pressure (left-hand curve) and the temperature (right-hand curve)*

and plants including phytoplankton, shielding all from damaging ultraviolet radiation from the Sun. It also largely determines the thermal structure of the stratosphere in which temperatures increase slowly with height.

The first serious indication that human activities could perturb the ozone layer came in 1971 when H S Johnson pointed out the danger of a large fleet of supersonic transport aircraft, emitting considerable amounts of nitric oxide into or just beneath the ozone layer. This followed scientific research by M Nicolet and others which showed that oxides of nitrogen could deplete stratospheric ozone. A few years later, the first indications emerged from the studies of Rowland and Molina that chlorofluorocarbons (CFCs), when transported to the stratosphere, were becoming a source of active chlorine atoms which through catalytic cycles could destroy ozone.

Assessments found a steady but increasingly steep decline of the total atmospheric ozone of 2.5% in the 1980s between 65°N and 65°S (Figure 2). The losses are mainly concentrated pole-ward of 30 degrees latitude, in the lower stratosphere (13–23km), and in the winter-spring periods where losses are at least twice as large as during summer. For instance, the lower stratospheric ozone loss in mid-latitudes over the last two decades is calculated at over 20%. Similar or even larger declines can be expected in the next few decades as the chlorine and bromine loading of the stratosphere continue their steady increase.

The catastrophic collapse of the Antarctic stratospheric ozone layer, its practical disappearance between 13 and 20km with the total ozone reduced over the whole atmospheric depth by more than half each Austral spring, is caused by a combination of dynamic, chemical and micro-physical processes. These include the presence of a strong circumpolar vortex which prevents ozone rich air from middle latitudes from penetrating the Antarctic stratosphere and also leads to extremely low stratospheric temperatures (less than –80°C),and the presence of polar stratospheric clouds (PSCs). These conditions in turn facilitate heterogeneous reactions which all are unique to the Antarctic in spring. It is assumed that when in this environment the concentrations of chlorine exceed 2 parts per billion by volume, radiation from the returning Sun starts a series of

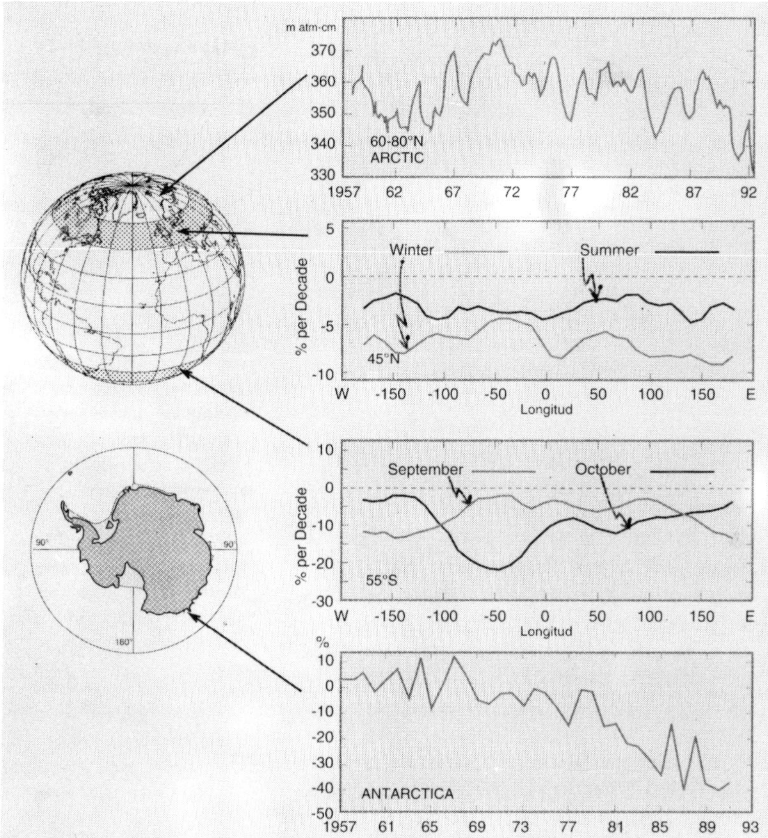



**Figure 2.** *The decline of the ozone content above the Arctic (top) since the late 1960s is more than 8%.* TOMS *satellite data (1978–91) shows that the winter decline is double than in summer for 45°N (above) and that the greatest decline is in October for 55°S (below). Over the Antarctic (bottom) the deviations in the October values from the 1957–78 mean show a drastic decline reaching about 40% in four of the last five years.*

catalytic reactions that result in rapid ozone depletion (Hofmann *et al.* 1992). Figure 3 shows schematically the process of depletion by, in this case, chlorofluorocarbon 12, whereby UV radiation frees a chlorine atom which immediately detaches atomic oxygen from ozone leaving chlorine monoxide and molecular oxygen. After further reactions, the chlorine atom is once again free to repeat the depletion cycle. The latest evidence on ozone depletion in this region also points to the crucial role that polar stratospheric clouds play in converting less reactive chlorine compounds such as hydrochloric acid into more reactive compounds (chlorine monoxide) capable of destroying ozone. Bromine which is even more destructive of ozone than chlorine is present in the stratosphere in relatively small amounts, but it is believed to be doubling every five years.

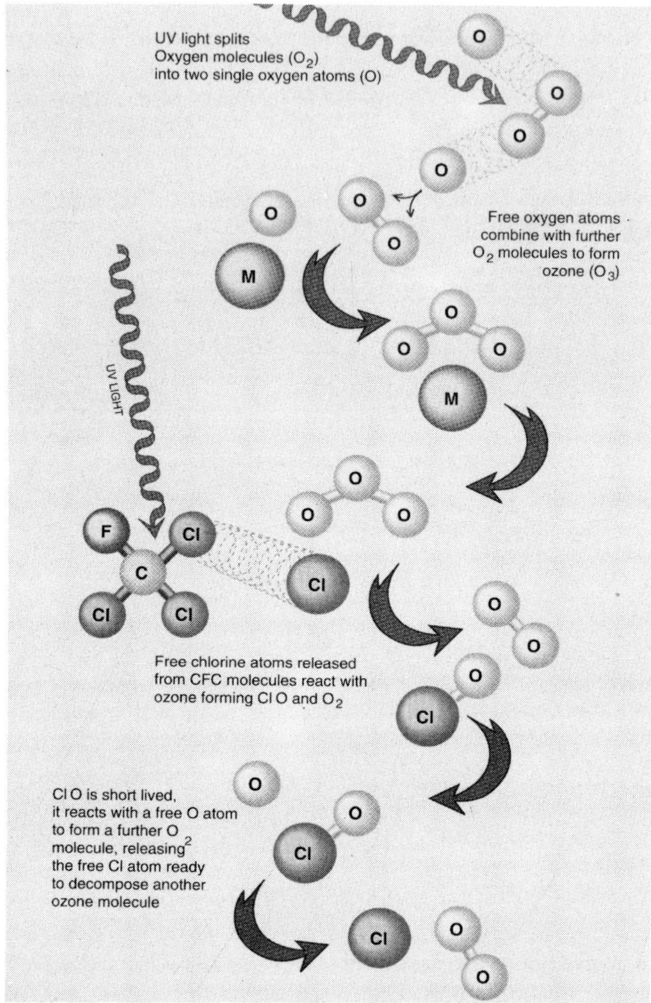



**Figure 3.** *Schematic sequence of the destruction of ozone by Cl released from a* CFC-12 *molecule*

At the same time that stratospheric ozone is decreasing, in the northern hemisphere at least, tropospheric ozone is increasing by 10% or more per decade. This tropospheric ozone increase is partly a consequence of the Sun's radiation on specific air pollutants, particularly oxides of nitrogen from aircraft and automobile exhausts, combined with the increasing concentrations of other precursors such as methane and carbon monoxide. Its increase poses threats to human health as well as a contributing to climate warming (Reeves and Penkett 1993).

Both ozone and halocarbons are greenhouse gases, much like carbon dioxide, that intercept and re-radiate the Earth's outgoing radiation in all directions, thereby helping to warm the lower atmosphere. The reported ozone loss in the lower stratosphere is, according to some studies, more important radiatively than the observed ozone increase in the troposphere, the net effect being a contribution to a reduction of surface temperature. For the 1980s the surface cooling due to stratospheric ozone loss is calculated to have been about one third of the surface warming calculated for $CO_2$ increases during the same period, at least in northern mid-latitudes. This cooling would be largely compensated by the tropospheric ozone increase, caused by the warming, however details of the climatic aspect of the ozone changes are inconclusive and are being studied intensively. Ozone depletion in the stratosphere is accompanied by an increase of harmful UV-B radiation (Berger 1976, Blumthaler and Ambach 1990). Each 1% reduction in ozone results in an increase of 1.3% to 1.8% in radiation reaching the Earth's surface. Its harmful consequences for human health, terrestrial plants, aquatic ecosystems including ocean plankton and the basic structure of the atmosphere prompted the international community to undertake regulatory actions in the mid-1980s; these actions are briefly described below.

# 2 Ozone depletion

## 2.1 Polar ozone

Although it was first proposed in the mid-1970s that stratospheric ozone could be depleted in chemical reactions involving the degradation products of CFCs, it was not until 1985 that unequivocal evidence of ozone loss was reported in the scientific literature. In that year, scientists from the British Antarctic Survey described the polar ozone depletion, now known as the ozone hole, in which during about six weeks in the spring the total ozone column falls by more than half, and at altitudes between 13 and 18km almost all the ozone is removed (Webb 1992). These observations attracted great interest and aroused considerable scientific debate. At that time, photochemical theories had predicted ozone loss in the upper stratosphere by catalytic cycles involving the chlorine monoxide radical (ClO). The Antarctic ozone loss occurs in the lower stratosphere and cannot be explained by this mechanism. It is now accepted that Antarctic ozone is destroyed by catalytic cycles, different from those first proposed, but still involving Cl O. These cycles are strongly favoured by the low temperatures present in the Antarctic lower stratosphere. The new theories had also to explain the production of ClO. We now believe that this arises from reactions on polar stratospheric clouds (PSCs) which convert less reactive chlorine compounds (HCl, $ClONO_2$) into more reactive forms capable of destroying ozone. Recent research has shown that reactions on sulphate aerosols can play a similar role to the PSCs in both polar and middle latitudes. These aerosol reactions have taken on an extra significance following the eruption of Mt. Pinatubo which injected large concentrations of sulphur dioxide into the stratosphere. The increased size of ozone hole in 1993 (Figure 4) could be related to the extra sulphate aerosol (Bluth et al. 1992 Johnston *et al.* 1992 Solomon *et al.* 1993).

The distribution of ozone is influenced strongly by atmospheric transport on a variety of time scales. The annual cycle arises as a balance between photochemical production and destruction and transport. Large day-to-day variations in the ozone measured at a particular site are known to be strongly related to meteorological conditions with,

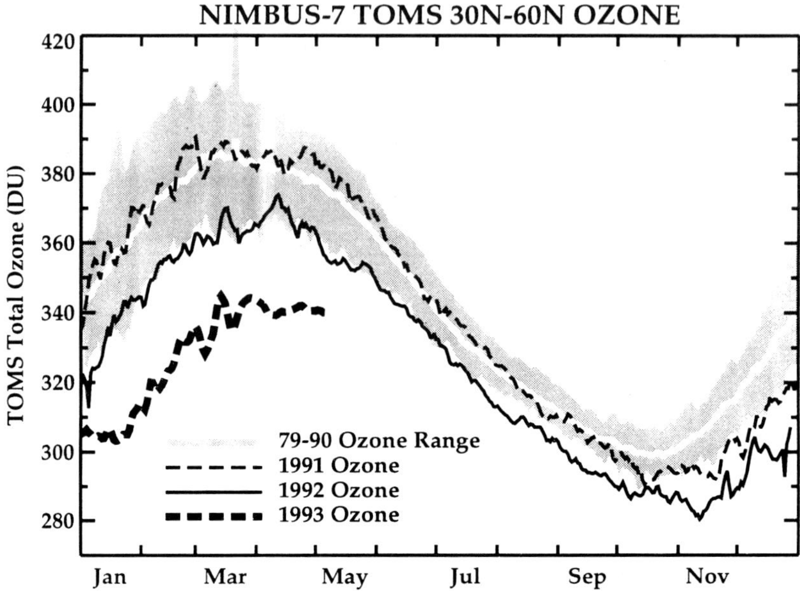



**Figure 4.** *TOMS measurements of zonally-averaged ozone, 30° N–60° N, for 1991,1992 and 1993, constrasted with the daily mean value and range for 1979–1990.*

for example, low ozone correlated with high pressure systems in the troposphere. The meteorology of the austral winter and spring is an important factor in producing the Antarctic ozone hole. Associated with the very cold polar temperatures during this period is a persistently strong circulation of westerly winds, or vortex, around the pole. This prevents substantial mixing of air between high and middle latitudes, thus preventing replenishment of ozone within the vortex by ozone-rich air from outside. These conditions also explain why severe loss of ozone, comparable to that in Antarctica, has not been reported in the Arctic (Austin *et al.* 1992). There, the mean winter temperatures are higher than in the south, the abundance of PSCs is lower and the vortex is more variable and breaks down earlier in the winter than its southern hemisphere counterpart. Nevertheless, observations taken during the 1991/1992 European Arctic Stratospheric Ozone Experiment, EASOE, have indicated that during the winter the composition of the lower stratosphere in northern latitudes is significantly perturbed chemically and this region appears to be primed for ozone destruction. Studies during the EASOE winter have demonstrated that the observed rate of ozone loss in the Arctic is consistent with the observed activation of chlorine compounds (Farman *et al.* Neuber *et al.* ).

## 2.2   Mid-latitude ozone

Analyses of global ozone records, from satellite and ground-based instruments, have shown that ozone loss is not confined to Antarctica. For example, data from the Total Ozone Mapping Spectrometer, TOMS, has shown that between 1979 and 1991 ozone amounts in mid-latitudes fell by about 4%, averaged over the year (Krueger *et al.* 1992). Over northern mid-latitudes the change was greatest in spring, varying from 6 to 8% with latitude. These mid-latitude changes, much slower than that seen in the Antarctic spring, proceeded more or less steadily, and could be adequately described as linear trends. Over the tropics there was no significant change. In 1992 and 1993 persistently low ozone values were observed, values below those expected from extrapolation of the linear trends. There were persistent large anomalies in flow patterns in the lower atmosphere in these years, and there were two volcanic eruptions, Mount Pinatubo in the Philippines and Cerro Hudson in Chile, that injected material directly into the stratosphere (Labitzke and McCormick 1993, Solomon *et al.* 1993 and Stowe *et al.* 1992). The last eighteen months of the record indicate a very significant shift in global average ozone, without precedent in the lifetime of TOMS. /indexTOMS

The causes of depletion of ozone outside the polar regions remain controversial. Three processes that certainly contribute to the observed trends are as follows.

1. The production of active chlorine by PSCs. PSC formation is not necessarily restricted to air masses strictly confined to the polar vortices, but can occur in, and equator-ward of the core of the stratospheric jet stream. Persistent PSCs in such locations can process very large volumes of air, that, as a result of jet stream meandering, may come from, and return to, latitudes as low as the sub-tropics.

2. Heterogeneous reactions on sulphuric acid aerosol. Reactions that have been demonstrated in laboratory studies are noted in next paragraph. Several atmospheric measurements have been reported that point to the significance of these reactions.

3. Transport of ozone-poor air from regions of severe depletion-dilution. In winter and spring vortex erosion is an on-going process, culminating in the final break-up of the vortex, that leads to eventual mixing of vortex air into middle latitudes. In the southern hemisphere this constitutes a significant flux of ozone-poor air, and of enhanced active chlorine, but nevertheless this dilution is not sufficient to explain the observed mid-latitude loss.

In a controversial interpretation of some observational data, some scientists have proposed that the flow of air through the vortex is much greater than can apparently be explained by current dynamical or radiative theory. The main issue is the rate at which air flows through the vortex (and so is chemically processed); this flux critically influences the amount of ozone-poor air of polar origin over the middle latitudes. To these processes must be added the changes in atmospheric circulation that may be associated with the large heating rates produced when there is a dense belt of aerosol in the tropics. Accelerated up-welling of air naturally poor in ozone will imprint an anti-correlation between aerosol and ozone that will survive poleward transport. The relative importance of these processes remains to be resolved. There may be no simple solution—their importance may vary from year to year, depending on the very variable climatology of the Arctic winter stratosphere.

## 2.3   Chemical processes leading to polar ozone depletion

The depletion of ozone in polar latitudes is attributed to a sequence of chemical reactions that starts with the conversion of compounds with long lifetimes into less stable forms. These are broken down by sunlight to give high concentrations of chlorine and chlorine monoxide radicals (Cl and ClO). This leads to catalytic cycles that can destroy ozone very rapidly, and in which a significant part is played by bromine and bromine monoxide radicals (Br and BrO) (MBSW 1992, Poulet *et al.* 1992 and Traub *et al.* 1992). The main long-lived inorganic carriers (reservoirs) of chlorine are hydrochloric acid (HCl) and chlorine nitrate ($ClONO_2$). Dinitrogen pentoxide ($N_2O_5$) is a reservoir of oxides of nitrogen. All the gas phase reactions of these reservoirs are relatively slow, but fast reactions can occur on a suitable surface or in solution (heterogeneous reactions). PSC particles are believed to support such reactions. PSCs can form in the lower stratosphere at temperatures below about 195K ($-78°$ C), and are believed to contain solid phases of mixed nitric acid and water. Laboratory experiments suggest that the most important reactions are:

$$
\begin{array}{lll}
HCl & + & ClONO_2 & \longrightarrow & HNO_3 & + & Cl2 \\
ClONO_2 & + & H_2O & \longrightarrow & HNO_3 & + & HOCl \\
HCl & + & HOCl & \longrightarrow & H_2O & + & Cl_2 \\
N_2O_5 & + & HCl & \longrightarrow & HNO_3 & + & ClONO \\
N_2O_5 & + & H_2O & \longrightarrow & 2HNO_3
\end{array}
$$

The chlorine compounds formed in these reactions are readily dissociated by sunlight, even in low light conditions releasing chlorine atoms (Cl). The nitric acid formed in these reactions remains in the particles, so that the gas phase concentrations of oxides of nitrogen are reduced. This reduction, "denoxification", slows down the rate of removal of ClO in the reaction:

$$
ClO \; + \; NO_2 + M \; \longrightarrow \; ClONO_2 \; + \; M
$$

where M is any air molecule, and so helps to maintain high levels of active chlorine. It is known that heterogeneous reactions can also take place on, or in, sulphate aerosol particles (droplets of sulphuric acid in solution), which are always present in the lower stratosphere. The droplets are formed following the oxidation of carbonyl sulphide (OCS), a naturally occurring sulphur compound that has a long enough lifetime to survive slow transport to the stratosphere. Another, highly spasmodic, source of aerosol is sulphur dioxide, injected directly into the stratosphere during some volcanic eruptions. This can raise aerosol concentration to many times the background level. Sulphate aerosol is known to support reactions (2) and (5), and possibly (3), leading both to the release of active chlorine, and to denoxification. Thus there is the potential for widespread ozone destruction (Fahey et al. 1993). However the temperature dependence of the heterogeneous reactions on the aerosol seems to be such that rapid destruction is possible only at temperatures close to those required to form PSCs, and the effects of the background aerosol are thought to be small. The production of active chlorine requires sunlight, and sunlight can drive the following catalytic cycles:

(I)

$$
\begin{array}{llllllll}
 & \mathrm{ClO} & + & \mathrm{ClO} & + & \mathrm{M} & \longrightarrow & \mathrm{Cl_2O_2} + \mathrm{M} \\
 & \mathrm{Cl_2O_2} & + & h\nu & \longrightarrow & \mathrm{Cl} & + & \mathrm{ClO_2} \\
 & \mathrm{ClO_2} & + & \mathrm{M} & \longrightarrow & \mathrm{Cl} & + & \mathrm{O_2} + \mathrm{M} \\
2\times & (\mathrm{Cl} & + & \mathrm{O_3} & \longrightarrow & \mathrm{ClO} & + & \mathrm{O_2}) \\
\hline
\text{net} & & & 2\mathrm{O_3} & \longrightarrow & 3\mathrm{O_2}
\end{array}
$$

(II)

$$
\begin{array}{lllllll}
\mathrm{ClO} & + & \mathrm{BrO} & \longrightarrow & \mathrm{Br} & + & \mathrm{Cl} + \mathrm{O_2} \\
\mathrm{Cl} & + & \mathrm{O_3} & \longrightarrow & \mathrm{ClO} & + & \mathrm{O_2} \\
\mathrm{Br} & + & \mathrm{O_3} & \longrightarrow & \mathrm{BrO} & + & \mathrm{O_3} \\
\hline
\text{net} & & 2\mathrm{O_3} & \longrightarrow & 3\mathrm{O_2}
\end{array}
$$

The dimer ($\mathrm{Cl_2O_2}$) of the chlorine monoxide radical involved in cycle I is thermally unstable, and the cycle is most effective at low temperatures. It is thought to be responsible for most (70%) of the ozone loss in Antarctica. In the warmer Arctic a large proposition of the loss may be driven by cycle II.

## 2.4  The role of halocarbons

The measured atmospheric concentrations of CFCs are still rising. The rate of increase has slowed because emissions have been reduced in response to legislation but concentrations will continue to rise until emissions of the compounds have been nearly eliminated (Wang 1993). The 1992 revision of the Montreal Protocol brought forward the phase out of CFCs, carbon tetrachloride and methyl chloroform. This will lead to a small reduction in the predicted peak chlorine loading arising from these compounds, and will reduce the time to return to 1990 levels by some 2 to 3 years. The 1992 revisions imposed controls on HCFCs for the first time. A cap was placed on consumption, with subsequent reductions leading to a complete phase out. If HCFCs were used at the full rate permitted by the cap, this could extend the time to return to 1990 levels of chlorine loading by up to 8 years. Bromine in the stratosphere is an important catalyst for ozone depletion, alone and in coupled reactions with chlorine. There is, however, still considerable uncertainty about both the natural and the man-made sources of bromine (Singth *et al.* 1993). The major uncertainty concerns methyl bromide, with estimates of the man-made contribution ranging from 30% to in excess of 50%. A reduction in the emissions of man-made methyl bromide would reduce atmospheric bromine loading significantly and rapidly (Singth and Kanakidou 1993). Slow release of halons from the "banks" will continue to contribute to atmospheric bromine loading until well into the next century. Following the recommendations in the World Meteorological Organisation's 1989 Scientific Assessment, the atmospheric chemistry of the alternative fluorocarbons has been studied (WMO 1989 1991). The results have not led to significant revision of the potentials of the substitutes, or their degradation products, to deplete ozone (AFEAS 1989).

The chlorine and bromine loadings are the calculated concentrations, in the lower atmosphere, of the chlorine or bromine contained in compounds which could be transported into the stratosphere, that chlorine and bromine could be released to enter into ozone depleting reactions. The relation between chlorine loading of the lower atmosphere and stratospheric ozone depletion is neither simple nor exact. In general, the higher the chlorine loading, the higher the potential concentration of chlorine in the air transported into the stratosphere. In addition, the longer the time for which the stratospheric chlorine concentration is above its current level, the longer ozone depletion might be expected to persist at or above the present rate. The consequences from bromine loading are similar. Furthermore bromine and chlorine interact in the stratosphere; an increase in the concentration of chlorine increases the effectiveness of bromine in ozone depletion and the effectiveness of both species is affected by stratospheric temperature. The 1992 (Copenhagen) revision of the Montreal Protocol limits the additional chlorine loading by curtailing the amount of HCFCs which are allowed to be used; setting a maximum amount and a time limit. The mechanisms to accomplish this is a cap on consumption which will be reduced stepwise, leading to phase out in the year 2030. The numerical value of the cap is the sum of the quantity of all HCFCs which were produced and used during 1989 and an amount equal to 3.1% of the calculated CFC consumption during 1989. Contributions to the 3.1are adjusted by their ODPs. For example, if the average ODP of the substitute HCFCs were 0.1, the 3.1% allowance would amount to about 300,000 tonnes / year, or approximately 30% of total CFC consumption in 1989. A similar mechanism has been proposed for the European Community but with a cap set at 2.5% and phase out by 2015; this regulation is still under negotiation (Fisher *et al.* 1990 and Fisher and Midgley 1993). There are now a large number of bromine compounds in the atmosphere, some of which are entirely anthropogenic in origin, some of which are entirely natural and some of which have both natural and anthropogenic sources. There has never been a systematic attempt to measure the complete set of bromine compounds simultaneously on a global scale. Measurements have been made spasmodically and these generally have been restricted to one class of compounds, such as the halons; even then the data has been incomplete, although they are consistent with estimates of halon emissions. In addition compounds such as $CHBr_3$, $CHBrCl_2$, $C_2H_4Br_2$ and $C_2H_5Br$ have been detected in the atmosphere, but their lifetimes are sufficiently short that their concentration close to the tropopause will be very low in most circumstances.

## 2.5 Effects of aircraft on the ozone layer

In the early 1970s it was realised that the fleets of supersonic passenger aircraft then being planned could pose a serious environmental threat. In particular, it was suggested that the exhaust material from the aircraft engines could possibly lead to a depletion of the stratospheric ozone layer. Supersonic aircraft cruise directly within the stratosphere so that even short-lived pollutants, which would be removed rapidly in the troposphere, could be important. One of the major exhaust products is NOX ($NO + NO_2$) produced as air is heated to very high temperatures in the aircraft engines. NOX could then destroy ozone by the following catalytic cycle:

$$\begin{array}{rcl}
\text{NO} \ + \ \text{O}_3 & \longrightarrow & \text{NO}_2 \ + \ \text{O}_2 \\
\text{NO}_2 \ + \ \text{O} & \longrightarrow & \text{NO} \ + \ \text{O}_2 \\
\hline
\text{net} \quad \text{O} \ + \ \text{O}_3 & \longrightarrow & 2\text{O}_2
\end{array}$$

In brief, increasing aircraft traffic, both sub-sonic and supersonic, represents a potential threat to atmospheric ozone in the early 21st century (CIAP 1974). Calculations currently suggest that ozone will decrease in the mid to upper troposphere. There are large uncertainties which make prediction difficult. These include our understanding of chemical and physical processes, stratosphere-troposphere exchange and possible radiative impacts.

# 3 Impacts of ozone depletion

## 3.1 On surface ultraviolet radiation

An important reason for concern about stratospheric ozone depletion is the consequent increase in ground level ultraviolet (UV ) radiation (Figure 5). Solar ultraviolet reaching the Earth's surface is known as UV-B (280–320nm) or UV-A (320–400nm) radiation. UV-B is absorbed by ozone but some UV-B photons reach the Earth with sufficient energy to initiate many chemical and biological reactions. UV-A radiation has less energy per photon and is not absorbed significantly by ozone. Ozone loss should lead, in the absence of other changes, to an increase in UV-B radiation reaching the ground. In the Antarctic a marked increase in surface UV-B radiation has been detected during the springtime ozone depletion. An increase has also been measured at Toronto from 1989 to 1993. These increases are consistent with theoretical calculations based on observed ozone changes (Kerr and McElroy 1993). Attempts to quantify long term changes in UV-B elsewhere have been greatly hindered, both by a lack of high quality data and by insufficient information on other atmospheric trends (*e.g.* pollution and cloudiness).

The factors which determine the amount of radiation reaching the surface are the following:

- The solar zenith angle, $Z$, depends on both time of day and time of year and can be calculated precisely. The solar zenith angle determines the path length of the radiation through the atmosphere (it is shortest when the Sun is high in the sky, *e.g.* at noon in summer, and longer at sunset). On a longer path through the atmosphere there is more attenuation due to the factors listed below. The zenith angle also determines the horizontal area over which the incident beam is spread; the area is larger with larger zenith angle. The rate at which energy reaches a unit horizontal area is reduced by a factor $\cos Z$.

- Scattering by air molecules (Rayleigh scattering) also affects the solar radiation reaching the surface by redirecting the radiation either back out to space or on towards the surface but at a different angle. This scattering process becomes much more effective as wavelengths decrease: mathematically the scattering is inversely proportional to the fourth power of the wavelength. It is very important

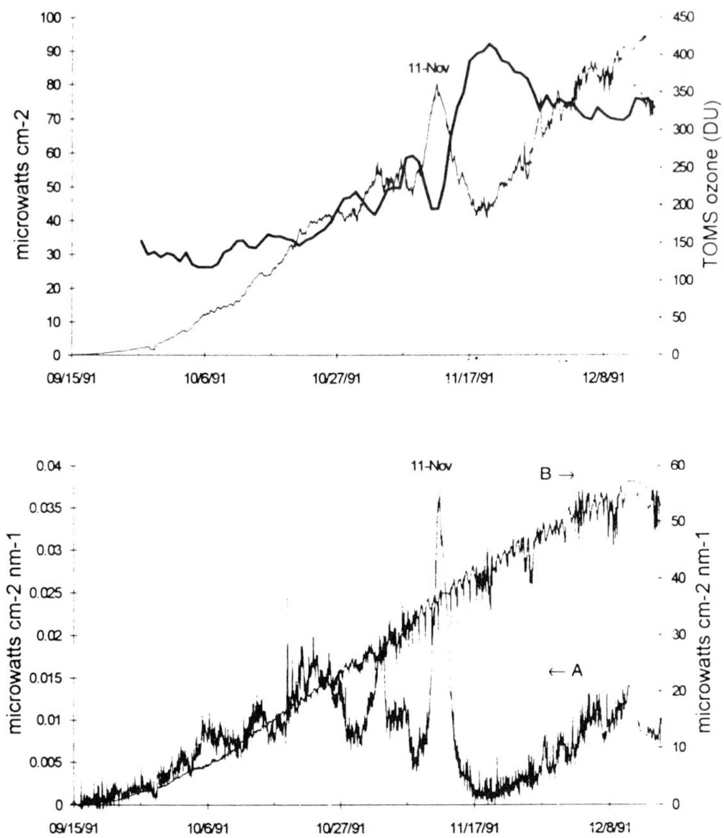



**Figure 5.** *Measurements of ozone and* UV *at the South Pole for a 3-month period in 1991. The top graph shows* TOMS *ozone data (bold curve) and* UV-B *(280–320nm) intensity. The bottom graph shows irradiance at wavelengths of 300nm (A) and 400nm (B). The anti-correlation at 300nm between ozone and* UV *is not evident at 400nm and is thus not an artefact of other atmospheric effects.*

at UV wavelengths where over half of the UV-B radiation reaching the surface is diffuse (*i.e.* has been scattered).

- Ozone is the major absorber of UV . It is usually expressed as the amount in a vertical column through the atmosphere, combining stratospheric and tropospheric ozone. Ozone in the troposphere has increased in some areas, with localised high ozone events dependent upon industrial activity and prevailing weather conditions. In the stratosphere attenuation of UV depends on the ozone concentration along the length of the radiation path. In the lower troposphere more than half the UV-B (*i.e.* radiation with wavelengths 280–320nm) has been scattered and is following an indirect path to the surface. Path lengths through a low altitude ozone layer can thus be increased and the apparent effectiveness of absorption by ozone increased (Bruhl and Crutzen, 1989). In addition to ozone trends, the ozone column can vary greatly with the passage of individual weather systems. Cyclonic weather systems (depressions) are associated with high column ozone content, whilst anticyclones are associated with low column ozone content.

- Clouds, sulphur dioxide and aerosols also affect the UV radiation reaching the surface. Cloud is highly variable, it normally has an attenuating effect, and can cause rapid local changes in incident UV .

**Surface UV-B radiation derived from satellite observations**

It is well known that the UV-B input into the troposphere can be estimated quite accurately because of the high quality record of stratospheric ozone measurements from the TOMS instrument on Nimbus 7. So the measured changes in stratospheric ozone can reliably be used to compute UV transmission through the atmosphere. Given that the relative accuracy of the ozone record from satellites has been maintained to $\approx 2\%$, the uncertainties in the calculation of surface UV-B from satellite observations will be the uncertainties involved in the transmission of UV through the troposphere (*i.e.* cloud cover, aerosol loading).

## 3.2   Impacts of ozone depletion on climate

Ozone is an important gas in the climate system; observed and predicted changes in ozone are calculated to have significant effects on the atmosphere (Oort and Liu 1993). Lower stratospheric temperatures have declined over the past few decades (Figure 6).

Much of this decline can probably be attributed to the effect of ozone loss. Stratospheric ozone is itself affected by temperature change. A cooling of the lower stratosphere, expected to result from increased concentrations of atmospheric $CO_2$, may increase ozone loss in the northern hemisphere. Decreases in lower stratospheric ozone are calculated to have partially offset the climatic effect of increases in other greenhouse gases over the past decade. The net greenhouse effect of hydrofluorocarbon replacements for CFCs may be comparable to, or exceed, the effect of equal emissions of CFCs, after account has been taken of chlorine-induced ozone loss. .The increase in UV penetrating to the troposphere, as a consequence of ozone depletion, has been calculated to alter the concentrations of some climatically important tropospheric contituents, such as methane.

Potential climate effects are characterised by a quantity called radiative forcing (which has units of $Wm^{-2}$). This is the perturbation to the planetary radiation budget; this forcing is expected to lead to a climate change, such as changes in temperature. Calculations of radiative forcing have associated uncertainties but the uncertainties in predicting surface temperature change are even greater; hence for a comparison of different climate change mechanisms, radiative forcing is a preferred yardstick (NOAA 1992, 1993).

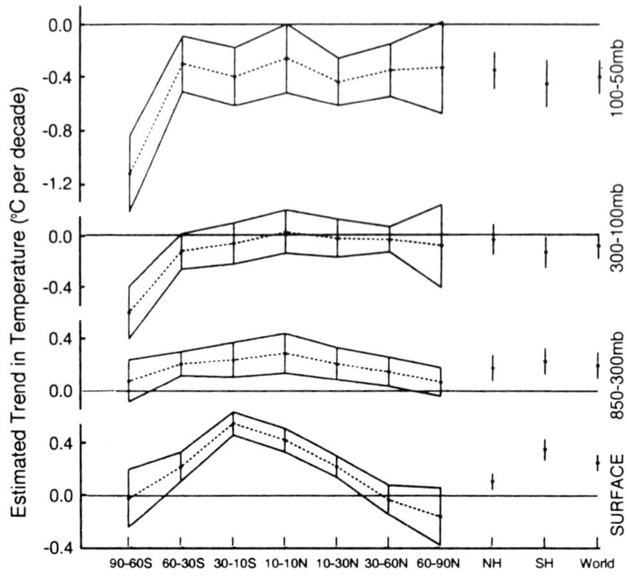

**Figure 6.** *Estimated trend in temperature (in °C per decade) between 1963 and 1989 based on an analysis of radiosonde data by Oort and Liu (1993). The main part of the figure shows the variation of this trend with latitude for four levels in the atmosphere- the surface (defined here as 950mb), the mid-troposphere (850–300mb), the upper troposphere/lower stratosphere (300–100mb) and the lower stratosphere (100–50mb). The dashed line indicates the best estimate of the trend, the outer solid line indicates the 95confidence limit. The hemisphere average (NH = northern hemisphere, SH = southern hemisphere) and global average trends are shown with associated confidence limits.*

# 4   Total ozone measuring instruments

## 4.1   Ground-based instruments

There are three ground-based instrument types that determine ozone column contents or densities by measuring scattered solar radiation in the Hartley-Huggins bands in the UV spectral region: the Dobson spectrophotometer, the M-83 filter ozontometer, and the Brewer spectrophotometer. There is also one that uses the Chappuis band in the visible region—a spectrometer with a diode array detector. These instruments use several wavelengths in order to separate out molecular (Rayleigh) and aerosol scattering from ozone absorption.

To understand how these instruments work, it is necessary to understand the vertical-profile nature of three properties of the atmosphere: molecular absorption, molecular scattering and aerosol scattering. To first order the vertical profile of ozone density affects molecular absorption of solar radiation as it descends through the atmosphere. The temperature dependence of the absorption is an important second-order effect. Molecular scattering is directly proportional to molecular density and is proportional to the inverse fourth power of wavelength (which causes the sky to appear blue). Molecular scattering directs the incoming radiation into various angles, including into the receiver of a zenith- pointed instrument. Aerosol scattering acts similarly to molecular scattering but with a different and variable wavelength dependence and much more spatial variability. If these three atmospheric properties are combined into a model that also considers solar angle and spectral characteristics, data obtained with these instruments over a several-hour period can be used to determine vertical profiles of ozone with approximately 5 km resolution (the so-called Umkehr technique). These instruments can also be used to determine ozone column content by looking directly at the Sun or Moon.

The primary advantages of the ground-based indirect-absorption-measuring instruments are that they can be used to measure ozone column content with fairly high accuracy routinely, as well as to provide vertical-profile data, and that they can make measurements for solar zenith angles greater than 90° (that is, pre-sunrise or post sunset). Because they are easy to operate, these types of instrument are deployed at more than 140 ground stations, forming the World Meteorological Organisation's Global Ozone Observing System (GOOS) network. This network is useful for monitoring ozone above the stations to give quasi-global values and for providing reference measurements for satellite-based ozone measuring instruments.

The disadvantages are that the instrument responses tend to drift with time, requiring frequent calibration for accurate work; they can suffer interference from aerosols or other molecular species that absorb in the same spectral region; that the ozone absorption cross sections used as a basis for measurements have changed with time, although data with reasonably high accuracy are now available; and that they are more sensitive to low-altitude ozone than they are to upper stratospheric ozone.

Note that these ground-based instruments are generally based on older technology that has been upgraded with time (the Dobson instrument dates from the 1920s and the M-83 and Brewer instruments date from the 1970s; the spectrometer mith diode array is a product of the 1980s). Although continued measurements with these instruments are invaluable in continuing the historical record so that we can determine trends, it is possible that a well-funded effort could suceed in developing an improved instrument that is based on more recent technological advances.

The four instruments are presented in historical order.

## Dobson spectrophotometer

The Dobson spectrophotometer has been an important instrument for measuring global ozone concentrations since its development by G M B Dobson around 1927. This development was based on earlier research with the Fabry-Buisson spectrophotometer that used photographic plates. The instrument, which has undergone some changes over the years, is based on making measurements of ozone in the Huggins bands by using direct or scattered sunlight. By using one pair of wavelengths, a differential measurement is made that includes the differential absorption of sunlight by ozone but also includes the effects of differential scattering by atmospheric molecules and aerosols. If measurements at two wavelength pairs are combined, the scattering effects can essentially be cancelled, because they are similar for different wavelength pairs. The instrument uses a double monochromator, with one monochromator used to reject interfering scattered radiation. A pair of wavelength-isolating slits within the instrument is coupled to an optical wedge that is adjusted at one slit to give a null difference reading. A number of pairs of wavelengths, with each member of a pair separated by approximately $2025 \mathrm{cm}^{-1}$, are used by Dobson instruments, with each given a letter designation. Thus there are the A pair (305.5 and 325.4nm), the B pair (308.8 and 329.1nm), the C pair (311.45 and 332.4 nm), the D pair (317.6 and 339.8nm), etc., as well as the C' pair (332.4 and 453.6 nm) with a different separation. Total ozone observations using direct sunlight or moonlight are usually made on AD and CD double-pair wavelengths. Similar, although less precise, ozone measurements can also be made by using clear or cloudy zenith skylight. Uncertainties arise from not knowing the ozone profiles involved in the diffuse scattering from the various directions. Additionally, zenith skylight observations can be made on CC' wavelengths; the C' wavelength pair measurement provides data on light scattering by aerosols. Measurements of ozone vertical distribution can also be made with the Dobson spectrophotometer (Frederic and Weatherhead 1992). The technique is called Umkehr-a German word meaning reversal, and it was first used by Goetz. Observations are made on the clear zenith sky as the Sun is rising or setting. Because the ozone is primarily concentrated in a shell centred between 20 and 40km above the surface of the Earth, the brightness of the zenith sky at a UV wavelength that is partially absorped by ozone will continuously decline as the Sun approaches the horizon and then, just before it sets, the brightness will increase briefly. Ozone vertical profile data are obtained through an inversion technique that takes into account ozone absorption and molecular and aerosol scattering of light that passes through the atmosphere from the Sun to the instrument. Long Umkehr measurements are made for solar zenith angles between 60° and 90° and short Umkehr measurements are made from A, C and D wavelengths at solar zenith angles between 80° and 90°.

A number of factors affect the measurement precision and accuracy of Dobson spectrophotometers, limiting the historical record to an accuracy of 5–10%. However, it should be possible for many instruments, for which adequate calibration records have been kept, to reduce the error to a 2% relative uncertainty and an additional 1–2% absolute uncertainty of the spectral parameters. A primary error source stems from absolute instrument calibrations, that is, from the determination for an instrument of extraterrestrial constants for the wavelengths at which observations are made. The procedure entails a series of measurements on the rising or setting Sun during clear-sky conditions when atmospheric ozone remains constant or is changing slowly. An alternative method to calibrate a Dobson instrument absolutely is to compare directly the results from the instrument with a reference-standard Dobson spectrophotometer. This method has been used extensively since 1977. Since the mid-1970s, virtually all Dobson spectrophotometers in the global Dobson instrument network have been calibrated several times, either directly or indirectly to Instrument 83. (There are approximately 140 instruments in the network, of which 50–60 have good long-term calibration records.) Absolute calibrations of Instrument 83 were performed in 1962 at Mauna Loa Observatory, Hawaii. A detailed analysis of the calibration data indicated that the long-term (25 year) ozone-measurement precision for the instrument is known to within five tenths of a percent although the absolute values may be low by 2–4% owing to errors in the spectral values used.

Ozone absorption cross sections used to calculate densities from UV intensities measured by Dobson spectrophotometers from 1 July, 1957 to 31 December, 1967, were derived from laboratory measurements made by Vigroux. Because these cross sections yielded inconsistent ozone data from measurements on different wavelengths, observations on the AD double pair wavelength, with an absorption cross-section difference of $1.19 \times 10^{-19} \mathrm{cm}^2/\mathrm{molecule}$ at $-4°\mathrm{C}$, were adopted as standard. After additional research was conducted by several investigators, the International Ozone Commission recommended that beginning on 1 January, 1968, the set of internally consistent absorption cross sections be used. The newer data preserved the $1.19 \times 10^{-19} \mathrm{cm}^2/\mathrm{molecule}$ differential absorption cross section for AD wavelengths and yielded essentially consistent ozone values from different kinds of observations. However, there were still problems with the absorption cross sections used. More recently Paur and Bass redetermined ozone absorption cross sections at the Dobson instrument wavelengths. The differential absorption cross section obtained for the AD double-pair wavelengths was $1.23 \times 10^{-19} \mathrm{cm}^2$ /molecule, which yields lower ozone values by approximately 3% than do the standard cross section data. However, Molina and Molina report slightly different values for the absorption cross sections. These new absorption cross sections have not been adopted; this is in order to preserve the historical record for trend analysis.

Dobson spectrophotometers are also subject to errors because of the presence of aerosols and $SO_2$. Because of the simplicity and reasonable accuracy of the Dobson spectrophotometer approach, and its early development, a widespread network comprising approximately 90 Dobson-equipped and approximately 50 M-83 filter ozonometer-equipped stations has allowed us to make many important observations over many parts of the globe. Results up to 1959 were reviewed by London. There were more than 30 Dobson stations in the global network in 1957 and 50 in 1963. Dobson for example, summarised the measurements as of 1965 in the Antarctic by noting that the ozone increased each year in November by approximately 30% and then gradually decreased to pre-November values in February. In the northern hemisphere, the ozone peak occurs in March. Farman et al. were the first to report a gradual decline in the October ozone levels at Halley Bay, Antarctica, beginning in 1968; a rapid increase in the rate of decline started in 1976. Similar results were obtained by Komhyr *et al.* at South Pole Station. Analysis of Total Ozone Mapping Spectrometer (TOMS) satellite data subsequently confirmed the results for the Dobson instrument measurements. A considerable effort has gone into analysing Dobson network data to determine whether a corresponding decrease in global, non-polar ozone has also occurred. It has been found that there has been a small decrease in the past 20 years, with analysis of data by months being the key to finding the trend, and that the effects of solar radiation as a function of sunspot cycle are significant for ozone densities.

## 4.2   M-83 filter ozonometer

The Soviet M-83 filter ozonometer measures ozone column content by looking directly at the Sun and by using a 21nm bandpass channel with peak spectral sensitivity at 299nm (strong absorption by ozone) and a 15nm bandpass channel that has a peak response at 326nm (weak absorption by ozone), in the Hartley-Huggins bands. The M-83 can also be pointed in the zenith direction to measure ozone through the scattering of solar radiation. The instrument was developed in 1963. Improvements were made in 1972 by using the stated bandwidths (reduced from 40nm originally) when a better data reduction algorithm was incorporated that included variations of the atmospheric characteristics with solar angle (atmospheric mass), and by including a third filter to correct for aerosol effects. Earlier versions of the instrument suffered from changes of responses of the filters and the incomplete elimination of aerosol effects. The M-83 has a measurement precision of approximately 5% in comparison with TOMS instrument measurements. The accuracy in the absence of aerosol effects appears to be 2-4% if sufficient data are averaged, although overestimations of ozone in the zenith of 30% have been noted during cloudy conditions. At least 45 stations make up the former USSR's ozone network, and these stations have been in operation for more than 25 years. Some of the accomplishments of these instruments are the study of diurnal variations in the total ozone content, and the comparison with measurements from satellite instruments. An improved version of this instrument, the M-124 ozonometer, was developed in 1981. Changes included a compact solar reflector, the elimination of the photomultiplier tube, increased leak resistance, and the placement of the optical filters in hermetically sealed cells with quartz windows to increase filter lifetime. The M-124 ozonometer was introduced into the USSR's ozone monitoring network in 1985–1986. Currently no data are available to evaluate its measurement precision and accuracy.

## 4.3   Brewer spectrophotometer

In the early 1970s a grating spectrophotometer was developed as an alternative instrument along with the Dobson spectrophotometer. This instrument was based on the Ebert-type spectrometer. The five wavelengths normally used for ozone measurements are 306.3, 310.1, 313.5, 316.8, and 320.0nm. Radiances at these wavelengths are monitored sequentially in 1.6 s. The main differences between the Brewer spectrophotometer and the Dobson spectrophotometer are: (1) the Brewer instrument can be used to observe several wavelengths simultaneously, whereas the Dobson instrument is limited to one pair of wavelengths at a time, (2) the Brewer instrument measures absolute intensity of light at each wavelength by using a single detector so that there are no wedges or other components to calibrate and (3) a different set of wavelengths is used. The recently revised ozone absorption cross sections are used with the data-processing algorithm. Also, the Brewer spectrophotometer is generally supplied with an automatic Sun tracker so that human involvement and error is minimised and measurement time is increased.

A Brewer spectrophotometer has been being compared with the Dobson spectrophotometer in Toronto since 1983. The differences have been 3% or less during this period. The causes of these differences are not fully understood, but several things are being investigated: (1) the different dependencies on total air mass, (2) the differences in the assumed temperature dependencies of the ozone cross sections and (3) interference by $SO_2$. The Brewer spectrophotometer has seen increased use since 1982 when a fully automated unit was delivered to Greece. There it was used to demonstrate that in cloudy weather depolarised solar radiation multiply scattered by clouds gave a better measurement of total ozone than did the unpolarised component. It has also been of value as a travelling standard by which to recalibrate Dobson spectrophotometers. Although there were some problems with the Brewer spectrophotometer because the electronics were not made to be fully rugged, these problems have been overcome. One instrument in Edmonton (Canada) has been operating continuously in the open for 3 years at temperatures that range from $-40°C$ to $+40°C$. In addition, the accuracy of the ozone absorption cross sections used as reference for all UV ozone instruments was somewhat low in the early 1980s but is now much improved. Brewer spectrophotometers are now in operation in Canada and several other countries. The Brewer instrument has a number of attractive features that seem to qualify it as an instrument able to be used to augment the Dobson network, especially since higher measurement accuracy is now desired.

## 4.4  Space-based instruments

Several instruments have been developed to measure ozone vertical profiles and column contents from space by measuring backscattered UV solar radiation. These instruments use several wavelength channels in the Hartley-Huggins bands to help separate the ozone absorption from the molecular (Rayleigh) and aerosol scattering as well as to detect the presence of clouds and variations in Earth-surface reflectance. The shorter wavelengths (approximately 250–300nm) are used for the higher-altitude (up to 50–55 km) measurements, and the longer wavelengths (to 380nm) are used for lower altitudes (to the ground in some cases). The spectral distribution of radiation scattered back to space by the atmosphere of the Earth is influenced by the presence of gases with selective absorption bands as well as by molecular (Rayleigh) and aerosol scattering. Most of the ozone lies above the tropopause, whereas most of the measured radiance comes from below the tropopause. Thus detailed knowledge of the ozone profile a priori is not required. The brightness is reduced in a spectral region where absorption occurs as compared with the situation when no absorbers were present. The most accurate measurements are made for solar zenith angles less than 60° but measurements with solar zenith angles to 88° could be made, although the standard deviation of the measurement increases at the larger angles. This approach was suggested by Wentworth and Wentworth on the basis of the Umkehr technique for ground-based vertical-profile measurements using forward scattering developed by Goetz *et al.* . The instruments can use spectrometers or filters to select the spectral channels. They can look in the nadir direction, as for solar backscattered UV, or be scanned cross track, as with total ozone mapping spectrometry. The largest difficulty with these instruments is the spectral changes observed in the diffuser plate used to direct exoatmospheric solar radiation into the instrument occasionally for reference purposes. These diffuser plates become contaminated, which slowly reduces the reflectance of the plate in a spectrally unknown way. In addition, some of the instruments have been adversely affected by scattered light inside the spectrometer. The instruments operate best in the portion of the atmosphere above 15–20km, primarily because of the strong absorption by ozone in the Hartley band, which prevents most UV radiation in the band from getting below the ozone density peak. These instruments continue to be quite useful in providing global coverage of ozone, both in column content and vertical profile with moderate (several kilometres) resolution.

## Total ozone mapping spectrometer

The Total Ozone Mapping Spectrometer (TOMS) is similar to the BUV and SBUV instruments but differs in that it observes in only six wavelength regions from 312.5 to 380nm, uses a single monochromator, and scans cross track over 105°. In addition, it uses an IR cloud-cover photometer to avoid problems from clouds in the measurements. As the name implies, it provides intergrated ozone burdens rather than profiles.

The algorithms used for data analysis are derived from those used for the BUV measurements and add features to account for off-nadir viewing angles. In addition, daily snow and ice maps are used to help eparate out the causes of high-reflectivity scenes, and the Temperature Humidity Infrared Radiometer (THIR) onboard the Nimbus-7 satellite is also used to determine cloud locations and heights. Such information is required in order to determine the altitude above which ozone is being measured. If clouds are present, the measured ozone amount has to be augmented by an estimate of the below-cloud amount to arrive at a total ozone column content.

The TOMS instrument measurement precision and accuracy has been verified by several comparisons with other instruments. The precision of the TOMS instrument is given as 2% or better, and there is a small drift in the data (underestimate) of approximately 0.4% per year, owing to changes in the reflectance properties of the diffuser plate. The drift in diffuser-plate reflectance is being corrected in the data processing by using an algorithm that takes measurements at five pairs of wavelengths and assumes that the drift is proportional to wavelength separation. It was determined after processing 5 years of data that the ozone absorption cross sections used in the data analysis were incorrect and that the new absorption cross sections yielded total ozone values 6% higher than the old cross sections.

In the Antarctic region the unusual shape of the ozone profile can also lead to errors in using algorithms to process the data. The TOMS instrument has been used to measure global ozone levels since 1978. The data from 1979 to 1982 were used to calculate a global climatology of total ozone. More recently, TOMS data were used to deduce global changes in total ozone from 1979/1980 to 1986/1987. A graph of latitude versus time in the year clearly shows the Antarctic ozone hole in the austral spring, as well as several-percentage decreases at other times and latitudes. In the process of comparing the TOMS data with data from Dobson and M-83 ozonometer ground stations, the quality of the ground-station data was evaluated, leading to the finding that a number of the ground stations were not producing high-quality data.

The TOMS instrument has also been used to study the ozone hole over Antarctica. Although these data are quite colourful and detailed, they were initially unable to discover the ozone hole because "in the first five years all TOMS total ozone values less than $180\mu$Watts cm$^{-2}$ were flagged (and rejected) as being climatologically invalid". TOMS data are also being used to study ozone in the troposphere. While only 10–15% of the ozone column amount resides in the troposphere, TOMS data are able to detect 5–10D.U. changes in total ozone, which, by comparison with ground monitoring station values or UV differential absorption lidar measurements are shown to be related to lower altitude changes in ozone.

# 5  References

AFEAS, 1989, Published as an Appendix to Scientific Asessment of Stratospheric Ozone, WMO Global Ozone Research and Monitoring Project—Report No 20.

AFEAS, 1992, Production sales and atmospheric release of CFC-12 (1992a) CFC-113, CFC-114 and CFC-115 (1992b) and HCFC-22 (1992c) through 1991, Alternative Fluorocarbons Enviromental Acceptability Study (Washington, DC).

Austin J , Butchart N, and Shine K P, 1992, Possibility of an Arctic ozone hole in a doubled-CO2 climate, Nature 360 221-225.

Berger D S, 1976, The sunburning ultraviolet meter design and performance, Photochem Photobiol 24 587-593.

Blutmthaler M, and Ambach W, 1990, Indication of increasing solar ultraviolet-B radiation in Alpine regions, Science 248 206-208.

Bluth G J S, Doiron S D, Schnetzler C S, Krueger A J, and Walter L S, 1992, Global Tracking of the SO2 Clouds from the June 1991 Mount Pinatubo Eruptions, GRL 19 151-154.

CIAP, 1974, The effects of stratospheric pollution by aircraft, U S Dept. of Transportation (Washington, DC).

Climate Diagnostics Bulletin, July 1993, Climate Analysis Centre, U. S. Department of Commerce, NOAA/NWS/NMC, World Weather Building, Room 605, 5200 Auth Road (Washington, DC).

COMESA, 1975, Report of the Committee on meteorological effects of stratospheric aircraft, Meteorological Office (Bracknell, U.K).

Fahey D W, Kawa S R, Woodbridge E L, Tin P, Wilson J C, Jonsson H H, Dye J E, Baumgarten D, Borrmann S, Toohey D W, Avallone L M, et al., 1993, In situ measurements constraining the role of sulphate aerosols in mid-latitude ozone depletion, Nature 363 509-514.

Farman J C , O'Neil A, and Swinbank R S, The dynamics of the Arctic polar vortex during the Easoe campaign, GRL.

Fisher D A, Hales C H, Wang W C, Ko M K W, and Sze N D, 1990, Model calculations of the relative effects of CFCs and their replacements on global warming, Nature 344 513-516.

Frederick J E, and Weatherhead E C, 1992, Temporal changes in surface ultraviolet radiation: a study of the Robertson-Berger meter and Dobson data records, Photochem. Photobiol. 56 123-131.

Hofmann D J, Oltmans S J, Harris J M, Solomon S, Deshler T, and Johnson B J, 1992, Observation and possible causes of new ozone depletion in Antarctica in 1991, Nature 359 283-287.

Johnston P V, McKenzie R L, Keys J G, and Matthews W A, 1992, Observations of depleted stratospheric NO2 following the Pinatubo volcanic eruption, Geophys. Res. Lett. 19 211-213.

Kerr J B, and McElroy C T, 1993, Evidence for large upward trends of ultraviolet-B radiation linked to ozone depletion, Science 262 1032-1034.

Khalil M A K, Rasmussen R A, and Gunawardena R, 1993, Atmospheric methyl bromide: Trends and global mass balance, J. Geophys. Res. 98 2887-2896.

Krueger A J, Schoeberl M, Newman P, and Stolarski R, 1992, The 1991 Antarctic ozone hole: TOMS observations, Geophys. Res. Lett. 19 1215-1218.

Labitzke K, and McCormick M P, 1993, Stratospheric temperature increases due to Pinatubo aerosols, Geophys. Res. Lett. 19 207-210.

MBSW, June 2-3 1992, Methyl Bromide Science Workshop, proceedings (available from Atmospheric Environmental Research Inc., Cambridge, M A) (Washington DC).

Neuber R, Beyerle G, Fiocco G, di Sarra A, Fricke K H, David Ch., Godin S, Knudsen B M, Stefanutti L, Vaughan G, and Wolf J P, Latitudinal distribution of stratospheric aerosols during the EASOE winter 1991/92, Geophys. Res. Lett. to appear.

NOAA, 1992, Southern Hemisphere Winter Summary 1992: Selected indicators of stratospheric climate, NOAA Climate Analysis Centre.

NOAA, 1993, Northern Hemisphere Winter Summary 1992/93: Selected indicators of stratospheric climate, NOAA Climate Analysis Centre.

Oort A H, and Liu H, 1993, Upper-air temperature trends over the globe, 1958-1989, J. Climate 6 292-307.

Penkett S A, Jones B M R, Rycroft M J, and Simmons D A, 1985, An interhemispheric comparison of concentrations of bromine compounds in the atmosphere, Nature 318 550-553.

Perner D A, Roth A, and Klupfel T, Ground-based measurements of stratospheric OClO, NO2 and O3 at Sondre Stromfjord in winter 1991/92, Geophys. Res. Lett., to appear.

Poulet G, Pirre M, Maguin F, Ramaroson R, and LeBras G, 1992, Role of the BrO + HO2 reaction in the stratospheric chemistry of bromine, Geophys. Res. Lett. 19 2305-2308.

Reeves C E, and Penkett S A, 1993, An estimate of the anthropogenic contribution to atmospheric methyl bromide, Geophys. Res. Lett. 20 1563-1566.

Singh H B, Kanakidou M, 1993, An investigation of tth atmospheric sources and sinks of methyl bromide, Geophys. Res. Lett. 20 133-136. Singh H B, Salas L J, and Stiles R E, 1993, Methyl halides in and over the Eastern Pacific (40oN—32oS), J. Geophys. Res. 98 3684-3690.

Solomon S, Saunders R W, Garcia R R, and Keys J G, 1993, Increased chlorine dioxide over Antarctica caused by volcanic aerosols from Mount Pinatubo, Nature 245-248.

Stowe L L, Carey R M, and Pellegrino P P, 1992, Monitoring the Mt. Pinatubo aerosol layer with NOAA/11 AVHRR data, Geophys. Res. Lett. 19 159-162.

Traub W A, Johnson D G, Jucks K W, and Chance K V, 1992, Upper limit for stratospheric HBr using far-infrared thermal emission spectroscopy, Geophys. Res. Lett. 1651-1654.

Wang J - L, 1993, Global Concentrations of Selected Halocarbons, 1988-1992, dissertation, University of California (Irvine CA).

Webb A R, 1992, Spectral measurements of solar ultraviolet-B radiation in southeast England, J. Applied Meteorology 31 212-216.

WMO, 1989, Scientific Assessment of Stratospheric Ozone: 1989, WMO Global Ozone Research and Monitoring Project, Report No 20.

WMO, 1991, Scientific Assessment of Ozone Depletion: 1991, WMO, Global Ozone Research and Monitoring Project, Report No 25.

# Atmospheric Acoustic Sounding


C G Helmis

University of Athens
Greece


## 1 Atmospheric acoustic sounders

The development of acoustic sounding of the lower atmosphere proceeded rapidly after the early work of Little(1969). The principle of the technique of monostatic acoustic sounding is straightforward. A burst of sound is projected into the atmosphere in a well-collimated beam using a suitable acoustic antenna. Antenna commonly used include an array of loudspeakers, a single loudspeaker at the focus of a reflecting dish or sometimes a simple horn loudspeaker. The antenna is switched to the listening mode after transmission when the loudspeakers have ceased ringing significantly. Received echoes are amplified and displayed, for an immediate visual record, in height-time form on a facsimile chart recorder, the intensity of the record giving an indication of the strength of the received echo. Since the same antenna is used for both transmission and reception only the back-scattered sound is detected. The transmitted frequency is normally chosen as a compromise between the effect of the atmospheric absorption of sound, which increases at higher frequencies, and ambient noise.

Facsimile records have been produced by different sounders operating at four frequencies between 820Hz and 6.5kHz (Asimakopoulos and Cole 1977). Transmitted acoustic powers from these systems range between 30W at 820Hz and 1–5W at 6.5kHz with pulse lengths varying between 5ms and 250ms corresponding to height resolutions from about 1m to 35m.

Acoustic sounding can offer a method of remotely estimating important turbulence quantities, such as the structure parameters for wind ($C_V{}^2$) and temperature ($C_T{}^2$) at any chosen height (Tatarski 1961), since the eddies producing the scattered sound are of a length scale found within the inertial sub-range of atmospheric turbulence. Additionally the echoes provide information on the wind structure in the boundary layer through the magnitude of the Doppler shift. In general terms the monostatic sounder will record any atmospheric structure having an associated fine scale turbulence which generates significant refractive index fluctuations at acoustic wavelengths, on a scale approximately equal to that of the transmitted sound half wavelength, *i.e.* between about 2.5cm and 20cm for the range of sounders used in this work. Thus it will readily indicate the presence of ground-based thermal activity, synoptic and nocturnal inversions, breaking waves and the turbulent thermal structure associated with fog. Examples of these features are discussed in the following sections.

Certain factors must be borne in mind when one attempts to relate the structure indicated on a monostatic sounder facsimile chart to the turbulent activity in the atmospheric boundary layer. The chart illustrates only that part of the atmosphere being advected above the sounder antenna, hence the sounder provides only a 'height-time' section through any structure and so inferences about the three-dimensional shape cannot be attempted nor can the Lagrangian evolution of any features of interest be usefully discussed. The correspondence between the chart record and the actual structure will depend on the gain and the dynamic range of the system. For example, the location of the edges of a thermal plume may be ill defined. Recent experimental work has shown that the intensity of small scale temperature fluctuations falls off rapidly with height. This means, for example, that a monostatic sounder, typical of those in general use, will significantly underestimate the upper limit of convection in deep boundary layers. Indeed, an examination of the charts reveals an apparent 'fading' of turbulent thermal activity with height (Kaimal *et al.* 1976).

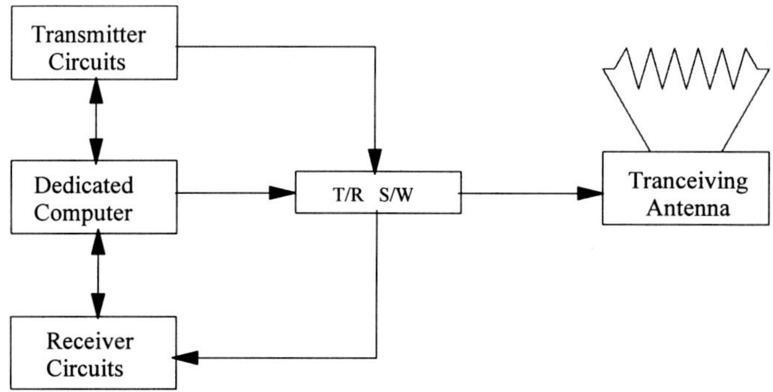


**Figure 1.** *Simplified block diagram of a typical acoustic radar system*

## 1.1   Acoustic sounder design

The considerations involved in the design of acoustic sounders have been discussed by a number of authors (*e.g.* Asimakopoulos and Cole 1977). A simplified block diagram of a typical acoustic radar (AR) system and almost of any active radar system is shown in Figure 1.

- The transmitter unit involves all the power systems part of the digital circuitry.

- The receiver unit incorporates all filtering, mixing and gating systems.

- The transmit/receive (T/R) switch is over simplified here but incorporates all gating and other protection hardware of the complete radar.

The transmitted frequency and all receiver control signals are derived by the dedicated computer. The transmit gate which is also produced is used to generate a tone burst of the required duration which is amplified and applied to the antenna which produces a narrow beam of high acoustic intensity.

In the case of the monostatic sounder, a common transmit/receive antenna is used and a passive T/R switch is employed to protect the sensitive receiver circuits from the high power transmit signal. The receiver also incorporates an active gate which opens after the 'dead time' resulting from the antenna ringing after transmission.

The received signal is amplified and filtered to remove out-of-band noise components and is detected using a precision rectifier having a dynamic range of about 50dB. The output of the rectifier can be recorded for quantitative analysis and can also be used to provide a visual display, in this case on a facsimile recorder. A time varying gain is used to compensate for the loss of amplitude resulting from spherical divergence of the acoustic echoes as they are received from longer ranges. The facsimile recorder helix drive is also derived from the crystal oscillator.

In order to produce detectable echo signals a high acoustic transmit power is desirable. This can be achieved with high electrical power to the antenna, implying high power handling capability and high antenna efficiency. For a high receive power it is necessary to use an antenna with a large collecting area and to operate at frequencies at which the atmospheric absorption is small, acoustic absorption increasing rapidly with frequency. However, if the antenna is too large the beam width becomes too narrow resulting in significant sound energy being lost due to refraction by the wind.

It is usually possible to construct amplifiers with noise levels well below the environmental noise contribution, *e.g.* wind, traffic, aircraft, etc. However, in very quiet environments, the thermal noise in the receiver transducer may impose the limit on sensitivity. The intrusion of environmental noise can be minimised by good antenna design, *e.g.* low sidelobes and screening. Often a low noise pre-amplifier is used in close proximity to the antenna to eliminate the problems caused by electrical pick-up when long cables (50–200m) are used to connect the antenna to the receiver. The pre-amplifier incorporates a step-up transformer to match the antenna impedance of a few ohms to the input impedance of the receiver, a few kilo-ohms.

In the design of the receiver electronics the most important requirement is to separate the wanted narrow band acoustic signal from strong, usually broad-band, noise signals and thus a narrow band filter is necessary. However, a filter bandwidth of at least a few per cent of the carrier frequency is usually required because of wind induced Doppler shifts and turbulent spectral broadening.

Most conventional AR use a carrier frequency in the region of 1–2kHz with maximum useful ranges from a few hundred metres to a few kilometres. Operation at higher frequencies, greater than about 4kHz, has a number of advantages. For example, external noise sources, such as traffic, often have little acoustic output at these frequencies which allows operation in a noisy town environment with a minimum of acoustic screening. Also the antenna is physically small which, besides the obvious advantage of being easily portable, means it has a short reverberation time, *i.e.* a short 'dead time' after transmission which allows sounding down to a few metres from the ground. However, these advantages are at the expense of reduced range because of the greater atmospheric acoustic attenuation at these frequencies.

## 1.2 Acoustic sounder antenna

The main limitations in the performance of an AR often depend to a great extent on the characteristics of the antenna, since present-day electronic techniques are capable of achieving almost ideal performance in a sounding system. However, in typical sounders, there remains considerable scope for improvement in antenna design. The basic requirements for an effective monostatic antenna are as follows.

- Narrow main beam to enable probing of a well defined atmospheric volume.

- Low side lobe levels to minimise reflections from nearby solid objects and to reduce reception from external noise sources.

- High transmitting efficiency to allow high acoustic outputs using acceptable electrical input powers.

- High receiving efficiency (in many cases this is equal to the transmitting efficiency) to avoid degradation of the received signal to noise ratio by receiver noise.

In quantitative sounding it is essential to know the polar diagram and the efficiency of the antenna. For Doppler wind measuring systems it is desirable that the antenna characteristics are known in order that measurement errors may be estimated (Owens 1975).

# 2  Applications of acoustic sounding

To understand better the atmospheric properties near to the ground, we will present the most representative thermal structure cases which are also detected by conventional AR systems.

## 2.1 Thermal plumes

Thermal plumes are features present in convective conditions and have been observed by *in-situ* sensors for some time. They originate in the super adiabatic region near the ground and normally extend to the top of the convective boundary layer. Direct observations show that they are of great importance in the diurnal development of the boundary layer over land and are responsible for carrying aloft significant heat and moisture from the surface. Of particular importance in, for example, pollution studies, is the time of onset of convection and the rate of change of the depth of the convective layer. Similarly, in the evenings when convection dies out, turbulent mixing reduces dramatically and the ability of the atmosphere to disperse pollutants is much affected.

Considerations such as these illustrate that a system which can remotely record the state of convection over long periods of time is likely to be of considerable usefulness.

The first atmospheric acoustic sounders showed strong echo regions in the lowest few hundred meters which were associated with thermal plumes. Since the scattering cross-section involves the intensity of small scale temperature fluctuations which fall off rapidly with height, most sounders do not show the full height of a plume. Normally, however, the upper limit of convection is marked by an inversion and this can be taken as the depth of convection. Sounders can give a good indication of the strength of turbulence within a plume, its size in the horizontal plane as well as its speed and direction of movement, although rather complicated multi-antenna sounding systems are required to obtain all of this information. The determination of the plume diameter from the facsimile recording of a single sounder is not straightforward since it depends not only on the wind speed but also the direction of the plume motion relative to the sounder. However, plume diameters ranging from about 50 meters up to many hundreds of meters are observed. Within the plume boundaries, localised regions of greater temperature fluctuations sometimes occur which indicates a distinct plume substructure. Assemblies of plumes may be several kilometres in extent and take 10 to 15 minutes to pass through the sounder beam.

An important parameter in an acoustic sounding system is the transmit frequency. This determines not only the atmospheric absorption of the acoustic signal and hence the range of the system but is also related to the scale of the atmospheric turbulence which the sounder will see. Within the range of frequencies commonly used in acoustic systems this scale is only marginally different. The plume record depends on the individual characteristics of the sensing remote device, *i.e.* frequency, power transmitted, antenna efficiencies and receiver gains etc.

Normally AR systems are used in the monostatic or bistatic (*i.e.* two separated antenna are used and off-axis sound is detected) configurations, or more recently in the triple monostatic (tristatic) mode; with the antennas being directed vertically or at some fairly large angle to the horizontal. This is necessary since spurious echoes from nearby objects would be received and mask the atmospheric returns. However the recently developed mini-sounders operating at higher frequencies and using small antenna arrays are suitable for mounting on a tower and can be directed at angles close to the horizontal, so providing the opportunity to study the plumes in two different planes, *i.e.* horizontally and vertically. These systems are also lightweight and can operate under heavy noise conditions, (near airports, close to wind turbines, in city centres etc).

## 2.2  Temperature inversions

One of the most useful applications of acoustic sounding may well prove to be the monitoring of inversions within the lowest few kilometres of the atmosphere. These are regions in which the temperature increases with height contrary to the normal decrease typical of the troposphere. They are generally associated with anticyclones (regions of subsiding air) and may extend over many hundred or thousands of kilometres in the horizontal plane whilst their vertical scale is usually of the order of several hundred metres, although sharp layers may exist in which the temperature can increase by as much as 5K in less than one metre.

Inversions are of great meteorological significance since their presence may indicate preferred regions for strong wind shear which can be of importance to aviation. They can also sometimes limit the growth of the convective boundary layer and hence influence such parameters as maximum daytime temperature, visibility etc. In some cases the height of the lowest inversion may be taken as the depth of the convective boundary layer which is an important scaling parameter in the description of turbulence in the lower atmosphere and is relevant to the dispersal of pollutants. Further, they can effectively impose a barrier to the transfer of heat, moisture and pollutants from the surface to the free atmosphere and thus may critically affect the formation of cloud and the air quality of the near surface layer. Inversions also frequently contain strong gradients of humidity and can therefore seriously affect radio wave propagation. Clearly the ability to monitor the height and characteristics of inversions remotely with surface based instruments would be of great value.

Much work has been carried out to investigate the mechanisms by which entrainment may occur, but the resultant effect is that entrained air is carried well down into the boundary layer by the return flow between convective plumes. The acoustic record provides direct qualitative confirmation of this process. Entrainment usually results in a lifting of the inversion base and hence an increase in the boundary layer depth. However, in the presence of strong subsidence, the two may be in approximate balance, in which case no net increase in depth occurs.

Clearly, the sounder is capable of yielding very useful information on the presence and nature of temperature inversions and of confirming the occurrence of entrainment. As well as these important aspects, the sounder can be used to monitor the break up of inversions. This process can occur when the boundary layer continues to warm through daytime surface heating in which case the inversion temperature may be reached or exceeded. Convection then breaks through the inversion layer and continues upward to some higher level inversion or more generally until its buoyancy is lost. Acoustic sounding has considerable potential for monitoring the evolution and character of synoptic scale and local temperature inversions in the lower atmosphere. This capability will doubtless prove of benefit to meteorological and air pollution scientists.

## 2.3 Diurnal variation of the convective boundary layer

The depth of the surface based convective boundary layer ($Z_i$) is an important parameter for studies of atmospheric diffusion. A great deal of recently published work presents models which predict $Z_i$ and the structure of the convectively mixed layer (Tennekes 1973). However, little experimental data to verify this work is available. In general it is necessary to obtain information from heights greater than a kilometre in order to determine the depth and investigate the structure of the higher regions of the convective boundary layer and this implies the use of a very high power sounder. In some circumstances a lower power sounder can be used on any available high ground. Recent work in Greece used a typical commercial acoustic sounder, operating at 1600Hz with an effective range of 1000m, on the top of a 1000m high mountain overlooking

Athens. Operating from such a height, the AR has the range to monitor the real depth of the atmospheric boundary layer (ABL) and to identify the onset and cessation of convection as well as the diurnal variation in the mixed layer depth. This information would be invaluable in, for example, estimating the diffusion of pollutants in the lower atmosphere.

## 2.4 Stable conditions and waves

With a stable stratification the atmosphere can support wave motion and the interpretation of atmospheric data may well be complicated by the coexistence of waves and turbulence. This topic is particularly relevant to studies of the nocturnal boundary layer and, in order to understand the structure of the turbulence field in these conditions, it is necessary to eliminate the wave contributions from the data. Acoustic sounding offers an excellent method for remotely assessing the significance of wave motion in particular situations. Furthermore, the waves in many instances may become unstable and form patches of turbulence which may be the major contributors to turbulent transfer on these occasions. This is particularly true of areas which are nearly continuously stable such as the polar regions in winter. The most common form of wave-like activity in the boundary layer is undoubtedly Kelvin-Helmholtz instability. This occurs in layers where the static stability is much larger than average, *i.e.* where there are large gradients of temperature and humidity. They are produced predominantly by tilting of these very stable layers. Following initiation the billows develop and 'roll-up', then secondary instabilities form and the initial region of billows may collapse to form irregular turbulent motions containing a cascade of smaller eddies. The mixture of billows and turbulence is detected by the acoustic sounder since it generates easily significant temperature fluctuations on a scale of one half the acoustic wavelength.

A second type of wave motion which is common in the lower atmosphere is the so-called gravity wave (Eymard 1978). These waves may be triggered by disturbance within the stable environment. An important difference between gravity and Kelvin-Helmholz waves is that the former propagate through the air with significant phase velocity. Gravity waves may also become unstable and generate local patches of turbulence and hence it is difficult to distinguish between the two by simple inspection of an acoustic record. It is, however possible to deduce wave speeds and directions from a network of sounders and in this way the generating mechanisms could be distinguished.

Although turbulence levels are generally much lower in stable conditions the small vertical fluxes may produce large net transfers when maintained for long periods of time. Acoustic sounding provides a method for immediately assessing the state of turbulent motion in the stable boundary layer.

## 2.5 Acoustic sounding of clouds.

A potentially very important application of acoustic sounding within the ABL is to monitor the height of low level layer cloud or fog top. This could be obtained since a stratocumulus cloud formed much higher in the atmosphere is usually accompanied by strong gradients of temperature and humidity at its upper boundary. These changes can occur in remarkably short distances, *e.g.* in some recent studies it was found that the mean temperature increased abruptly by 5°K in less than 1m at the top of nocturnal stratocumulus. Since the cloud layer is turbulent, intense small scale variations in temperature are generated at cloud top and hence generate acoustic returns. Acoustic scattering has also been observed to occur at the boundaries of cumulus clouds. Shaw, 1971 convincingly demonstrated that convective clouds within the range of his sounder (1500m) were detected. Normally direct observations of cloud base are available and hence sounding can offer a simple and effective method of monitoring cloud depth (Gaynor *et al.* 1976).

The long range to the cloud top clearly limits the resolution with which the motion occurring may be resolved. This is because in order to achieve a reasonable echo level a very long acoustic pulse must be used (0.5s) and hence the vertical resolution is rather poor (84m). In this way even with low spatial resolution the highly contorted and variable cloud top is resolved.

## 2.6  Acoustic sounding of radiation fog.

A possibly important practical application of acoustic sounding is to assist in the measurement of the depth of radiation fog and the prediction of the time of its clearance. Present methods for estimating fog depth involve the measurement of temperature and humidity profiles using tethered balloon instrument packages. Profiles of potential temperature usually show a strong inversion associated with the fog top which is formed by the lifting of the surface-based nocturnal inversion following fog formation and continued radiation cooling from the fog top. With the ground radiatively shielded, a small super-adiabatic lapse rate can develop which may result in the establishment of a weakly convective regime within the fog, thereby transferring heat flux upwards from the ground and enhancing vertical mixing. Echo patterns obtained in radiation fog have typically demonstrated a strong echo layer overlying a region usually exhibiting evidence of weak convective activity. It has been suggested that the echo layer is generated by the interaction of the convective cells with sharp temperature gradients found in the vicinity of the fog top and also by fluctuations in temperature resulting from unstable internal waves forming in the inversion layer at the fog top.

## 2.7  Structure parameters and turbulence dissipation

**Monostatic sounding.**

Early work sought to demonstrate a correlation between the intensity of differential temperature fluctuations ($\Delta T^2$), measured across a distance of 0–12m, *i.e.* an acoustic half wavelength, and the intensity of back scattered sound. More detailed studies required the accurate calibration of the acoustic system in the field and also the verification that the turbulence field was such that the simplified version of the scattering cross-section involving only $C_T{}^2$ and $C_V{}^2$ was justified, (Haughan and Kaimal 1978).

It is simplest first to consider a simple vertically directed monostatic sounder detecting back-scattered sound and hence the contribution to the scattering cross-section from wind fluctuations is zero and the precise measurement of echo strength may be used to estimate $C_T{}^2$ . Measurements were made in two distinct atmospheric conditions:

1. Stable conditions at night with moderate wind speeds.

2. Strongly convective conditions with well defined thermal plumes.

## Bistatic sounding.

More recent studies investigated the feasibility of using off-axis scattered sound, *i.e.* a bistatic sounder configuration, to estimate $C_V{}^2$ and hence the rate of dissipation of turbulent kinetic energy ($\epsilon$). These experiments have been described in detail by Caughey *et al.* (1978). The vertically directed array was used in both transmit and receive modes while a second array was used only to receive.

## Tristatic sounding

When combining three monostatic systems synchronised together or a genuine tristatic system with three tranceiving antennae, we have the benefit of interrogating the atmospheric properties in three dimensions. It is typical then to obtain the three components of the wind and three dimensional 'pictures' of the target area. Naturally in this case there are very many applications but significantly higher manufacturing costs.

# References


Asimakopoulos D N and Cole R S, 1977, An acoustic sounder for the remote probing of the lower atmosphere. *J Phys E (Sci Instrum)*, **10** 47

Asimakopoulos D N, Helmis C G and Stephanou G J, 1987, Atmospheric acoustic minisounder, *J Atmos Ocean Technol* **4** 345

Asimakopoulos D N, Moulsley T J, Helmis C G, Lalas D P and Gaynor J, 1983, Quantitative low-level acoustic sounding and comparison with direct measurements, *Boundary Layer Met,* **27** 1

Barrett E W and Ben-Dov 0, 1967, Application of the LIDAR to air pollution measurements, *J Appl Met,* **6** 500

Caughey S J, Crease B A, Asimakopoulos D N and Cole R S, 1978, Quantitative acoustic bistatic sounding of the atmospheric boundary layer, *Q J R Met Soc,* **104** 147

Deardorff J W, 1972, Numerical investigation of neutral and unstable planetary boundary layers, *J Atmos Sci* **29** 91

Eymard L, 1988, Ondes de gravité dans la couche limité planetaire étude experimentale par sondage acoustique. Note 54, Centre National d'études des Télécommunications, France

Francis R, 1992, Radar altimeter measurements, *Earth Observation Quarterly,* Nos. 35-36

Gaynor J E, Mandics P A, Wahr A B and Hall F F, 1976, Studies of the tropical marine boundary layer using acoustic backscattering during GATE, In Reprints of the 17th Radar Meteorology Conference, 303-6, American Meteorological Society, Boston, Mass

Grams G W, 1974-75, LIDAR - some current uses and potential applications in the atmospheric sciences, *Atmos Technol,* **6** 61

Hogstrom U, Asimakopoulos D N, Kambezidis H, Helmis C G and Smedman A, 1988, A field study of the wake behind a 2 MW wind turbine, *Atmospheric Environment,* **22** 803

Kaimal J C, 1978, NOAA instrumentation at the Boulder atmospheric observatory. Presented at 4th Symposium, Meteorological Observations and Instrumentation, Denver, Co, American Meteorological Society

Kaimal J C, Wyngaard J C, Haugen D A, Coté O R, Izumi Y, Caughey S J and Readings C J, 1976, Turbulence structure in the convective boundary layer, *J Atmos Sci,* **33** 2152

Little C G, 1969, Acoustic methods for the remote probing of the lower atmosphere, *Proc IEEE,* **57** 571

Lutz H and Armandillo E, 1991, Laser-based remote sensing from space, *ESA Bulletin,* 73-78

May P T, Koran K P and Strauch R G, 1989, The accuracy of RASS temperature measurements, *J Appl Met,* **28** 1329

Owens E J, 1975, NOAA Mk VII acoustic echo sounder. Tech. Memo. ERL WPL-12, NOAA, Boulder, Colorado

Peters G, Hinzpeter H and Baumann G, 1985, Measurements of heat flux in the atmospheric boundary layer by sodar and RASS: a first attempt, *Radio Sci,* **20** 1555

Peters G, Timmerman H and Hinzpeter H, 1983, Temperature sounding in the planetary boundary layer by RASS, system analysis and results, *Int J Remote Sensing,* **4** 49

Readings C J and Butler H E, 1972, The measurement of atmospheric turbulence from a captive balloon, *Met Mag,* **101** 286

Shaw N A, 1971, Acoustic sounding of the atmosphere. Ph.D. dissertation, Roy Aust Air Force Acad, University of Melbourne, Melbourne, Australia

Shermann J W III, 1991, The near-term ensemble of satellite remote sensors for earth systems monitoring, *Proceedings of the 5th AVHRR Data Users Meeting,* Tromso, Norway, 25-28 June, p.161-168

Tatarski, V I, 1961, Wave Propagation in a turbulent medium. Translated by R.A. Silver, McGraw-Hill, N.Y.

Tennekes H, 1973, A model for the dynamics of the inversion above a convective boundary layer, *J Atmos Sci* **30** 558

Weill A, Klapisz C and Baudin F, 1986, The CRPE mini sodar: applications in micrometeorology and in the physics of precipitations, *Atmos Res,* **20** 317

Wyngaard J C, Cote' O R and Rao K S, 1974, Modelling the atmospheric boundary layer. *Advances in Geophysics,* Aca- demic Press, **18A** 193

# Meteo Disc: An Interactive Meteorological Videodisc

Guenter Warnecke, Christian Zick
and Brian Toussaint

Freie Universität and Technische Universität
Berlin

## 1 Introduction

Planet Earth is a highly complex dynamic system. The atmosphere and, to some extent the oceans, are its most dynamic subsystems. It is appropriate, therefore, to search for ways and means of visualising their dynamics. Viewing the earth from space, remote sensing from orbiting space platforms provides a favourable perspective for the observation of dynamic processes. Besides their reflection in certain characteristic structural features, such as wave patterns, vortices, spirals, *etc.* in single pictures, information on the dynamics is, of course, predominantly contained in time series of various data sets. Such image sequences become noticeable as pattern changes or object displacements. Many of these displacements and pattern changes are usually related to motion. Others, however, are related to growth, expansion, *etc.* , depending on the nature of exhibited objects or processes .

Since satellite data are usually characterised by their tremendous volume, one way to present such data for visual inspection is achieved economically by displaying the data as images. Because, in addition, many satellites provide images of an area at appropriate repetition rates, cinematographic methods can be applied to visualise the dynamics as movie-like motion scenes. Well-known examples include television presentations of animated half-hourly images from geostationary weather satellites. This technique was first applied to satellite image data and by Suomi and Fujita (1967). Since then, motion scenes from satellite image sequences have become rather common; they are well accepted even in public as a regular component of most televised daily weather presentations.

Thus, motion scenes from satellite image sequences represent a new quality of visual

meteorological and oceanographic observation. This fact is due to
  - the extraordinary perspective (viewing from space),
  - being four-dimensional (space and time)
  - exhibiting most scales of motion superimposed (from mesoscale to hemispheric or global scales).

These three points lead to the visualisation of *dynamic processes*, as characterised by *motions* and *pattern changes*, in their real natural appearance. In particular, motion scenes from satellite image sequences exhibit the
  - formation,
  - growth,
  - displacement,
  - decay,
  - dissipation of textures, patterns and dynamic structures.

In our case, this means  motion and  structural changes of
  - clouds and cloud systems,
  - sea surface temperature fields,
  - snow and sea ice coverage,
  - changes of land surface characteristics, *etc.*

These facts thus imply that satellite image sequences provide, by far, much more information than on mere displacement, *i.e.* horisontal motions, only.

For scientific purposes, quite a number of existing 'classical' cinematographic animation techniques have been adapted, or have been further developed, into a variety of useful methods of motion display manipulation (Warnecke and Zick 1981, Warnecke 1987, 1991). These techniques take into account that the viewed atmospheric scenes are composed of a large variety of atmospheric-oceanic processes covering a wide spectrum of temporal and spatial scales. Motion display manipulation can then be used to filter time and space scales. It is possible to suppress or to emphasize certain scales, in order to optimize the perception of selected phenomena. Without such filtering, the complexity of motion patterns and processes makes analysis by visual observation rather difficult in many cases, and sometimes even impossible.


## 2   Animation and manipulation of image sequences

As mentioned above, a well-known method to display image sequences, and thus to visualize dynamical processes, is provided by the cinematographic principle. If the contents of consecutive images in an image sequence are well correlated, a display rate of approximately 20 images per second yields the illusion of continuous motion to a human observer. Correlation, in this context, means rather continuous optical flow, *i.e.* 'smooth changes' from image to image. In the case of image sequences from earth oriented satellites, this display rate usually introduces long time lapse. Considering the natural time scales involved, such time lapse favours the perception of atmospheric and oceanic changes, even in the case of extremely complex dynamics.

Three basic abilities of the human visual system favour visual observation of complex dynamic scenes.

**Figure 1.** *Letter 'A' in various printing styles ( from Letraset Catalogue)*

1. Since we are able to position objects at a high degree of accuracy, we are very sensitive to displacements of objects within animated time sequences of images.

2. We recognize defined patterns, even if they are distorted or spurious if the context of the scene is defined (Figure 1). Knowing that a capital letter of the Roman alphabet is reproduced, everybody familiar with this alphabet will immediately recognize letter 'A' in each of the figures printed in Figure 1. This recognition of a capital 'A' occurs although there are obviously no common rules for it. We are able to perform instantly the required quick 'coding' of patterns, textures and events. Coding, in this context, refers to the build-up of 'signs' and 'super signs', in terms of pre-described models. It allows the recognition and the pursuit of complex change patterns and dynamic processes in complex dynamic scenes.

3. Unconsciously, and very effectively, we are able to interpolate optical flow integratively in space and time , even across a large scene. This ability allows the perception of complete, continuous flow patterns.

Beyond this well-accepted, common empirical knowledge, specific anthropo-technical research has recently provided experimental evidence for the superiority of motion scenes over still pictures in the comprehension and understanding of dynamic processes. The superiority increases with growing complexity of observed dynamic phenomena (Faber *et al.* 1990, 1991, 1992 and Ruschkin 1992).

Thus, we seem to have good and sufficient reasons to consider motion scenes from image sequences as a powerful tool of visual observation, particularly in earth sciences. This tool is even more useful for displaying image data remotely sensed aboard space platforms.

Visualizing dynamic scenes for visual observation by displaying image sequences requires considering two kinds of boundary conditions. This requirement is due to the extremely high degree of complexity of most satellite scenes. Each one of these conditions suggests a number of possible manipulations in order to allow extraction of useful visual information on the dynamical aspects of interest. Specifically we must take into account (a) structural characteristics, *i.e.* the specific time and space scales of natural objects and processes and (b) certain characteristics of the human vision system.

Let us consider the second item first. According to a model suggested by Frank 1969 (see Figure 2), the data received by the eye at a bit rate of 106 bits per second

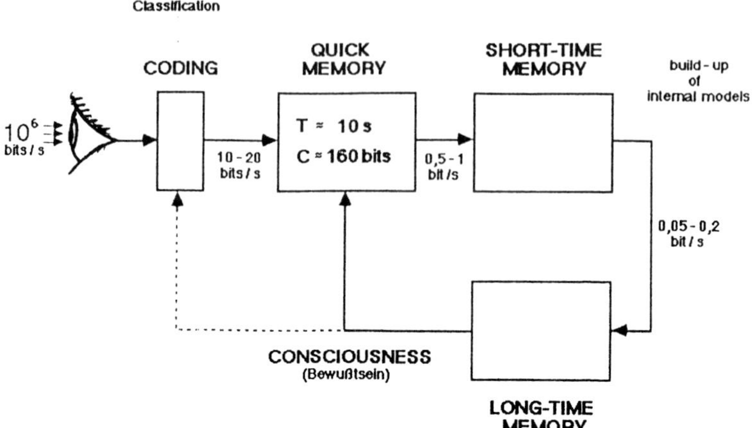



**Figure 2.** *Simplified model of data reception, data flow, data storage and information processing within the human vision system. T = storage time, C = storage capacity. (After Frank 1969)*

is internally transmitted into a hierarchy of three serial memories in our brain. Each one is defined by different capacities and functions. The significantly lower level of data flow capacities of the provided internal transfer channels is the main, surprising characteristics of this transfer system, if compared with the high initial optical inflow rate. Although a pronounced bottleneck obviously exists at the very beginning, transfer rates decrease even further with each further transfer step. The indicated 'coding' is an important, selective step for accomplishing the required high, initial data reduction without much loss of relevant information potential. This coding is commonly achieved by transforming selected meaningful or interesting groups of data into more easily transferable data, *i.e.* 'signs' and 'super-signs'. From this model, we can derive and understand three practical consequences:

- First, one well established method of displaying motion scenes is through loops. They permit an unlimited repetition of scenes. Applying the loop display mode can be interpreted as a plausible measure to accomplish the transfer of sufficient amounts of bits from our 'quick memory' into a 'short-time-memory', and finally into the long-term-memory. The effect of repetition will compensate for the low and, with each step, further decreasing bit rates.

- Second, still frames at the beginning or at the end of a highly complex motion scene allow for instantaneous visual comprehension of the image contents. Thus, these still frames favour a reasonable and effective coding, *i.e.* the necessary build-up of signs and super-signs.

- Third, particularly in the case of animated display of meteorological or oceanographic satellite image sequences, *time lapse* will favour the detection of changes. The limited storage time for coded image data of our 'quick memory' allows us to perceive dynamic objects and patterns only at a sufficiently high rate of apparent changes.

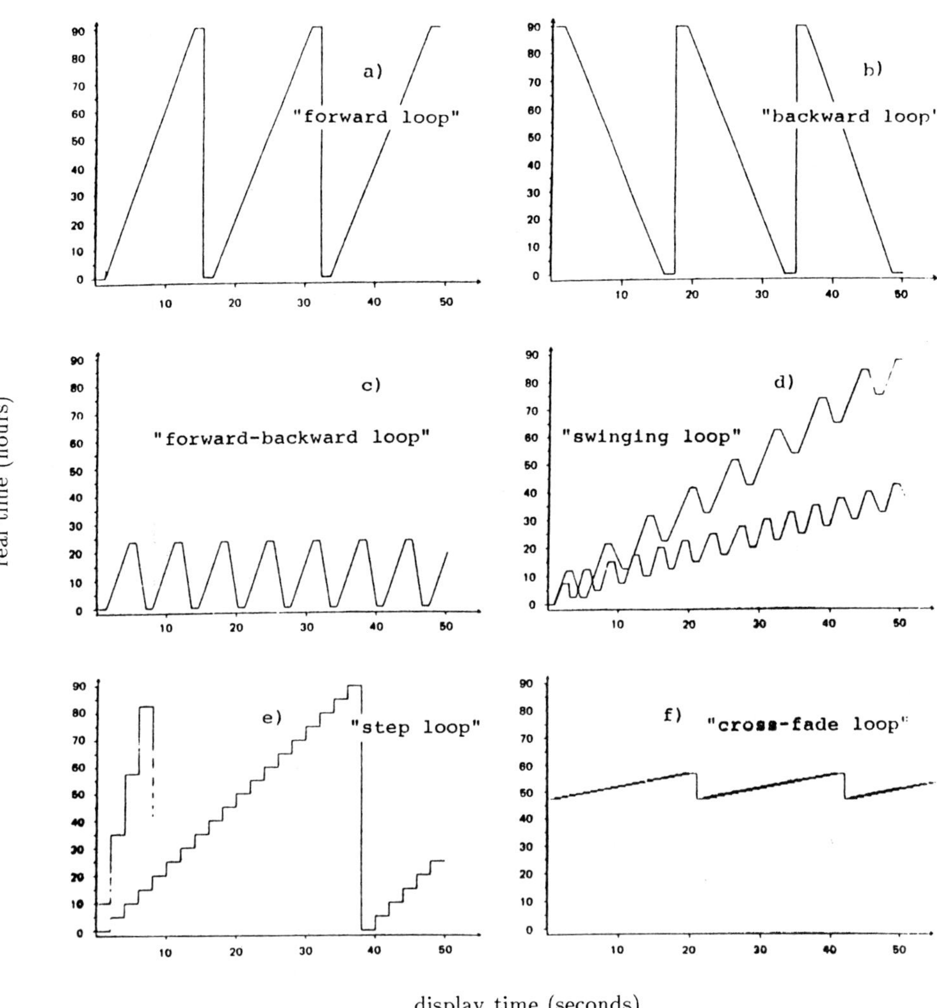


**Figure 3.** *Real time versus display time for different loop modes*

Frequently, it may happen that pronounced structures within an image sequence develop from rather fuzzy or completely unorganised patterns. When motion scenes are displayed in original sequence (*forward mode*, see Figure 3a), it is very hard to detect and to follow the early stages of pattern organisation. This fact suggests displaying the scene with time reversed (*backward mode*, see Figure 3b). The sudden jump from the end of the scene to the beginning of a looped image sequence usually appears as rather brutal. Within the early parts of a repeated sequence, it is also often hard to locate and to re-identify previously viewed objects or patterns of interest. The consequence requires using a *forward-backward mode* (see Figure 3c) which allows continuous eye tracking of individual features.

At times, the observed data stream of an image sequence appears too superfluous

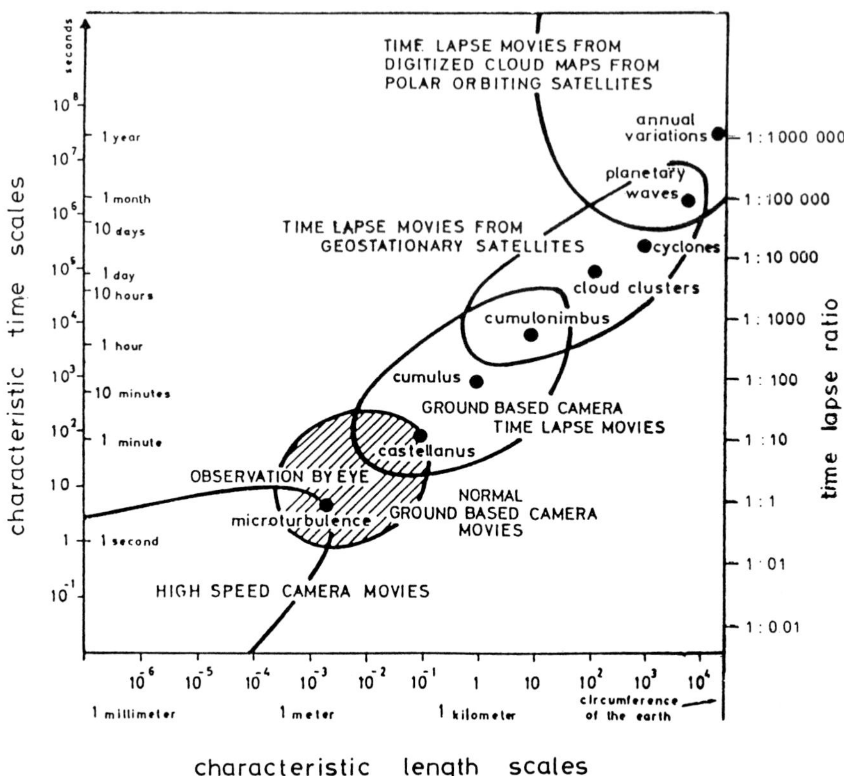


**Figure 4.** *Relation between atmospheric scales of motion, appropriate technical systems for visualisation by motion pictures and appropriate time lapse ratios.*

to allow a real gain of useful information. Mostly this happens due to the length of the sequence and a simultaneously high degree of complexity in the dynamic scenery. For appropriate visual perception such situations will also require specific manipulations. In these cases one can take advantage of considering the other kind of boundary conditions mentioned before. They are introduced by the well-known fact that atmospheric and oceanic dynamics cover a wide spectrum of temporal and spatial scales (see Figure 4). As known from experience, considerable improvement of visual perception can be achieved by proper attention to the specific time and space scales of observed dynamic objects and events. One useful step, if applicable, can be a reduction of the covered area. More remarkable gains, however, can usually be observed by carefully tuning the time lapse ratio of display to the time scales of the phenomena of interest (see, for example, Figure 1 and Figure 5).

In cases of a longer life-span of exhibited processes, the wealth of detailed events frequently exceeds the observer's capacity for visual perception. Here also a special manipulation techniques is available. The so-called *swinging loop* mode (see Figure 3d) allows one to observe shorter time-spans of development repeatedly whilst proceeding through the entire sequence. This is achieved by a modified forward-backward-mode with the

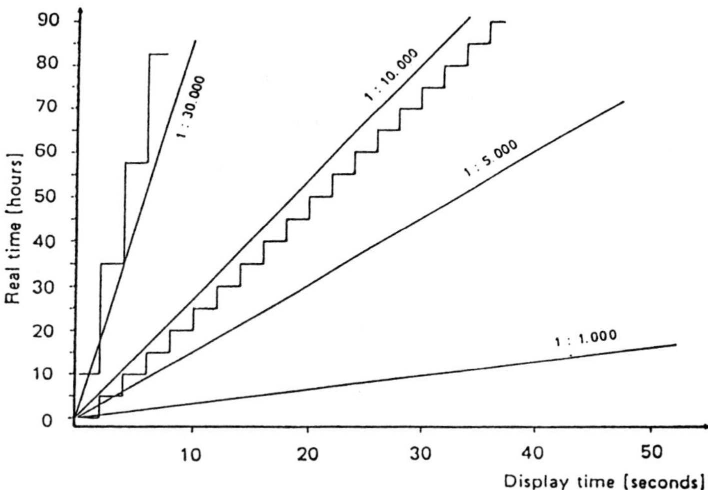


**Figure 5.** *Real time versus display time for displayed smooth and stepwise motions at different time lapse ratios.*

number of backward steps being always less than those forward. This loop mode allows the observation of developments in detail and of their modifications throughout the entire life time of a process as well.

Occasionally, the horizontal migration of atmospheric objects may exceed by far its own dimensions. If, in such a case, the image size had initially been tuned to the spatial dimensions of the studied object, it may soon move out of the displayed geographic area. Thus, it becomes useful to make the image position following the migrating object. In this so-called 'relative motion loop', which conserves the image size, the exhibited motions apply relative to the moving object. By this manipulation, the translation velocity of the observed object is, at least to a far extent, eliminated. This important loop mode usually provides a remarkable visual enhancement of all non-translatory kinematic properties of the exhibited motion fields, thus visually emphasizing predominant effects of divergence, rotation and deformation on cloud motions.

Further cinematographic manipulations are applicable. For example, the cross-fade technique slows down and smoothes image sequences (see Figure 3f). This mode is particularly useful when the time interval between images is too coarse in relation to the time scale of exhibited dynamic phenomena. For example, this mode has been successfully applied to sea ice drift from 12 hourly NOAA images or to local thunderstorm developments in GOES or METEOSAT imagery.

For long, interactive motion manipulation of satellite image sequences has usually been confined to rather costly image processing workstations and was mainly restricted to research applications. In the past, distribution of motion scenes generated for teaching and training purposes has had to rely on movie-film or video tape as data carriers. Producing quality motion scenes from meteorological satellite data on these carriers is rather costly and time consuming. For this reason, only a marginal number of such scenes has been distributed, compared to the enormous wealth of images available day

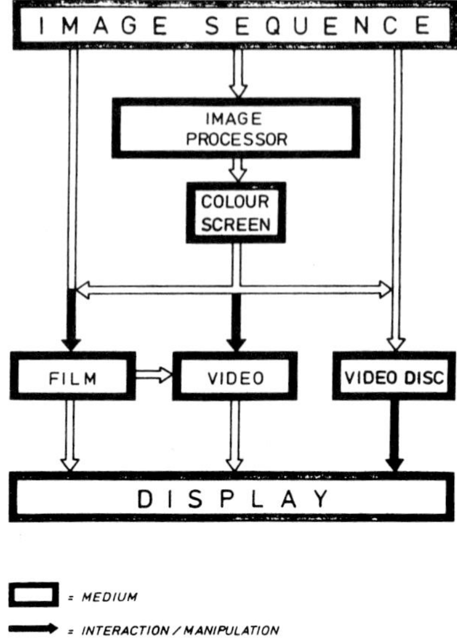


**Figure 6.** *Present procedures of motion scene manipulation for satellite image sequences. Heavy boxes indicate media, heavy arrows the crucial phases of interactive motion manipulation*

by day. Optimum display requires investigating the appropriate manipulation procedure, designing a film script, and generating the final image sequence to be filmed or to be copied onto video tape. Once the film or video has been made, however, it can only be displayed in its fixed linear form. Since motion manipulation and film script production are highly subjective processes, film and video products are unavoidably subjective, too. Interactive laser videodiscs, however, offer a valuable alternative.

## 3   Interactive videodiscs: METEO DISC

Interactive laser videodiscs are, in the first place, very potent carriers, of large amounts of analogue image data. Their capacity amounts to about 55,000 images on each side. Thus, laser videodiscs not only allow a much wider distribution of animated satellite image sequences than provided by film and video, they also utilise a much easier system of display. Secondly, interactive laser videodiscs allow the manipulation of linear image sequences, even after the discs have been produced. Freedom of motion scene manipulation is achieved by remote control or by computer linkage. Image sequences on a videodisc can even be generated and simultaneously manipulated during display (Figure 6). This possibility opens a new world of satellite image distribution and presentation. METEO DISC has been produced in order to demonstrate and prove these various possibilities, and to explore their widespread application.

METEO DISC has been produced by ZEAM/FUB, supported by EUMETSAT. METEO DISC is a CAV, *i.e.* constant angular velocity, videodisc in PAL standard, designed as a so-called resource disc. Besides a 10 minutes introductory sound movie 'Cloud Motions:

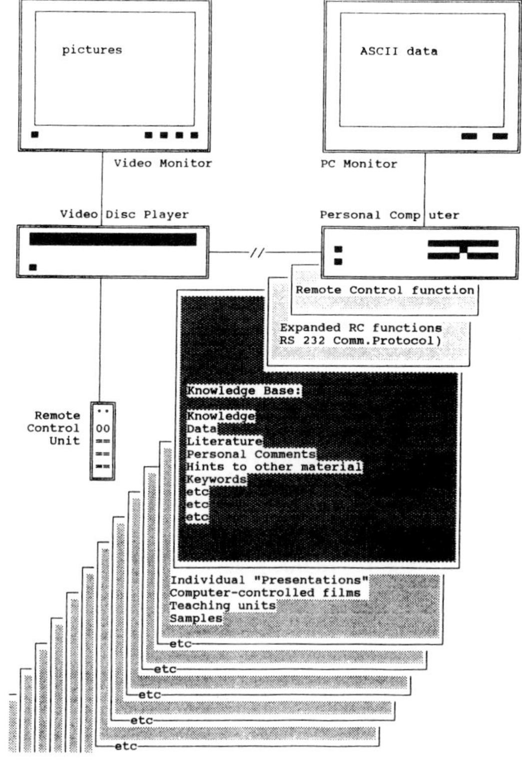



**Figure 7.** *The interactive video-disc display system*

Dynamic Scales and Display' (English or German by choice), METEO DISC contains about 42,000 further analogue satellite images. These pictures represent more than 600 satellite image sequences of primarily meteorological content. The total image content of about 55,000 is equivalent to a linear playing time of 37 minutes at normal video speed. Several scenes are also of oceanographic, others of geomorphological interest. Each image is only once on the disc. When some image sequences appear repeatedly, each one has been differently processed. Some of these differences pertain to contrast enhancement and pseudo-colour coding. Others refer to either absolute or relative motion modes. Images have been mainly chosen from northern hemispheric METEOSAT data. A number of scenes, however, have also been selected from ATS-3, SMS-1, GEOS-E, NOAA-7 and NOAA-9 (Arctic), and GMS-1 depicting New Zealand. A selection of the images available is illustrated in Plate 1 in the frontispiece of this book.

A standard remote control unit allows displaying scenes from videodiscs interactively. Using a computer to drive the videodisc player provides even more interaction. (Figure 7).

METEO DISC is divided into 40 chapters (Table 1). Its CAV format allows momentary access to each single frame or chapter. METEO DISC, therefore, can best be compared to a hypermedia document. One can easily and quickly navigate through the disc, picture

| METEO DISC INHALTSVERZEICHNIS METEO DISC CONTENTS |
| --- |

| Daten<br>Dates | | Ausgewählte Stichwörter<br>Selected key words | Kapitel<br>Chapter |
| --- | --- | --- | --- |
| Inhalt/ Contents | | | 01 |
| Stichwörter/ Key words | | | 02 |
| Film: WOLKENBEWEGUNG UND SCALES/ CLOUD MOTION | | | |
| AND SCALES | | von/by Christian Zick, Günter Warnecke | 03 |
| Satellitenbildsequenzen/ Satellite image sequences: | | | |
| 18 NOV + | 19 NOV 1967 | ATS-3 full disk scenes | 04 |
| 15 MAY | 1969 | Florida sea breeze circulation | 05 |
| 27 JUL - | 04 AUG 1974 | Sahara dust cloud crossing Atlantic | 06 |
| 11 AUG- | 14 AUG 1974 | N-Atlantic comma cyclone | 07 |
| 30 JUN | 1976 | USA, subsynoptic convective systems | 08 |
| 20 MAR- | 25 MAR 1978 | Dust, Africa, VIS/IR/WV comparison | 09 |
| 25 APR - | 26 APR 1978 | N-Africa, convective systems | 10 |
| 31 MAY- | 02 JUN 1979 | VIS/IR/WV, Atlantic cyclone, E-Europe | 11 |
| 27 SEP | 1979 | VIS/IR/WV single frame comparison | 12 |
| 27 FEB - | 28 FEB 1982 | Comma clouds, condensation trails | 13 |
| 04 SEP - | 07 SEP 1982 | IR/WV, Atlantic cyclone, cut-off | 14 |
| 20 JAN - | 25 JAN 1983 | Atlantic cyclone, trailing waves, dust | 15 |
| 24 MAR- | 31 MAR 1983 | NOAA-7, ice motion off Novaya Semlya | 16 |
| 04 AUG | 1983 | Kàrmàn vortex streets, relative motions | 17 |
| 05 AUG | 1983 | Kàrmàn vortex streets, relative motions | 18 |
| 06 AUG | 1983 | Kàrmàn vortex streets, convection | 19 |
| 22 AUG | 1983 | Kàrmàn vortex streets, gravity waves | 20 |
| 27 SEP - | 02 OCT 1983 | Mediterranean warm core cyclone | 21 |
| 05 NOV- | 10 NOV 1983 | Comma cyclone, Iberia wave clouds | 22 |
| 23 JAN - | 26 JAN 1984 | North Atlantic cyclone, relative motions | 23 |
| 25 FEB - | 02 MAR 1984 | Polar lows, cyclones over Europe | 24 |
| 12 MAY- | 15 MAY 1984 | North Sea surface temperatures | 25 |
| 12 JUL | 1984 | The Munich hailstorm, relative motions | 26 |
| 17 SEP - | 21 SEP 1984 | IR/WV, general circulation, cyclones | 27 |
| 27 MAR- | 31 MAR 1985 | VIS and IR, Sahara dust | 28 |
| 30 SEP - | 01 OCT 1987 | I.C.E., dissolving front over North Sea | 29 |
| 15 OCT- | 17 OCT 1987 | FRONTEX cyclone over Europe | 30 |
| 12 NOV | 1987 | FRONTEX cyclone over Europe | 31 |
| 13 MAY | 1988 | NOAA-9, ice/cloud motion off Greenland | 32 |
| 20 NOV- | 21 NOV 1988 | VIS/IR/WV, Egypt, Near East | 33 |
| 07 APR - | 09 APR 1989 | Atlantic cyclone, dust, Baltic Sea | 34 |
| 17 OCT- | 19 OCT 1989 | VIS/IR, I.C.E., condensation trails | 35 |
| 05 NOV- | 07 NOV 1989 | Cyclones over Europe | 36 |
| 09 NOV- | 16 NOV 1989 | Cyclones over Europe | 37 |
| 01 JUL - | 30 SEP 1990 | New Zealand and Tasman Sea winter | 38 |
| 25 JAN - | 12 MAR 1991 | VIS/IR Near East, smoke plumes | 39 |
| Danksagungen und Information/ credits and information | | | 40 |

**Table 1.** *Table of Contents of the interactive videodisc* METEO DISC

by picture, chapter by chapter, scene by scene, or in any of their combinations. Motion scenes from METEO DISC exhibit predominantly synoptic and sub-synoptic scale atmospheric structures and processes such as cyclones, cyclogeneses, jet streams, comma cloud developments and polar lows. Other scenes, for example, show Sahara dust storms, long-range transports of Sahara dust clouds, mesoscale circulations such as mountain and sea breezes, and intense convective systems like thunderstorms, particularly the famous Munich Hailstorm of July 1984. Special phenomena shown include solitary and gravity waves which, in most cases, can be identified and monitored by specific motion manipulation only. Other phenomena concern sea ice motions and oil soot plumes over the Near East. Each chapter contains a quick-look map survey and additional relevant information such as date, spectral channel, applied enhancement functions, and data source.

METEO DISC is accompanied by a booklet containing relevant weather maps for most of the scenes and by a set of diskettes containing PC user interfaces.

As mentioned before, videodiscs differ from film and video tape because the discs can be controlled interactively. For example, display of motion scenes from videodisc image sequences can be manipulated with respect to length, time lapse ratio (speed), still frame inclusions, repetition rate, loop mode, *etc.* This manipulation occurs simultaneously during display. It also allows optimum adaptation of display to the time scales of exhibited dynamic phenomena and processes, or to individual requirements of the observer. Some useful basic manipulations can already be achieved through remote control devices for common videodisc players. In addition, recommended PC-linkage offers even more possibilities for scene access, motion manipulation and presentation (Toussaint 1992). Relative motion and cross-fade loop modes, however, can not be produced by videodisc manipulation. Due to their specific nature, such image sequences have to be imported, already pre-processed, during the production of the disc.

The specific display modes of motion scenes selected for presentation, can be programmed in advance. The PC further allows the development and introduction of a scene-oriented knowledge base, which can be structured for dates, key words, literature, comments, and supplementary data concerning weather. Actually, the knowledge-base can store any kind of information as text. The individual user can edit the knowledge base, and exchange it on diskette, with other users. In this way, users can accumulate knowledge bases for archiving.

All these characteristics make the videodisc a data carrier of versatile applicability. It is an extraordinary tool for atmospheric, oceanic and terrestrial observation, research, education and training. METEO DISC provides access to such a large amount of satellite image sequences for the very first time, and at rather low cost. This disc supports the widespread use of large amounts of satellite image data, even in motion, at relatively little costs. The possibilities of interactive control are unprecedented. The system has a high potential for visualizing complex dynamic processes as a basis for visual image data and scene analysis.

During the Ninth METEOSAT Scientific Users Meeting (Locarno, 1992) the sponsors set aside a room for demonstrations of METEO DISC. Members of the meeting were able to see demonstrations of METEO DISC, discuss the advantages and disadvantages of laser videodisc technology, and actually use the system for scientific analysis. One guest speaker from Nigeria spontaneously incorporated METEO DISC into his talk concerning dust outbreaks in West Africa. He pre-programmed four motion scenes of Sahara dust storms within a very short time before his talk, and played them back during his presentation. The system was connected to an overhead projector that allowed the selected motion scenes to be displayed on the main screen in the auditorium. This process demonstrated the flexibility of the METEO DISC laser videodisc system.

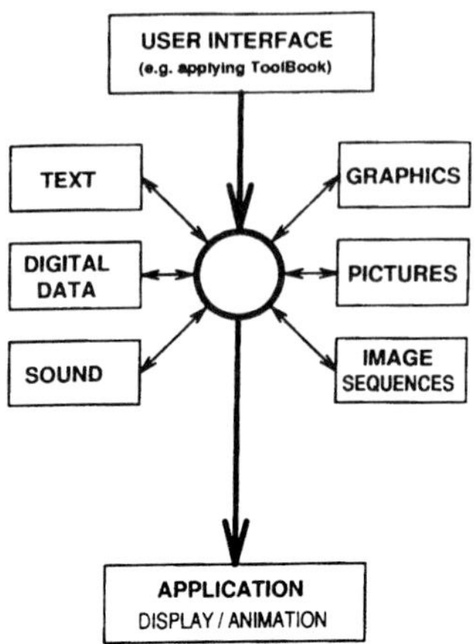


**Figure 8.** *User interface in a hypermedia environment*

# 4  Interactive use in a hypermedia environment

We have seen that the videodisc is an efficient and cost-effective device for storing analogue images, and permitting fast and easy access. This enables the widespread distribution of satellite image sequences and opens up many novel ways of displaying them. The videodiscs 'METEO DISC' (ZEAN/FU and EUMETSAT 1991) from Freie Universität Berlin and 'MetVUW' (RSES/VUW, 1993) from Victoria University of Wellington are first examples to demonstrate these capabilities of distributing satellite image sequences. METEO DISC is a constant angular velocity (CAV) laser videodisc. It has purposeley been designed as a resource disc for a very large number of satellite images. CAV laser disc technology provides fast, almost momentary ($\lesssim$ 2s) access to every individual image. A standard remote control device already allows basic manual motion display manipulations interactively. The enormous number of images on METEO DISC, however, require computer assistance for further display manipulations and for linkage to other analogue or digital data systems. One pre-requisite, of course, is the development of appropriate computer user interfaces.

As stated above, fast access to every image is, due to the CAV disc technology, already a common pre-condition of the laser videodisc hardware system. The software for computer controlled access, however, has to match the internal structure and the content of images and image sequences on the disc. This correspondence demands the construction of individual user interfaces for each disc. They are strictly object-oriented.

Such computer control not only considerably helps to improve the visualization of dynamic processes from satellite image sequences, it also allows the introduction of hypermedia applications. In our sense, 'hypermedia' means a non-sequential access to, or non-sequential presentation of multimedia data. These data can be textual, graphical, pictorial, still or animated, and aural (see Figure 8). The hypermedia environment provides the potential to gain a maximum of information from multimedia. The essential element 'that distinguishes hypermedia from other mixed media is the user's ability to choose information paths and segments' (Grice and Ridgway 1993), *i.e.* the power of interaction. Thus, 'hypermedia is fundamentally more natural than the printed page. The electronic link in hypermedia simulates the connections in the mind' (Conklin 1987 and Slatin 1987, after Wickliff 1993).

In the beginning, METEO DISC utilized the PC prototype user interface ZEAMDISC (Zeam 1991). First experiences with ZEAMDISC lead to further research and development (Toussaint 1992) and the introduction of COMPASS (Zeam, 1994). Released by Zeam, COMPASS is an advanced and much more convenient PC user interface for METEO DISC. Like ZEAMDISC, COMPASS also runs on MS-DOS. The use of the programmimg language PROLOG allows most versatile interactive motion display manipulation, the integration of data-bank knowledge bases, and pre-programming of presentations.

A further approach has produced the user interface HYPERMET. It incorporates METEO DISC into hypermedia. The programming environment in this case is 'ToolBook', a software construction set including the OpenScript language, running on MS-Windows, versions 3.0 or higher. COMPASS and HYPERMET will be briefly described.

## 4.1 COMPASS

The COMPASS user interface has replaced ZEAMDISC; COMPASS has been providing user interaction with METEO DISC since 1991. A recent COMPASS update is available as shareware through Internet 'Anonymous FTP' via the CALMET server. Although COMPASS still provides the best means for data retrieval with METEO DISC, further interface development has taken place. Future interfaces will run entirely under MS-Windows, and will include access to data stored in a relational database.

For the meantime, Windows can display COMPASS if the user so desires (by pressing alt-return keys).The recommended monitor resolution is $1024 \times 768$. This high resolution allows the user to open a second window for word processing knowledge bases, or presentations files. In fact, there is still room on the screen for a third window displaying video through an overlay card. The video card displays the METEO DISC images. A separate video monitor is, therefore, no longer necessary.

COMPASS comes with a set of knowledge base files and allows the user to search images for time or for keyword strings included in the knowledge base. The user can make his own individual knowledge bases by writing them in a PROLOG-type syntax. Experience has shown that it is convenient to divide knowledge-bases into sequence-oriented and frame-oriented bases. This division enhances loading and search operations. For example, when searching for phenomena via key-words in a sequence knowledge-base, dates are usually irrrelevant. When searching for dates and times in a frame knowledge-base, key-words are usually irrelevant. In general, experience has lead to the definition of three types of data or knowledge bases:

1. sequence knowledge-bases, containing actual meteorological data;

2. data-bases, containing date and time of still-frames;

3. knowledge-bases written for specific, personal purposes.

COMPASS allows recording presentations. A presentation is a sequence of commands that the program plays back, automatically. Users have already developed a number of presentations that have been useful during lectures and demonstrations. Users have also begun to share and trade their presentation files. Such cooperation benefits all.

## 4.2   HYPERMET

HYPERMET 1.1 is an object-oriented electronic book developed with ToolBook, version 1.5. The author designed HYPERMET originally as a rather personal user interface. It was aimed to serve him as his personal notebook, 'Guenter's World', to apply selected METEO DISC image sequences in his teaching courses. It became obvious, however, that HYPERMET could also be useful for other METEO DISC appliers. For them, HYPERMET can serve at least as a motivation and an exemplary basic skeleton for the construction of other individual user interfaces. Thus, though in a stage of permanent development, HYPERMET 1.0 has recently been released as public shareware for the already mentioned CALMET international Anonymous FTP communication link. It can be procured, hence, from the TOOLBOOK subdirectory of the CALMET FTP server. Version 2.0, presented here, differs from Version 1.0 by a more comfortable motion display handler, newly developed by Christian Zick, ( ZEAM). It also allows use of the complete image reference data base provided by COMPASS.

HYPERMET is an electronic book of 21 pages as shown in the following Table.

| 1. | Front Page | 11. | Dust Clouds & Dust Storms No 1 |
|----|-----------|-----|-------------------------------|
| 2. | Table of Contents | 12. | Dust Clouds & Dust Storms No 2 |
| 3. | Introduction to Motion Display | 13. | Gravity Waves |
| 4. | What can be seen by animation | 14. | Water Vapour Channel, Meteosat |
|    | only | 15. | Gulf War Oil Plumes, 1991 |
| 5. | Mountain Effects No 1 | 16. | New Zealand & Tasman Sea |
| 6. | Mountain Effects No 2 | 17. | Sea ice & fog |
| 7. | Mountain Effects No 3 | 18. | Miscellaneous |
| 8. | Cyclones & Cyclogenesis No 1 | 19. | Special Areas |
| 9. | Cyclones & Cyclogenesis No 2 | 20. | Long Sequences |
| 10. | Thunderstorms & MCCs | 21. | Sea Surface Temperature Fields |

**Table 2.** *Table of Contents of the* HYPERMET *electronic book.*

Besides Front Page and Table of Contents the remaining 19 pages are deserved to specific topics. Each of these pages contains, next to a number of handling buttons, a template which exhibits a list of selected, topic-related image sequences (see, for example, Table 3 and Figure 9. Each line of the listing has been defined as a so-called 'hot word'. A 'hot word' is automatically indicated by a change of the shape of the mouse cursor to a rectangle. By ToolBook definition, it signals the possibility of initiating pre-programmed, follow-on commands by a simple left mouse click.

In addition, clicking the right mouse button on objects, like hot words and buttons, in most cases automatically opens information fields. These fields usually contain image samples, necessary image-related alpha-numeric source and background information and, in many cases, a specific knowledge base. Usually, this knowledge base contains an interpretation of the image sequence and supplementary information such as 'lecture notes', references *etc.* — in many cases still incomplete.

Table 3 is an example of the data field for page 11 of HYPERMET. The numbers represent selected start and end frame (index) numbers of METEO DISC. From the chosen example the field provides four special sequences of the May 1985 Sahara dust storm. They have been selected for display out of a larger number actually available from the disc. All are produced from one and the same original data but each one exhibiting different aspects of the dust cloud.

| Page 11, Dust Clouds & Dust Storms No 1 | |
|---|---|
| Sahara dust storm, 27 - 31 May, 1985 | Transatl. dust transp. 27 July - 4 Aug, '74 |
|  | VIS 15119 - 15320 |
| Original scene, 32679 - 32822 | IR 14743 - 15116 |
| with colour and enh., 33656 - 33669 | Further examples: |
| with topography, 33781 - 33820 | 'Flip-flap-flop' case, 15797 ff. |
| 'the curly protrusion', 33759 - 33770 | 'Second curl ?' case, 15798 ff. |

**Table 3.** *A specific page of the* HYPERMET *electronic book.*

Clicking the left mouse button momentarily displays the first image of the selected image sequence. Clicking then one of the handler buttons makes immediately display the sequence in the chosen motion mode, from single step to sophisticated loops.

In Figure 9 we see the result of opening pages 8 and 9. In both of these the 'date' button allows one to import date and time of displayed image into the bright information field. The 'index' button toggles the index display within the image on and off.

In the upper part of Figure 9 the motion control handler exhibits the selected *Automatic Loop Status*: backward, forward-backward, forward or swinging loops (from left to right) can be chosen; the indicated loop bounds can be altered stepwise by clicking 'upward arrow' or 'downward arrow' buttons; the index number of each actually displayed image can be imported automaticly by clicking the 'entrance' button. Present parameters of motion dispay can be called into the information window, dwell time of first and last images, display speed and swinging loop characteristics be altered by clicking corresponding 'hot words'.

In the lower part of Figure 9 the motion control handler exhibits the selected *Manual Motion Control Status*: The exhibited control bar allows controlling of linear motion display in a one-step, standard video speed or a fast scan mode, each one forward or backward, according to pictographs. Clicking the 'automatic' or 'manual' hot words toggles the loop control bar and the manual control bar, alternatively.

As stated above, HYPERMET is in a state of continuous development. Besides filling some gaps within existing sections, it is planned to (a) to implement further hyper-text capabilities, and (b) to investigate the integration of supplementary object-related videotape and CD-ROM image data.

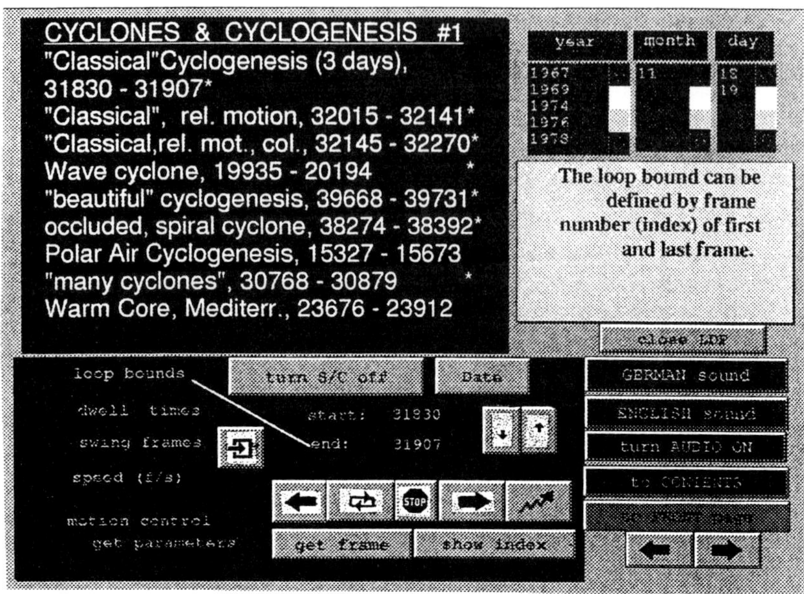


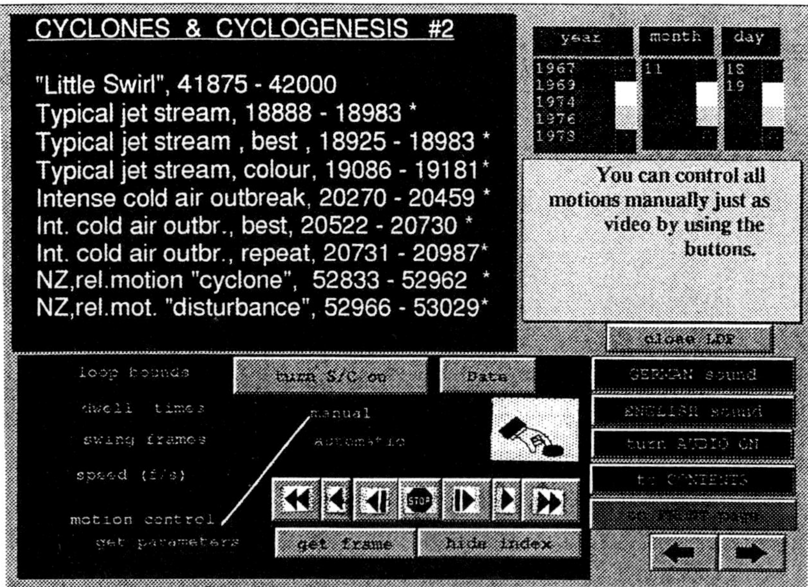


**Figure 9.** *Two full pages from* HYPERMET *2.0. showing Cyclones & Cyclogenisis Number 1 (upper) and Number 2 (lower). See text for discussion*

# 5 Conclusions

COMPASS has already proved itself a useful user interface for METEO DISC. New programming tools like ToolBook and Multimedia ToolBook, however, opened new paths into the hypermedia world. Smaller electronic books for METEO DISC, developed for specific topics like Karman vortices and gravity waves (Cassells 1993) and sub-synoptic scale vortices (Zick 1994) represent one line of approach. A second line is indicated by the more voluminous HYPERMET. It covers a larger amount of topics which can be extended easily into a more general user interface. In principle, there are, of course, no objections to integrate independently developed smaller electronic books into HYPERMET, as long as certain hardware and software conventions are observed.

Different laser videodiscs each require, of course, their own user interfaces. This statement is particularly true when the basic structure of a videodisc is very different, as in the case of METEO DISC and 'MetVUW' (RSES/VUW 1992). METEO DISC, as stated, is a collection of a large number of different image sequences, covering a large variety of selected phenomena from all across the globe and gathered over almost 25 years. MetVUW, in contrast, although also a resource disc, is made of a continuous series of hourly GMS images over a period of two years from the New Zealand and Tasman Sea region, including the concurrent series of supporting weather maps on the disc. However, this does not exclude cross-referencing between such discs in further stages of constructing teaching and training material. On the contrary, modern hypermedia environments invite such developments and provide the proper means.

In summary, experience with the development and application of METEO DISC (ZEAM/FU and EUMETSAT, 1991) and MetVUW (RSES/VUW, 1993) proved the laser videodisc as a very useful medium to provide easy and immediate access to a very large amount of satellite image data and to allow most versatile motion display and manipulation. Available PC linked user interfaces raise the power of the system by allowing hypermedia applications.

The combination of METEO DISC and hypermedia has still to reach its full potential. User interfaces like HYPERMET for laser videodiscs, like METEO DISC, can help to explore the utilization of these abilities of hypermedia and to favour and encourage similar developments in other disciplines. This appears particularly suggestive where ever interactive access to large amounts of image data is helpful or required. The presented results, at least, are quite promising and asking for continuous effort in exploring further this interesting field.

# 6 Glossary

**CALMET** is the acronym of 'Computer Aided Learning in Meteorological Education and Training' and the name of a communication mailbox on the 'anonymous ftp' computer network available through Internet. CALMET serves international exchange of ideas, data and programs related to this growing field of research and development. CALMET was initiated, and is managed, by the Department of Meteorology, University of Edinburgh (Dr Charles Duncan).

**ToolBook** is roughly the equivalent of 'HyperCard' for MacIntosh, or of 'IconAuthor'. ToolBook allows to import bmp images, dBase data banks and ClipArt graphics, in its recent version 3.0 also formatted text.

**ZEAM**, Zentraleinrichtung fuer Audiovisuelle Medien, Freie Universität Berlin, is the Audio-Visual Centre of the Free University of Berlin, Germany.

# References


Cassells C, 1993, Development of a 'Toolbook' for the display and control of 'METEODISC' image sequences. MSc Dissertation, University of Dundee

Conckin J, 1987, Hypertext: A survey and introduction. *Institute of Electrical and Eqlectronics Engineers Computer* **20**, No. 9, 465-472

Frank H G, 1969: Kybernetische Grundlagen der Paedagogik; Vol.2, Agis, Baden-Baden

Grice, R A and Ridgway L S, 1993, Usability and Hypermedia: Toward a Set of Usability Criteria and Measures. Technical Communication, 3rd Quarter 1993, pp.429- 437. RSES/VUW, 1993: MetVUW - CAV laser videodisc. Research School for Earth Sciences (RSES), Victoria University of Wellignton (VUW), New Zealand

Slatin J M, 1989, Reading hypertext: Order and coherence in a new medium. College English 52, No. 8, pp. 870-883

Soumi V, Fujita T, 1967: Film (9 Min.): Detailed Views of Mesoscale Cloud Patterns. The Walter A. Bohan Comp., Park Ridge, Ill., USA.

Toussaint B C, 1992, Bildplatten und wissensbasierte Systeme zur interaktiven Szenenbeobachtung und -analyse. PhD Dissertation, Fachbereich Kommunikations- Wissenschaften, Freie Universität Berlin, 247

Warnecke G and Zick C, 1981: The Use of Cinematographic Methods for the Presentation of Atmospheric Motions as revealed by Remote Sensing Techniques from Satellites. In: Remote Sensing in Meteorology, Oceanography and Hydrology, (ed: Cracknell A P) pp452–473 Ellis Horwood Ltd., Chichester.

Warnecke G, 1987: The Visualisation of the Ceaseless Atmosphere. In: Remote Sensing Applications in Meteorology and Climatology (ed: Vaughan R A), pp245–257, Reidel

Warnecke G, 1992: Beobachtung von dynamischen Prozessen aus dem Weltraum in Zeitraffung: Eine neue Wahrnehmungsdimension. In: Warnecke G., M. Huch, K. Germann (Ed.): Tatort Erde - Menschliche Eingriffe in Naturraum Und Klima, 2nd Edition, pp.222-235, Springer, Berlin Heidelberg. ZEAM/FU, EUMETSAT, 1991: METEO DISC - Laser Vision (PAL 625/50, CAV active play). Zentraleinrichtung fuer Audiovisuelle Medien (ZEAM), Freie Universität Berlin

Wickliff G A, 1993, Special Section: Designing, Testing, and Distributing Hypermedia Introduction. Technical Communication, 3rd Quarter 1993, 410-413

ZEAM/FU, EUMETSAT, 1991: METEO DISC - Laser Vision (PAL 625/50, CAV active play)

ZEAM 1991, ZEAMDISC - A PC based user interface for the laser videodisc METEO DISC. Zentraleinrichtung fuer Audiovisuelle Medien (ZEAM), Freie Universität Berlin, Germany

Zeam 1994, COMPASS - An advanced PC based user interface for the laser videodisc METEO DISC. Zentraleinrichtung fuer Audiovisuelle Medien (ZEAM), Freie Universität Berlin, Germany

Zick C, 1994, Sub-synoptic Scale Vortices - A ToolBook application for METEO DISC. Zentraleinrichtung fuer Audiovisuelle Medien (ZEAM), Freie University Berlin, Germany

# The Design and Implementation of MetVUW Workbench Version 1.0

James McGregor

Victoria University of Wellington
New Zealand

## 1 Introduction

MetVUW Workbench is a multimedia database of historical meteorological data. The system is designed to assist weather forecasters and other meteorologists, and to serve as a test bed for research into intelligent retrieval mechanisms. Version 1.0 of the system was completed in April 1993.

MetVUW Workbench allows retrieval and display of the following kinds of data: laser disc video imagery and corresponding digital satellite imagery, time-tagged text descriptions, hydrological data, and numeric fields from the European Centre for Medium-Range Weather Forecasting (ECMWF). Examples of numeric fields include pressure, temperature, relative humidity, wind speed, and relative vorticity. The data currently covers a fixed region including New Zealand, the south-west Pacific, the Tasman Sea, and Eastern Australia. Data can be retrieved by date and time, by full text search, or by scanning the laser disc. A number of mechanisms are provided for manipulating and displaying data once they have been retrieved, including false-colouring of satellite imagery, calculating contour overlays, scrolling text windows, and animating laser disc video. An experimental audio output device is also provided.

The current system accomplishes the following design objectives.

1. *Seamless access to multiple sources of data*

   It is possible to use any of the several search mechanisms provided to retrieve all data available at a given time. For example, a user can locate an image of interest using the laser disc player and then create contour displays of numerical fields at the time of that image. Alternatively, a user can locate a document of interest using full text search and then have instant access to the numerical data and laser disc imagery for the time period to which the document refers. A user can also retrieve images or numerical data by date and time and then start a laser disc animation at that point.

2. *Ease of Use*

   MetVUW Workbench is intended to provide straightforward access to the operations most commonly required by meteorologists. We therefore provide a graphical point and click interface that makes it very easy for users to carry out searches and to specify options for displaying data.

   This approach contrasts radically with the design philosophy of other existing meteorological databases such as the McIdas system employed at the Australian Bureau of Meteorology. McIdas provides a powerful command line interface with much greater functionality than MetVUW Workbench, but forecasters apparently find it very difficult to use in practice.

3. *Speed of Retrieval*

   We intend later.versions of MetVUW Workbench to be actively used by operational forecasters. Fast retrieval is vital to these users, as they operate under severe time pressure. The current version of MetVUW Workbench runs on a SparcStation 10 model 30 and takes at most two seconds to execute any command from a user to retrieve or display data

Version 1.0 of MetVUW Workbench serves several groups of users. First, research students in the Institute of Geophysics at Victoria University of Wellington use MetVUW Workbench as a tool for meteorological research. Second, within the Department of Computer Science, the system is used as a test bed for research and development of intelligent mechanisms for retrieving historical meteorological data. Two of us are actively engaged in the design of a mechanism for retrieving past weather situations from descriptions of high-level features of the meteorological processes and systems (Jones and Roydhouse, 1993).

# 2   Overview

A number of interesting problems and issues arose in the design and implementation of MetVUW Workbench. These divide into five broad categories:

1. *User Interface*

   MetVUW Workbench is operated by an easy-to-use graphical user interface that provides seam-less access to multiple-datatypes (satellite imagery, text, numerical fields, and so forth), multiple search mechanisms (date and time, full text search, laser disc), and multiple devices (computer screen, laser disc player, audio). Managing this complexity gives rise to several interesting issues of interface design, two of which are as follows:

(a) Coordination of multiple data types

After retrieving data of one type, a user can immediately access all other types of data occurring at the same time. For example, having located a satellite image by browsing the laser disc, pressure and wind data for that date and time are immediately available. The user interface provides 'synchronisation' buttons for this purpose that align different data types by their time of occurrence; see Section 4.7.

(b) Control of the laser disc player

The laser disc player is operated by means of the same point and click interface used to drive the rest of the system, as opposed to employing an external device such as a remote control. MetVUW Workbench provides a control panel that relies heavily on the analogy to a video tape player, although it provides a greater range of functions; see Section 4.3.

2. *Memory Organisation and Data Storage*

To facilitate access to multiple types of data using a variety of search mechanisms, an appropriate memory organisation is required. In particular, given data of one type occurring at some point in time, it must be possible to retrieve all other types of data for the same time period. For example, given a satellite image, it must be possible to construct contour overlays for numerical fields such as mean sea level pressure. The indexing and retrieval strategies that MetVUW Workbench employs for this purpose are described in Section 4.10. MetWUW Workbench also provides a number of data compression mechanisms, one for each data type, and allows compressed and uncompressed data to be freely intermixed; see Section 5.1.

3. *Communication with the Laser Disc Player*

No Unix-based device driver for the laser disc player was available, so we designed our own. Two problems arose in its design. The first problem was to design a high-level interface to the device driver that is more useful than the raw device commands. Our command set is described in Section 6.1

Second, certain laser disc commands such as searches and repeats generate asynchronous return values, so the device driver has to be able to accept these return values at any time and process them in a sensible fashion. The complications this gives rise to, and our solution to them are described in Section 6.2.

4. *Audio presentation of data*

In co-operation with the Electricity Corporation of New Zealand, we implemented an experimental audio output device that outputs audio tones of varying frequencies corresponding to times series of numeric data. Whilst browsing the laser disc, the user can simultaneously listen to an audio track representing time series of numeric data such as lake inflow rates. The audio driver is discussed in Section 7.

5. *Use of publicly available software*

A common fault of many software developments is to try and re-invent the wheel by writing all components of the project from scratch. Designing a new text search package, a database system, or a data compression algorithm would be interesting

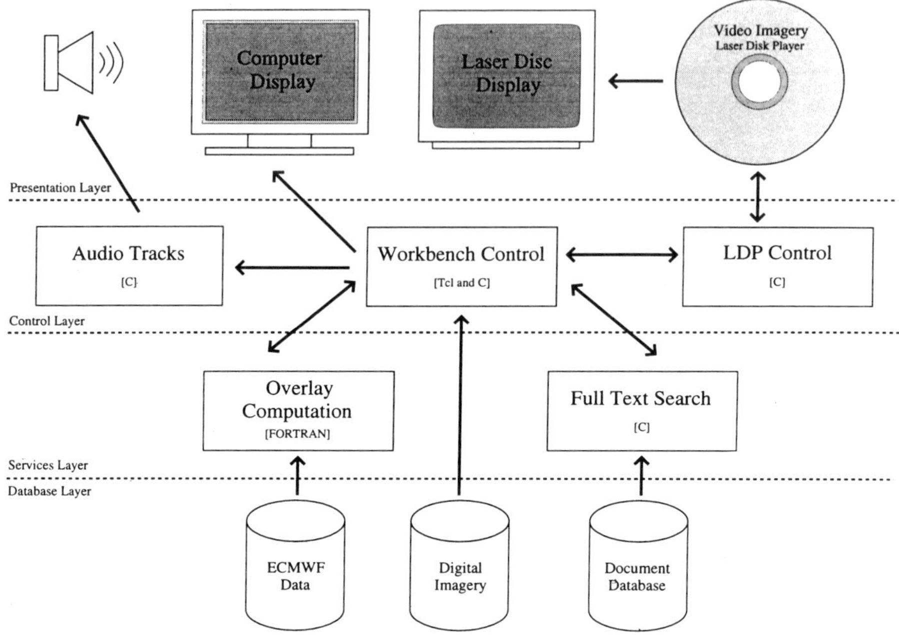


**Figure 1.** *MetVUW Workbench Organisation*

tasks in themselves, but they were not central to our goals. Instead, we used existing packages wherever possible to achieve speedy development and allow us to concentrate on the novel components and aims of MetVUW Workbench. The packages we used are freely available on the Internet. We discuss the contribution of these packages in Section 8

# 3   Architecture of MetVUW Workbench

This section describes the internal architecture of MetVUW Workbench. We first describe the organisation of the elements of the system, then provide brief details on the implementation.

The MetVUW Workbench architecture consists of four layers, as shown (Figure 1):

- The **Presentation Layer** is the user's view of the system. A high-resolution computer monitor displays digital imagery, overlays of contours and wind barbs, and a graphical user interface. Laser-disc video imagery is displayed on a second monitor. Audio is generated on the host computer's speaker.

- The **Control Layer** supervises the laser disc player and audio devices, and brings together requested digital satellite images, text documents, and overlays for presentation on the computer display.

- The **Services Layer** computes contours and wind barbs and carries out searches of the document database.

- The **Database Layer** manages large collections of numerical data, digital images, and text documents.

Because of the scope of tasks required for the MetVUW Workbench implementation, no one language is suitable. Version 1.0 of MetVUW Workbench is implemented in Tcl/Tk, C, and FORTRAN. The graphical user interface and core of MetVUW Workbench is implemented in Tcl/Tk. The drivers for the laser disc player and audio output are implemented in C, the overlay computation library is derived from code written by Dr Crispin Marks and implemented in FORTRAN. This code includes a contouring algorithm which uses the St Andrews Cells method and thus correctly contours saddle points (Sabin, 1986). Other packages for image compression (JPEG), full text search (LQ-text), database management (NDBM), and Tcl/Tk itself are implemented in C. The Tcl code acts as the glue that brings these components of the implementation together to form a complete application, as discussed in Section 8, below.

The satellite images available to the system are stored in two different formats: on hard disk in a digital format, and on laser disc in an analogue video format for purposes of animation. The satellite images are available for each hour and show a region covering New Zealand, the south-west Pacific, the Tasman Sea, and eastern Australia. ECMWF data are available for every twelve hours and at 2.5 degree grid-points. Some of the data are available for 14 atmospheric levels from surface to the 10mb level, including pressure, temperature, potential energy, geopotential, vertical wind velocity, horizontal wind velocity, and relative humidity. Other ECMWF data are by their nature available at only one specific level, including two-metre dew-point, two-metre temperature, mean sea level pressure, surface pressure, surface geopotential, and ten-metre wind. A number of "derived" fields are pre-calculated from the ECMWF data, including wind speed, surface wind speed, 300mb thickness, thickness, potential vorticity, relative vorticity, residual force, and heating rate. The text document collection includes brief synopses of a subset of the satellite images, and articles from the newsletters of the Meteorology Society of New Zealand.

The development platform for MetVUW Workbench is a SparcStation 10 model 30 running SunOS 4.1.3. The laser disc player is a Sony 3600. The languages and compilers used were Tcl 6.7, gcc 2.3.3, Sun FORTRAN 77 1.4p6, and Texinfo 2.16. The library versions used were Xl lR5, Tk 3.2, JPEG 4.0, SunOS 4.1 NDBM, LQ-text 1.12gamma, and TkInfo 0.3.

# 4   Interface design

The user interface of MetVUW Workbench integrates a number of different mechanisms for retrieving and displaying meteorological data. The system currently supports retrieval and display of laser disc video imagery, digital satellite imagery, hydrological data, ECMWF data, derived numeric parameters, and time-tagged text descriptions. These resources are all accessible using the graphical user interface shown in Figure 2. We next discuss each component of the interface.

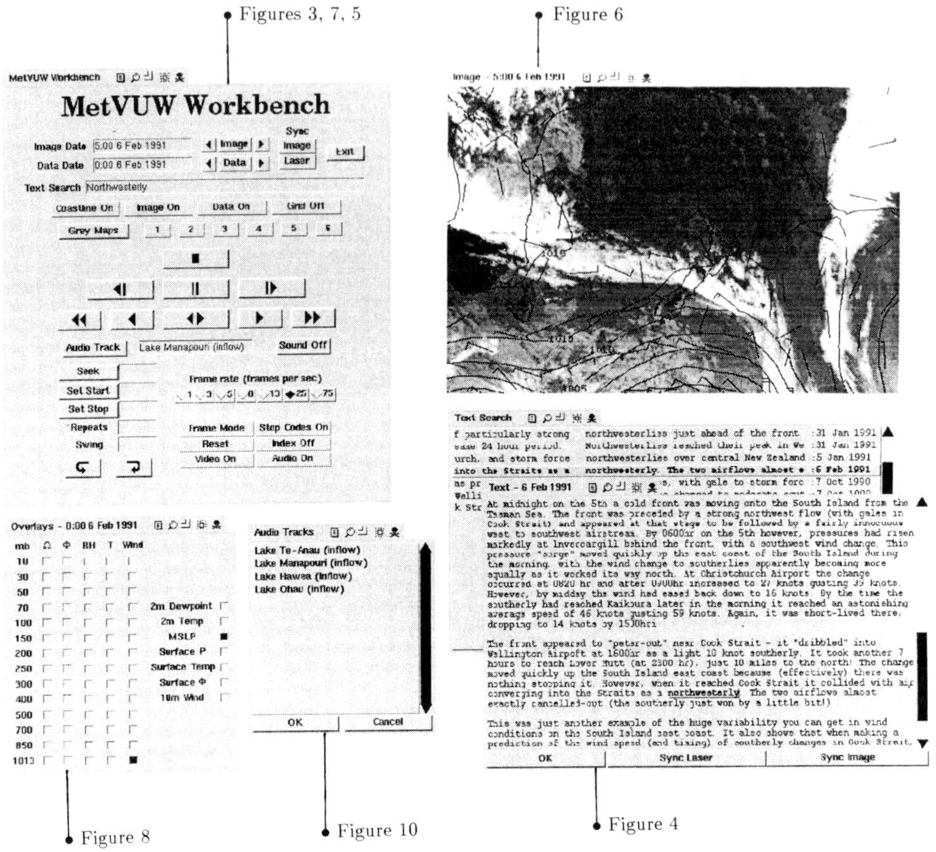

Figures 3, 7, 5

Figure 6

Figure 8

Figure 10

Figure 4

**Figure 2.** *The MetVUW Workbench graphical user interface*

## 4.1   Retrieval by date

A date and time entered in to the 'Image Date' entry box, shown in Figure 3, will cause the corresponding satellite image to be retrieved. If the requested image is missing for some reason, then the image for the nearest available hour is retrieved. In this case the Image Date entry box is changed to reflect the actual date and time of the displayed image.

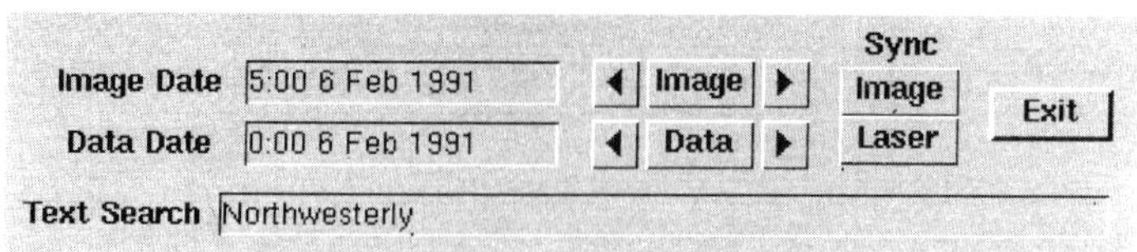


**Figure 3.** *Retrieval by date or key phrase*

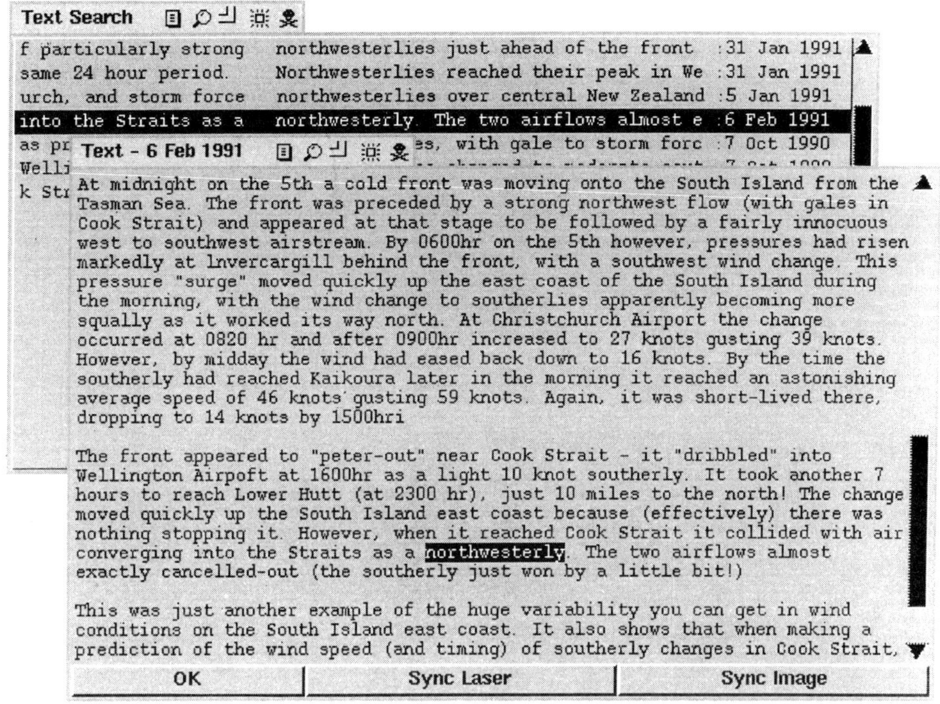

**Figure 4.** *Retrieval by text search*

Retrieval by image date also selects the closest ECMWF data set, providing it is available within six hours of the image date. The 'Data Date' entry box is changed to contain the date and time of the selected ECMWF data In a similar fashion, ECMWF data can be retrieved by entering a date and time in the 'Data Date' entry box; if available, an image within two hours of the data date will be retrieved.

Stepper buttons step forward and backward, image by image or data set by data set, synchronising the other type of data at each step.

## 4.2 Retrieval by text search

In Figure 3 we have entered the key phrase 'Northwesterley' in the 'Text Search' entry box This causes a window to pop up that lists all occurrences of this phrase in the on-line document archive (Figure 4). Each line of the text search window gives the context of an occurrence of the key phrase and the date of the document in which it occurs. Selecting any of these occurrences causes the whole document to be presented in a separate window which in Figure 4 partially obscures the text search window.

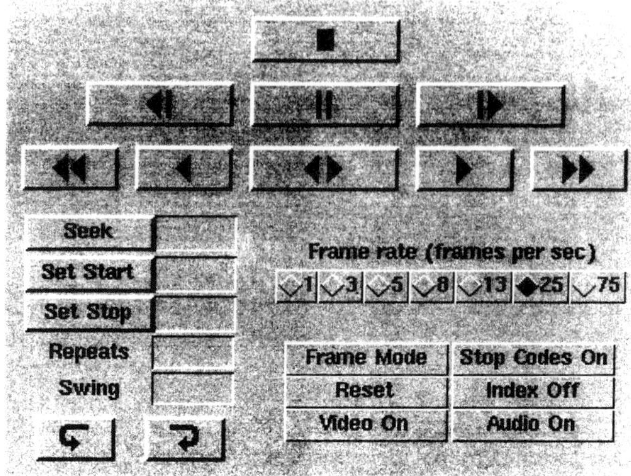

**Figure 5.** *Laser disc player controls*

Each window that contains a document also has buttons for retrieving laser disc or digital imagery occurring at that time. These buttons are described in Section 4.7.

## 4.3　Laser disc player controls

MetVUW Workbench provides a standard set of laser disc controls for playing and searching through the laser disc, as shown in Figure 5. The three rows of buttons at the top are as follows: stop, frame back, still frame, frame forward, rewind, play backward, play other direction, play forward, and fast forward. The right-hand middle row of buttons controls the frame rate for playback operations. At the bottom right is a set of toggle buttons for features of the laser disc player. These features are as follows: seek by frame or chapter, stop at stop codes, reset the player, display frame index on screen, display video, and play audio . On the middle left, the 'Seek' button displays the frame whose number is in the adjacent entry box. On the bottom left are entry boxes and buttons for controlling repeat and swing loops. Repeat loops play a sequence of satellite images a number of times in row. Swing loops are like repeat loops except that on each cycle, the sequence is played forwards then backwards, and a larger number of frames can be played in one direction than the other on each cycle. The 'Set Start' and 'Set Stop' buttons determine the frames to be repeated. The 'Repeats' box sets the number of times to play a repeat loop. The 'Swing' entry box determines the number of frames by which swing loop iterations are offset. The pair of buttons at the very bottom left initiate repeat and swing loops; the left-hand button plays loops backwards, and the right-hand button plays loops forwards.

## 4.4　The image window

The image window displays digital satellite imagery and overlays of contours and wind barbs. The region covered includes New Zealand, the south-west Pacific, the Tasman

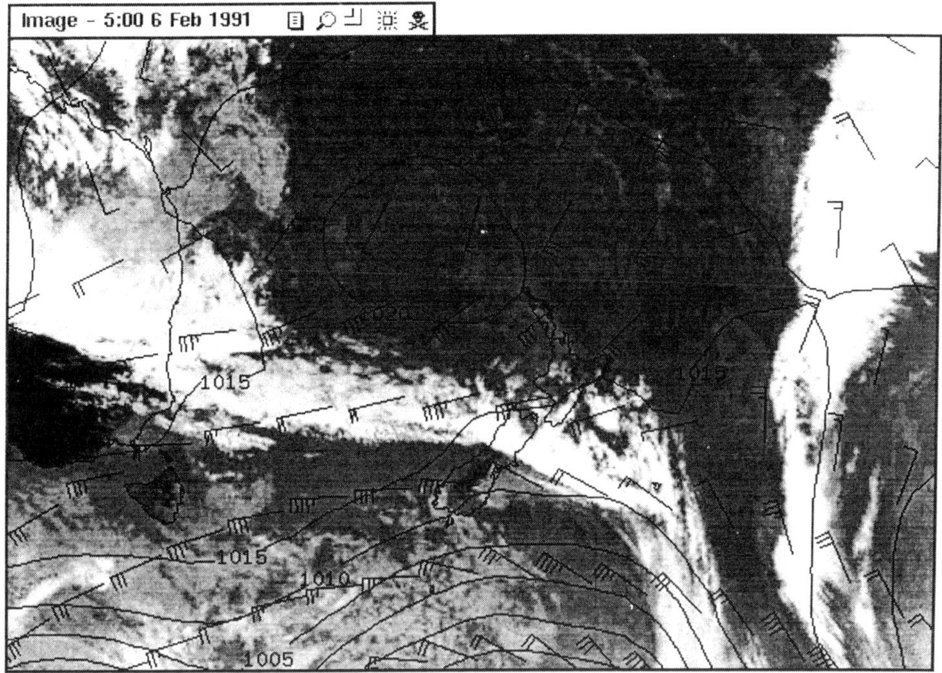


**Figure 6.** *Image window showing mean sea level pressure and wind barbs*

Sea, and Eastern Australia. Data for many atmospheric levels are available; see Section 4.6. For example, Figure 6 shows an infra-red satellite image overlayed by contours for the mean sea level pressure and wind barbs indicating the speed and direction of the surface winds.

## 4.5   Image controls

Figure 7 shows the interface for the image controls. Several toggles and options affecting the image window are available:

- **Coastline On/Off** toggles an overlay showing an outline of land masses.

- **Image On/Off** specifies whether or not digital satellite imagery is displayed.

- **Data On/Off** specifies whether or not data overlays are displayed.

- **Grid On/Off** toggles an overlay showing lines of latitude and longitude.

The lower row of buttons control the false-colouring facilities. Twelve options are available: six colour and six grey-scale. Each option is a piece-wise linear function that maps segments of data values to ranges of colours. A toggle button is provided to switch between colour and grey-scale option sets; buttons labelled '1' to '6' select an option.

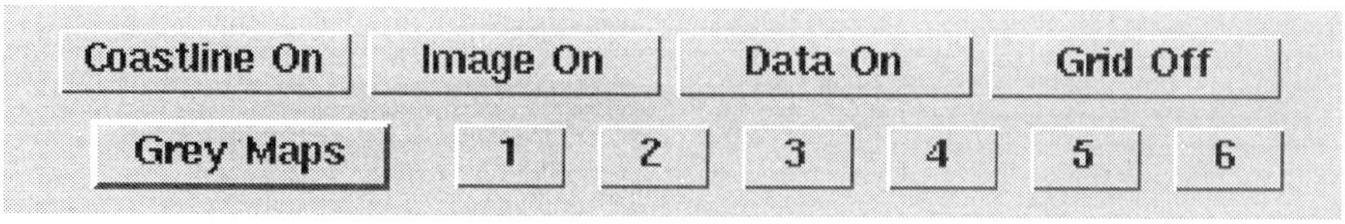

Figure 7. *Image Controls*

## 4.6 ECMWF Overlays

A separate window, shown in Figure 8, allows the user to specify contour and wind-barb overlays for the current data date. The rightmost column contains data fields that exist at only one atmospheric level. On the left is an array of data fields. Each column of the array refers to a different data field; each row refers to a different level of the atmosphere. Following standard meteorological practice, pressure is used as a measure of height.

In Figure 8, mean sea level pressure (MSLP) and wind speed and direction for 1013mb are currently selected. The overlays created for this selection are shown on the image in Figure 6.

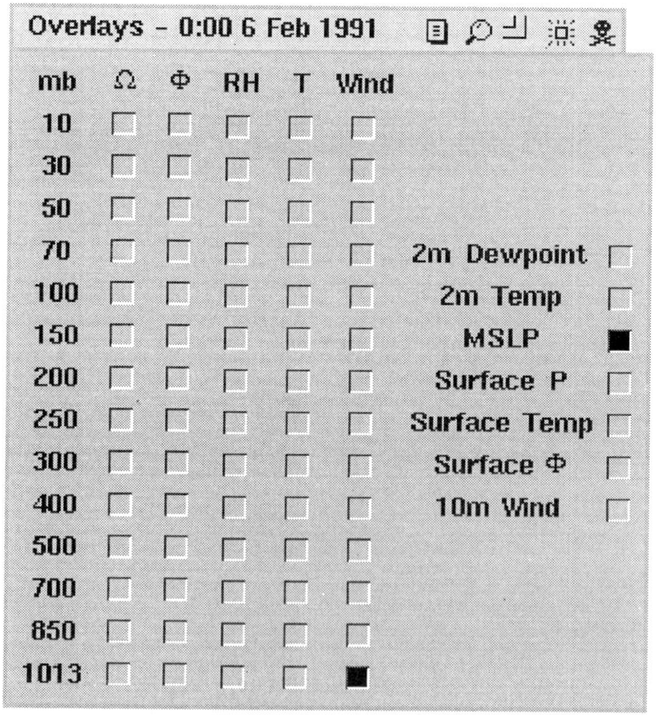

Figure 8. *ECMWF Overlays*

## 4.7   Synchronisation buttons

Synchronisation buttons, shown in Figure 9, align data for a given time with other types of data for the same time.

Image synchronisation buttons, shown at the top of Figure 9, align digital image data with laser disc data and vice versa. Selecting sync 'Laser' causes the laser disc player to display the video image corresponding to the currently selected digital image. These two images are identical except that one is stored in an analogue video format and the other in a digital format. Similarly, selecting sync 'Image' cause the image window to display the digital image corresponding to the video image currently displayed by the laser disc player

Figure 9 also shows a text window containing a document retrieved as the result of a full text search Synchronisation buttons at the base of the text window trigger retrieval of the laser disc and digital images occurring at the time of the document.

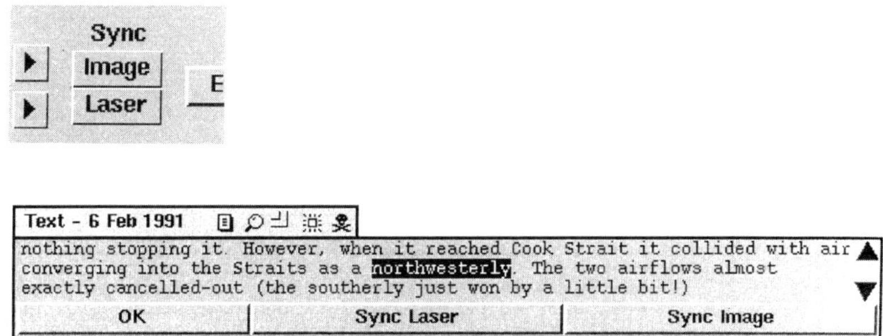


**Figure 9.** *Synchronisation buttons*

## 4.8   Audio tracks

Audio tracks of hydrological data can be played synchronously with the laser disc imagery as described in Section 7, below. A selection box allows the user to select the data series they wish to hear, whilst that generation of audio can be toggled on and off at any time; these controls are shown in Figure 10.

## 4.9   Help facility

MetVUW Workbench provides a hypertext help facility, shown in Figure 11. Help information is organised as a hierarchy of menus, with hypertext-style cross references. Clicking on any window or button of the computer display will bring up a new window displaying a hypertext node that contains appropriate help information. This information may include cross references and a menu of related topics. Clicking on a menu item or cross reference will cause that topic to be displayed.

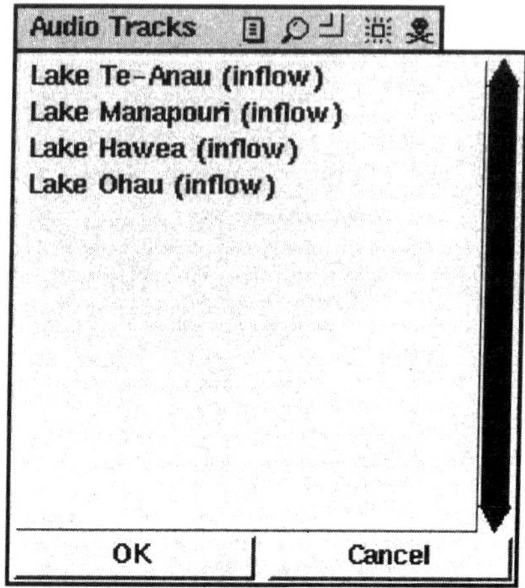


**Figure 10.** *Audio controls and track selection*

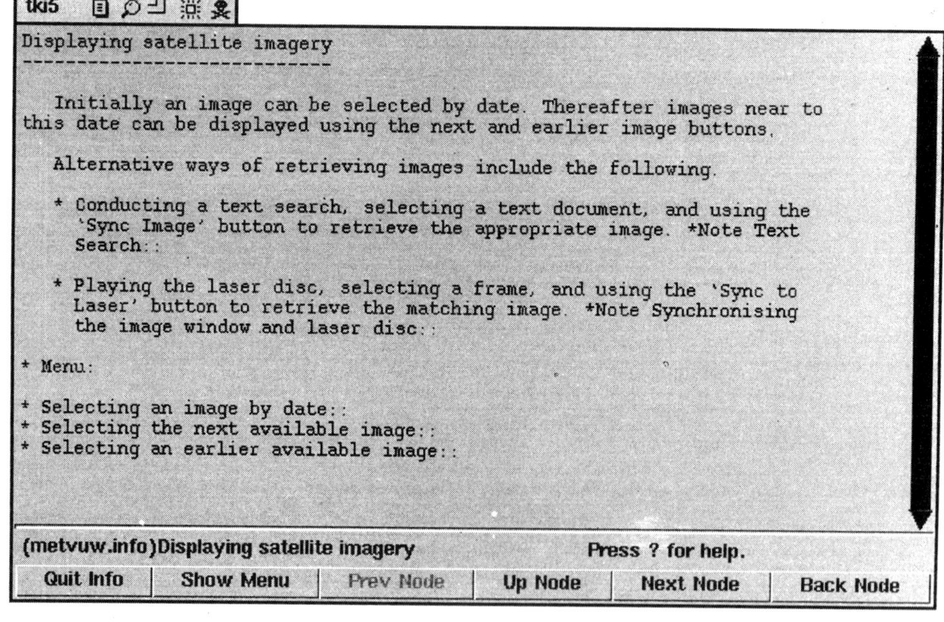


**Figure 11.** *Help facility*

## 4.10  Data indexing and synchronisation

To facilitate access to multiple types of data using a variety of search mechanisms, all data are indexed by the date and time of their occurrence. Most of these indices are bi-directional: not only can data be located by date and time, but given a datum, the date and time of the datum can be recovered. For example, a table is provided that records the date and time of each frame number of the laser disc player. Using this table, the current frame number displayed by the laser disc player can serve as in index to retrieve the date and time of the image. A similar table can then be used in the other direction to retrieve the corresponding digital image for that date and time.

In addition to these indices, every non-textual datum provides access to its immediate predecessor and successor in the historical record of data of that type. These 'relative' indices are used for stepping forwards and backwards through the record. Further indices to textual data are also compiled for full text search.

In more detail, the following indices are provided:

1. *Laser disc frame index*

   A table indexed by frame number contains a date and time for each frame. This table is used for synchronising other data sources with the image displayed by the laser disc player

2. *Date and time index*

   A table indexed by date and time permits access to a number of data sources. For each date and time, the table includes the following items

   (a) The *frame number* for the image on the laser disc that depicts the closest date and time available.

   (b) The *frame numbers* for the closest available laser disc images for later and earlier times.

   (c) An *image filepath* for the file that contains the digital image.

   (d) *Image filepaths* for the closest available images for later and earlier times.

   (e) A *data filepath* for the set of files containing numeric data.

   (f) *Data filepaths* for the closest available data sets for earlier and later times.

3. *Filename index*

   A table indexed by filename contains the date to which each file refers. This index is used to recover the date and time from a filename. For example, if the date and time index is used to obtain a file path for the closest earlier image for a given date and time, then the date and time of that earlier image can be identified using the filename index.

4. *Array of hydrological data*

   Hydrological data are coupled to laser disc imagery by storing it in a file that can be directly indexed by frame number (see Section 7).

5. *Text concordance*

A concordance records all occurrences of each word used in the collection of text documents and is constructed automatically by the full text search system. The concordance is used to locate text phrases in the document collection.

The freedom to move from one source of data to another that MetVUW Workbench provides is shown in Figure 12. Interconnectivity between the different data sources is nearly complete, with one important exception: the existing implementation does not permit access to text by date and time, so text cannot be retrieved by way of retrieving other kinds of data. This needs to be fixed.

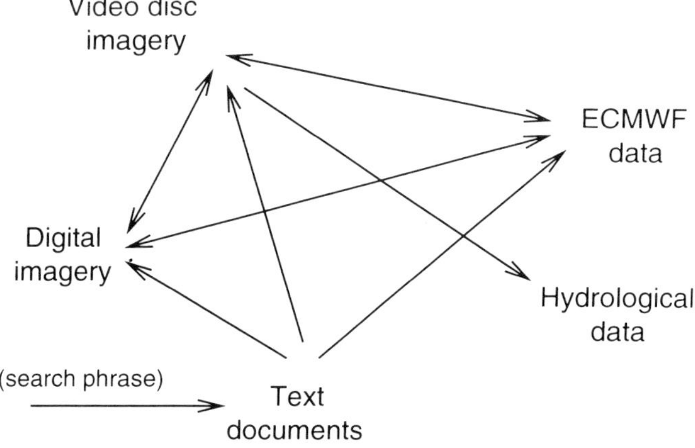


**Figure 12.** *Data connectivity provided by version 1.0 of MetVUW Workbench*

# 5  Memory Organisation and Data Storage

## 5.1  Data compression

To reduce storage requirements, meteorological data can be stored in compressed form. The appropriate form of compression varies according to data type. Presently we employ two compression algorithms, one for images, and another for numerical data. Images are compressed using an algorithm called lPEG, which is specifically designed for lossy compression of still images. Images compressed with lPEG appear almost indistinguishable from the original even when compressed to a tenth of their original.

Numeric data are compressed using Lempel-Ziv coding, a general-purpose loss-less compression algorithm. The effectiveness of Lempel-Ziv depends of the type of data being compressed. For our numeric meteorological data we achieved 10 - 20% compression. In the future we hope employ domain specific compression techniques for numeric data to improve on this figure.

# 6   Communication with the Laser Disc Player

## 6.1   High level interface

The laser disc driver provides a high-level, easily callable interface to all the features of the Sony 3600 laser disc player. The driver is naturally device-specific, but should convert with only minor modification to other models of Sony player.

The laser disc player is a device capable of displaying video frames and sound from a laser disc on a television monitor. Images can be displayed still frame, or played back at a variety of speeds, either forwards or backwards. A sequence of frames can be looped repeatedly. Frames are sequentially numbered, and the player can randomly access any frame from its number. The player can play both PAL and NTSC format discs.

Although the laser disc player can play back at a variety of speeds, each speed requires sending a different command to the player. Rather than providing a separate set of playback commands for each possible speed, the driver maintains a current frame rate for the player. The frame rate determines the speed of all playback commands. It is the job of the driver to translate a high-level playback call, in conjunction with the current frame rate, into the appropriate laser disc player command. Actual frame rates are determined by the format of the disc (PAL or NTSC).

The repeat loops provided by the player are only capable of looping a fixed cycle of frames forwards or backwards at normal playback speed. The driver enhances these loops to provide repeat loops or swing loops played at the current frame rate. Swing loops are like repeat loops except that, on each cycle, the sequence is played forwards then backwards, and a larger number of frames can be played in one direction than the other on each cycle.

Other capabilities of the driver include support for control of multiple players, full error handling, and the ability to connect to the laser disc player via TCP/IP or direct serial connection.

The complete command set of the driver is listed in Figure 13.

| **Connection** | **Playback** | **Control** |
|---|---|---|
| Open | Set frame rate | Reset |
| Close | Stop | Clear |
|  | Still | Video on/off |
|  | Frame advance | Audio on/off |
| **Status** | Start | Set frame mode |
| Get disc ID | Scan | Set chapter mode |
| Get frame address | Set memory position | Stop codes on/off |
| Get chapter address | Memory search | Index on/off |
| Get frame rate | Search | Audio channel 1/2 |
| Get status | Repeat | Motor on/off |
| Is playing | Swing | Eject disabled |
|  |  | Eject enabled |
|  |  | Noise reduction on/off |

**Figure 13.** *Function calls provided by the laser disc player driver*

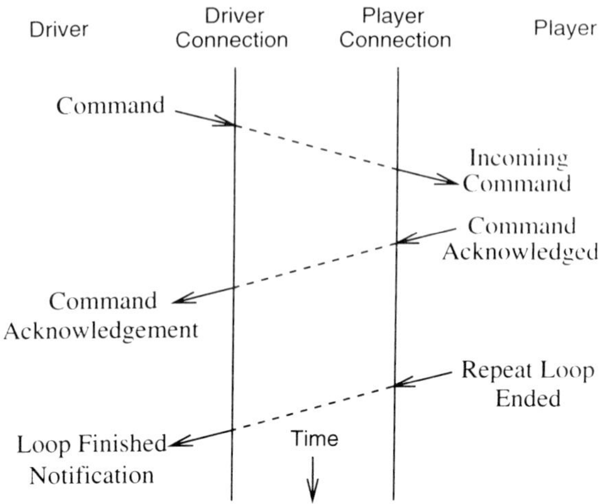



**Figure 14.** *Normal communication with the player*

## 6.2   Implementation

All the high level commands listed in Figure 13 are provided as a library of function calls that are linked with the main program. These function calls generate commands that are sent to the laser disc player. Communication with the laser disc player is via a 9600 baud serial line. All commands are single byte codes, possibly followed by multiple byte arguments. Normally, the player returns either an acknowledgement byte signalling that the command was received and is acceptable, or an error byte. On status queries, however, instead of returning an acknowledgement, the player returns a stream of bytes encoding the player status. The player also asynchronously sends bytes when the laser disc is ejected manually from the player or the end of a repeat loop is reached. These asynchronous bytes may arrive at any time except in the middle of a stream of bytes returned from a status query.

Whenever asynchronous bytes arrive, the operating system raises an interrupt to notify the driver. The driver then catches the interrupt, reads the asynchronous bytes and takes the appropriate action. It is possible that only one interrupt will be raised for a group of asynchronous bytes, so the driver must be able to deal with many outstanding bytes any time an interrupt is raised.

The arrival of asynchronous bytes can cause problems. Normal command processing happens as shown in Figure 14, where a command is sent, and the player response then read. However, it is possible that an asynchronous byte will be sent by the player at the same time as the driver is about to send a command to the player; see Figure 15. Thus the first byte on the serial line after the command is not the response from the command but the asynchronously received byte. The driver copes with this problem using the method shown in Figure 16.

A limitation of the driver is its inability to notice some changes in the state of the laser disc player. Currently there is no feedback from the driver to any caller when a

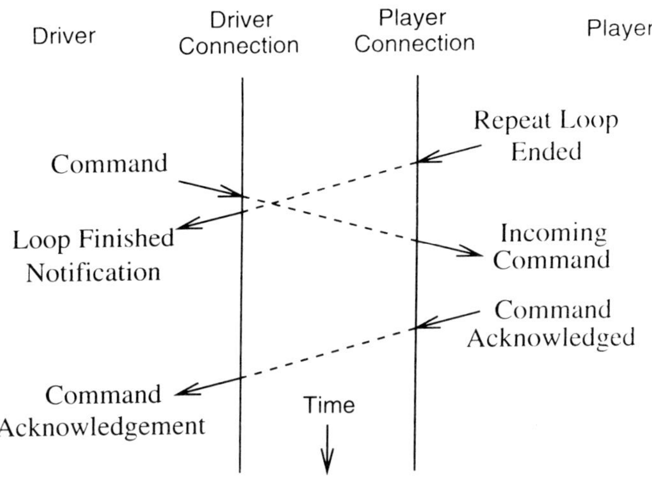


**Figure 15.** *Asynchronous return overlapping a player command*

```
• When an interrupt is received:
    For each outstanding asynchronous byte
        If it indicates the completion of an iteration of a loop,
            Decrease the number of iterations left, and if necessary,
            do another iteration.
        Else if it indicates that the disc has been ejected (manually)
            Reset player state to reflect the lack of a disc
• To send a command:
    Disable interrupts
    Send command
    Repeat
        Read 1 byte
        If it is an asynchronous byte, queue it to process later
    Until byte read is not asynchronous
    Read any more bytes expected as the result of the command
    Process queued asynchronous bytes
    Re-enable interrupts
```

**Figure 16.** *Handling asynchronous bytes*

repeat loop is completed, nor when the disc is ejected. There is also no feedback from the laser disc player to the driver when the player reaches the end of a disc and stops playing. The only way to detect that the player has reached the end of the disc is to constantly poll the status of the laser disc player. This is not a desirable option because of the very high overhead this incurs.

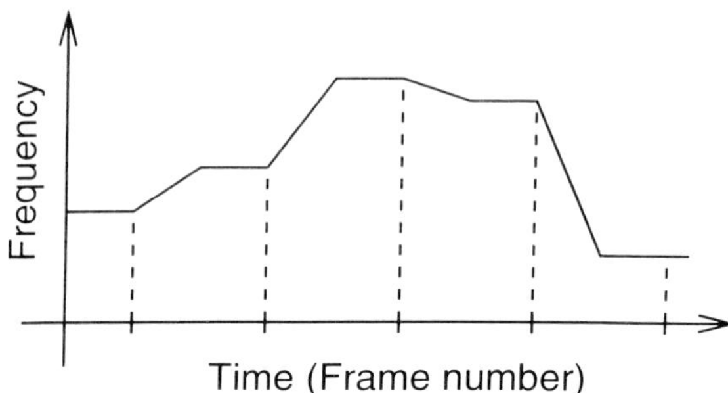



**Figure 17.** *Frequency versus time for the mapping of hydrological data to audio frequencies, showing frequency ramping between data.*

# 7   Audio Presentation of Data

The audio driver plays numeric data 'in sync' with the laser disc player imagery. The Electricity Corporation of New Zealand provided the MetVUW project with data series for lake inflow rates. While the laser disc player operates, the hydrological data are retrieved and mapped to an audio tone. Different data values give rise to an audio tone at a different frequency. As each laser disc frame triggers matching hydrological data the tone is ramped from the previous level to the new level, as shown in Figure 17.

# 8   Use of Publicly Available Software

Version 1.0 of MetVUW Workbench was constructed very quickly, and is a substantial application; nevertheless, it is already stable and relatively bug-free. Instrumental in this success was the use of innovative tools contributed to the Internet. We now discuss the role these packages play in MetVUW Workbench.

1. *Graphical User Interface*

   Tool Command Language (TCL) is an interpreted and extensible programming language (Ousterhout, 1991). The most popular set of extensions to Tcl is a graphics tool kit, called Tk. Tk is a library for implementing of graphical applications for X-windows. We used the Tcl/Tk combination to implement the graphical user interface for MetVUW Workbench. Because Tcl/Tk code controls both the user interface and the top-level program control, we are able to closely couple the state of the program and the interface. For example, the date and time of the current image displayed by the graphical user interface is directly linked to the program state variable storing the current date and time of the image. Whenever this variable is changed, the graphical user interface is automatically updated.

2. *Full Text Search*  MetVUW Workbench employs a package for full text search called LQ-text. LQ-text is able to quickly locate all instances that match a key phrase in a large corpus of indexed free-format text. LQ-text supports simple 'stemming', which means, for example, a search phrase containing 'north-westerly' will also match 'north-westerlies' in the text. MetVUW Workbench uses text to retrieve documents that match a key word or phrase from a collection of meteo-rological documents. After such a search the user can immediately retrieve other sources of data for the period to which a document refers.

3. *Image Compression*

One of the challenges of the MetVUW Workbench project is dealing with high volumes of satellite imagery. Hourly satellite images for one day requires 6Mb in raw form, while a year of imagery requires over 2Gb of storage. Ordinary loss-less compression algorithms such as Lempel-Ziv only reduce this requirement by about 30%. We therefore employ the lossy image compression algorithm specified by the Joint Photographic Experts Group or JPEG (Wallace, 1991; Pennebaker and Mitchell, 1993). Images compressed with JPEG appear almost indistinguishable from the original, even when compressed to a tenth of their storage requirements. Using JPEG, a year of imagery requires only 200Mb.

4. *Help Facility*

MetVUW Workbench Version 1.0 is already a large application that requires a manual and an effective on-line help facility. We wrote the manual using GNU'S Texinfo system (Stallman and Chassell, 1993). The Texinfo program automati-cally encodes the text of the manual in two ways: as a word-processed document, and in a hypertext format suitable for use by the GNU Info hypertext system. The hypertext version of the manual is built into the graphical user interface of MetVUW Workbench using a Tcl/Tk interface package called TkInfo designed for this purpose. Appropriate sections of the manual are indexed in terms of the corresponding windows and buttons of the user interface. Clicking on any win-dow or button of the user interface pops up a window containing the appropriate section of the manual. Users can then follow further cross references to retrieve related information. MetVUW Workbench uses the TkInfo user interface to the GNU Info hypertext system to display the on-line help.

# 9   Conclusions and Future Work

Our initial efforts in developing MetVUW Workbench have proved highly successful. Version 1.0 of MetVUW Workbench is already a powerful and easy-to-use package for retrieval and display of historical meteorological data. The system is in active use within the Institute of Geophysics at Victoria University of Wellington, and has attracted considerable interest from the Meteorological Service of New Zealand.

We plan to extend the present system in two ways. First, we intend to add an intelligent retrieval system that allows forecasters and other meteorologists to retrieve past weather situations by specifying high level descriptions of them. The system is to serve as a 'memory amplifier' that allows users to rapidly locate historical situations of interest. In particular, forecasters should be able to use the system to quickly retrieve situations that are similar to the current one in meteorologically significant respects, providing them with an additional source of information to supplement the output of sometimes equivocal numerical models

Second, we intend to extend and improve MetVUW Workbench's existing facilities for retrieval and display of data in a number of ways, including the following:

1. *Improve the database management system.*

   Presently, indices to data are pre-built and static, which make it difficult to add or remove data. We plan to construct facilities that make it easy to add or delete data. To solve this problem, we intend to switch to an extensible RDBMS called Postgres as a substrate to future versions of MetVUW Workbench (Stonebraker and Rowe, 1986). This change will also facilitate the development of intelligent retrieval techniques.

2. *Provide customisable false colouring and contouring.*

   Currently, false colouring schemes are presets defined in configuration files, contours can only be constructed at pre-configured intervals, and contouring is limited to the horizontal plane. We plan to allow users to specify their own false colouring schemes using the graphical user interface. Users should also be able to specify different intervals for contouring, and request contouring of vertical or oblique planes through the three dimensions of meteorological data

3. *Derive numerical fields on-line.*

   At present, all data fields must be pre-computed, including fields such as wind velocity that we derive ourselves from the component fields. These derived fields exacerbate the high data storage requirements of MetVUW Workbench. We plan to allow some of these fields to be calculated on demand, allowing storage requirements to be significantly reduced.

4. *Develop techniques for domain dependent compression of numeric data.*

   We already use JPEG, a domain dependent compression algorithm, to compress image data We plan to also introduce domain dependent techniques for compressing numeric data

5. *Improve the laser disc interface.*

   Most user interface controls for the laser disc player adhere closely to the analogy to a video player remote, but some advanced features do not fit this model. In particular, repeat loops and swing loops of frames are currently specified by frame number; it should also be possible to specify them by date and time. Also, the addition of a time-line view of the laser disc would enhance the user's model of the data making it possible to select animation periods and particular dates by point-and-click.

6. *Make use of frame grabbing technology.*

Satellite images on laser discs are stored in an analogue format, so they are not readily computer-readable. However, frame-grabber technology allows a video signal to be reinterpreted as a digital image. With the addition of frame-grabber hardware to the SparcStation, we could animate satellite imagery, directly on the computer display. This would eliminate the need for a separate video monitor and for storage of satellite imagery in a digital format if corresponding video imagery is available on laser disc. It will also be possible to false-colour these animations and to calculate contour overlays in real time. (Note that the slow data rate of the hard disk drive precludes rapid animation of digital imagery.)

7. *Improve organisation of document collections.*

The current system allows full text search of document collections, but provides only a primitive mechanism for organising and displaying the documents that are retrieved. We expect the size of the document collection to increase considerably in the near future, so this mechanism will soon be inadequate. We therefore propose to construct a more sophisticated facility for summarising and displaying the results of full text searches that is better suited to larger and more diverse document collections.

# Acknowledgements

This research was supported in part by IGC research grants and by a grant from the Electricity Corporation of New Zealand. Additional funds were supplied by the Department of Computer Science and the Institute of Geophysics at Victoria University of Wellington. Thanks to Dr. Crispin Marks for provision of contouring code. Kim Rutherford implemented a prototype for the audio presentation of hydrological data. Finally, we would like to thank staff at the Meteorological Service of New Zealand for helpful advice at various stages in the design of the system.

# References

Jones E K and Roydhouse A, 1993, Intelligent retrieval of historical meteorological data. In Proceedings of a Workshop on Artificial Intelligence and the Natural World, Melbourne, Australia.

McGregor J and Jones E K, 1993, MetVUW:A multi-media teaching aid,In the First International Conference on Computer-Aided Learning and Distance Learning in Meteorology, Hydrology, and Oceanography, Boulder, Colorado.

Ousterhout J K, 1991, An X11 toolkit based on the Tcl language. In Proceedings of the winter USENIX Conference, pages 105-115. USENIX Association.

Ousterhout J K, 1994, Tcl and the Tk Toolkit. Addison-Wesley. Forthcoming.

Pennebaker W B and Mitchell J L, 1993, JPEG Still Image Data Compression Standard. Van Nostrand Reinhold.

Sabin M, 1986, A survey of contouring methods. *Computer Graphics Forum,* **5** 325

Stallman R M and Chassell R J, 1993, The Texinfo Manual, Version 2.18. Online documentation.

Stonebraker M and Rowe L, 1986, The design of Postgres. In Proceedings of the 1986 SIGMOD Conference on Management of Data, Washington, D. C. ACM Press.

Wallace G K, 1991, The JPEG still picture compression standard, *Communications of the ACM*, **34**(4) 30

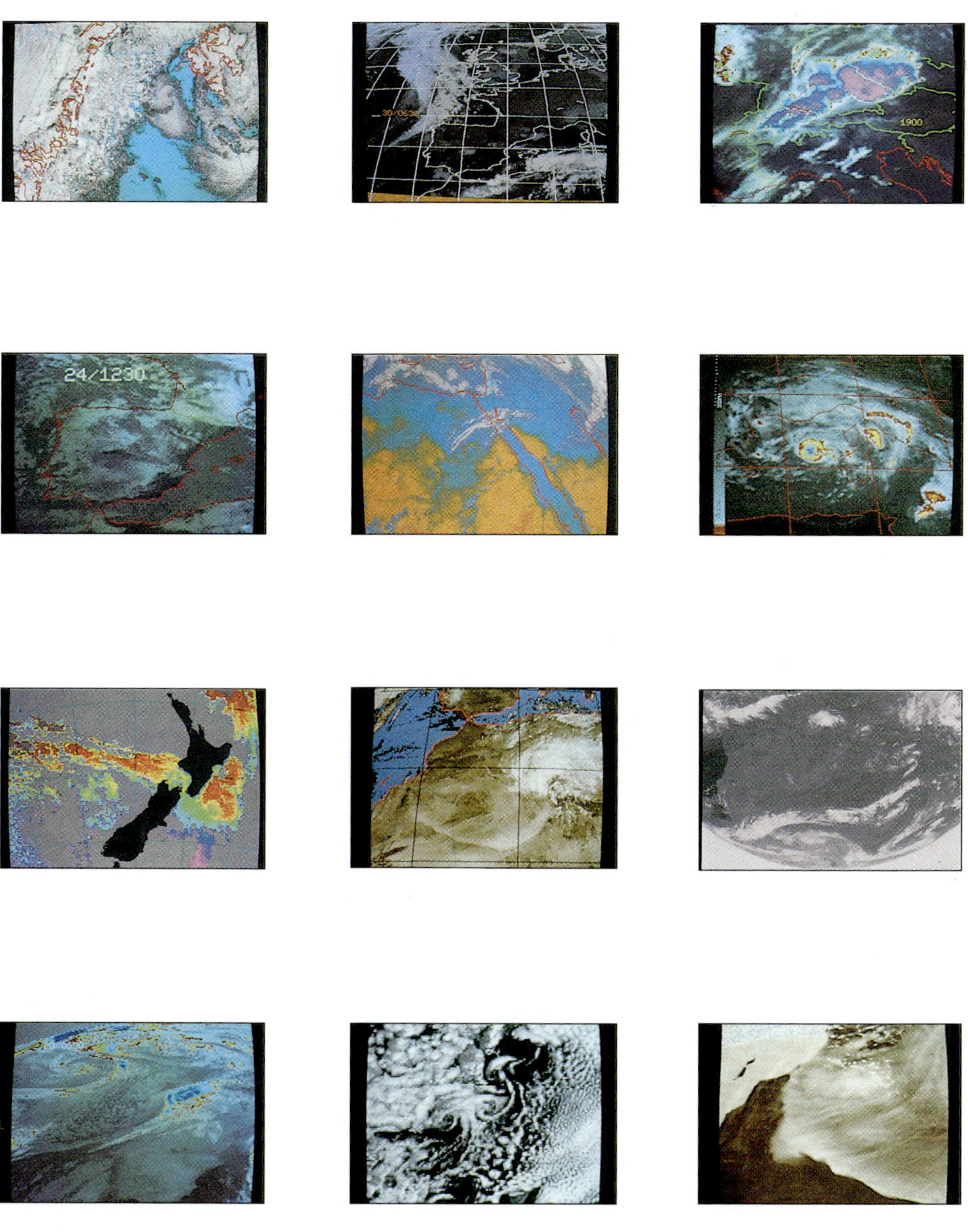

**Plate 1:** *METEO DISC contains some 650 cloud motion videos covering a wide range of weather situations, scales and special phenomena: From near-polar ice and cloud motions (1) to sub-tropical Sahara dust storms (8, 12). From short high-level cirrus cloud motions (2) to three continuous months of sea surface temperatures (7). From Kàrmàn vortices (11) and orographically induced waves (4) to tiny hurricane-like vortices over the Mediterranean (6) or hail storm super cells over Munich (3). From classical cyclone developments (10) to types of cyclogenesis (9) that were understood only after the availability of tools such as Meteo Disc. See Warneke, Zick and Toussaint pp293–310.*

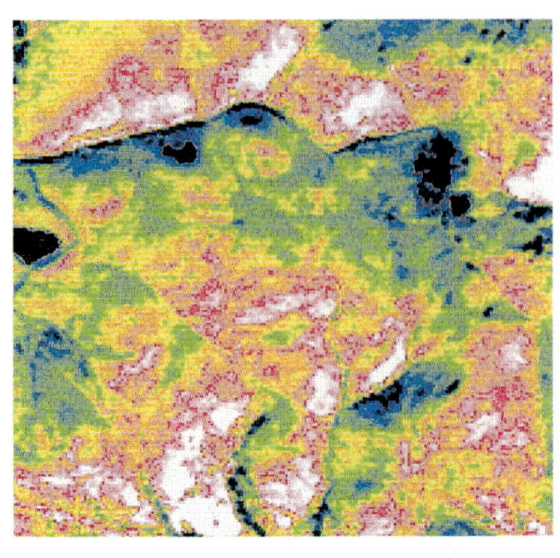

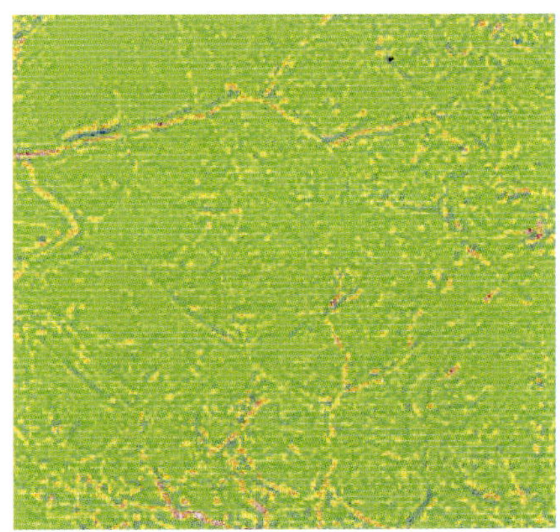

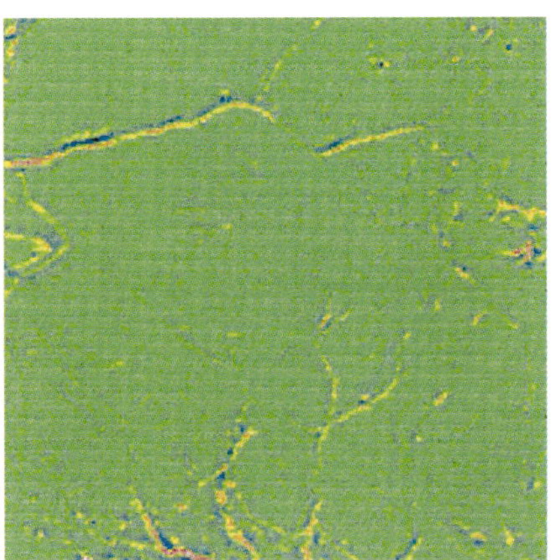

**Plate 2:** *(a) A SPOT Band 3 image of the Ochil Hills in Scotland and the result of applying 'rolling ball' filters of dimension 3 and 7 in (b) and (c) respectively. These methods are discussed by Watson on pages 241–247.*

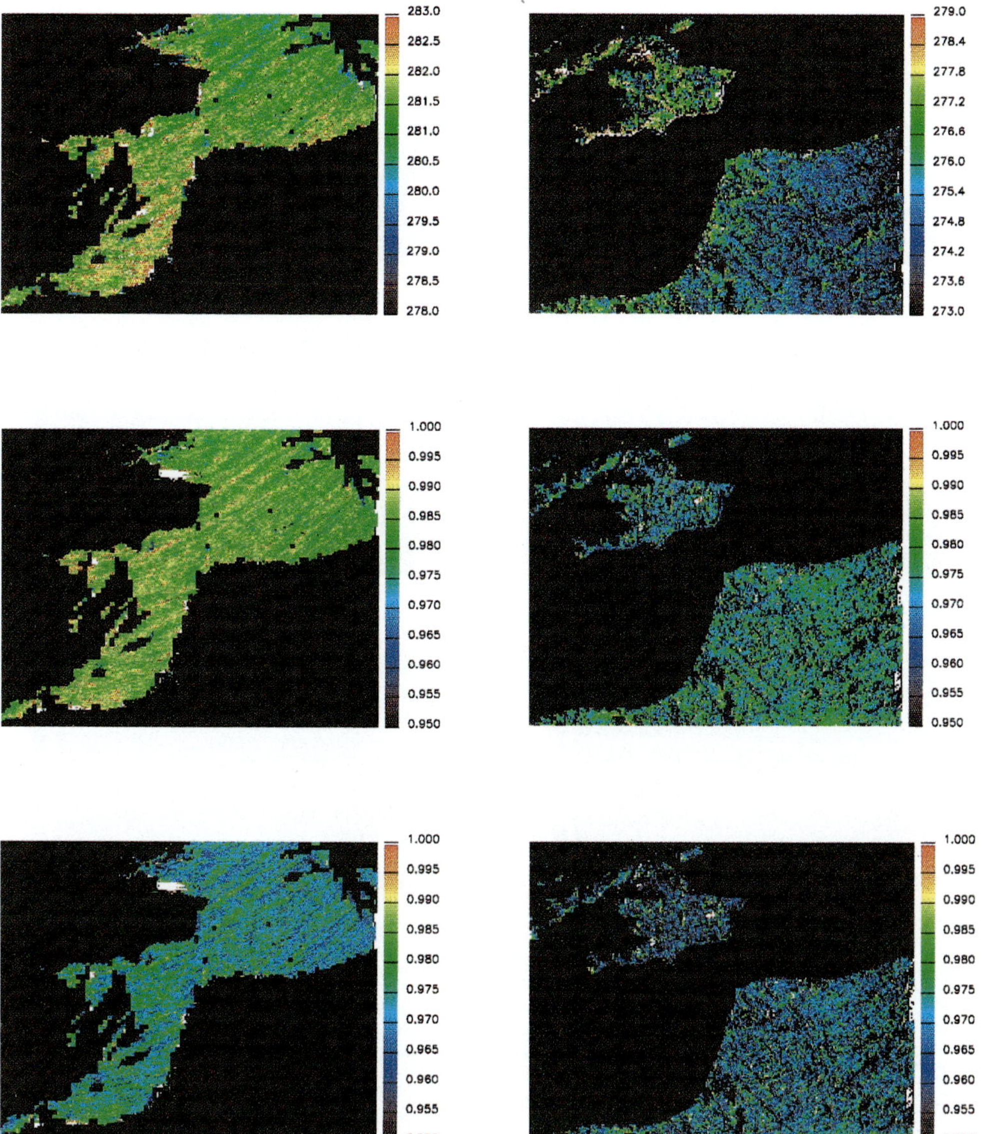

**Plate 3:** *The sea surface temperature (top left) and beneath it the the emissivities of the sea surface for the 11μm and 12μm channels. Top right is the the land surface temperature and beneath it the emissivities in the same channels—derived from* ATSR *thermal band data using the method described in Section 3.7 of the article by Xue and Cracknell (pages 209–240). The area is located at English Channel (0–3° E, 49.5°–52° N). Dark blue is cloud covered area and sea.*

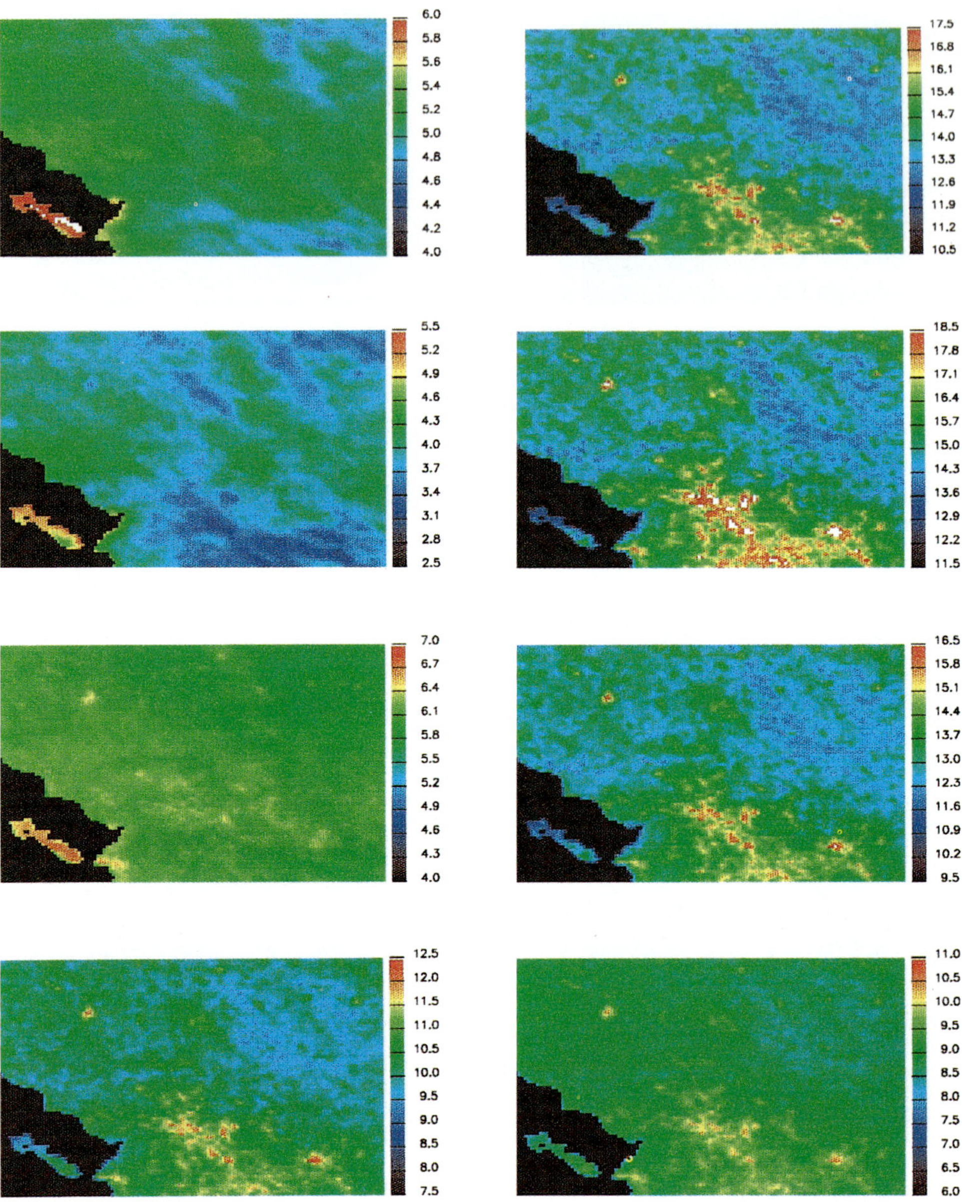

**Plate 4:** *Ground surface temperature predicted using the method developed by Cracknell and Xue (pages 209–240) with data from France (about 47–48° N, 0.8° E–0.4° W) over a one day period (9 April 1992). The temperatures were calculated at intervals of 3 hours: 0:00 GMT is at top left and increasing to 9:00 GMT down the first column; the second column covers 12:00 GMT (top) to 21:00 GMT.*

# Displacement Vector Fields of Oceanic Surface Structures from Satellite Image Sequences

L Bannehr and M Rohn

Institut für Meteorologie
Freie Universität Berlin

## 1 Introduction

Satellite image sequences provide detailed oceanographic information of the biological activity, the transportation of various water constituents and sea surface currents. More knowledge about the displacement of the sea surface structures such as algae, chlorophyll, yellow substance, sea surface currents etc. is needed in order to understand the biological cycle and protect our environment for future generations. For the modelling of our climate the interaction between atmosphere and ocean is currently a key issue in climate research.

With the development of the new generation of high resolution spectral satellite radiometers many new parameters can be inferred and the accuracy of the known ones can be improved by using passive remote sensing techniques. Therefore, new and better retrieval algorithms have to be developed by the scientific community to achieve more knowledge about our globe.

Over the past 20 years the Maximum Cross Correlation MCC method has been employed extensively to infer the wind fields in the troposphere from the infrared channels of Meteosat satellite. Image data from polar orbiting satellites such as NOAA and NIMBUS are used to infer the sea surface currents and ice motion from visible and infrared channels. Some weaknesses remain when employing this very well known statistical approach. The deficiencies of the MCC technique are the difficulties to reconstruct rotational and deformational motion patterns properly (Kamachi, 1989) as well as to retrieve information where the image sequence shows little contrast.

We adopted a new functional analytic approach from computer vision to estimate

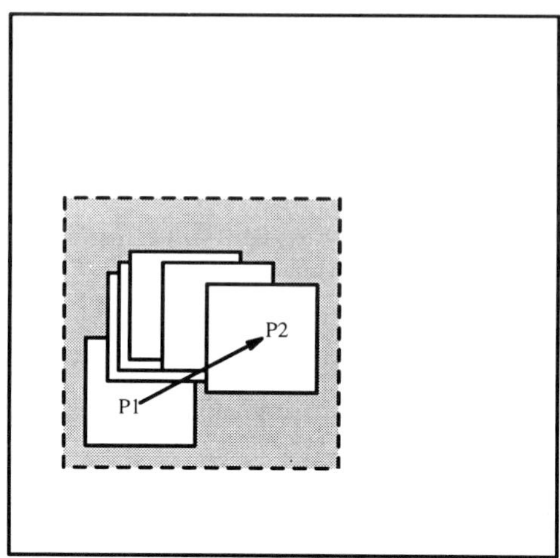


**Figure 1.** *Illustration of the* MCC *concept.*

the displacement vector fields from satellite image sequences, which overcomes some deficiencies of the MCC method. It has been found that this method is especially suitable for images where a fuzzy greyvalue structure is prevailent. Since this method is new to geophysical sciences, some problems are unresolved so that it remains an active area of research. For the sake of a better understanding concerning the differences in the results achieved by the MCC technique and the Functional Analytic Method (FAM), both concepts will be outlined.

## 2 Maximum cross correlation (MCC) technique

To estimate a displacement vector the MCC technique needs a sequence of at least two successive images. For this method it is assumed that the local intensity distribution is conserved. Figure 1 illustrates the principle of the maximum cross correlation technique whereby the black frame denotes the actual image size. Around the point P1 of the first image a square containing an invariant structure is selected. This square is called template. The template is moved pixel by pixel around a larger area (grey area) in the second image which is called the search window. The position of the displaced vector is found from the position of the maximum cross correlation.

For oceanographic applications the template size (first image) often chosen is between $16 \times 16$ and $22 \times 22$ with a search window size (second image) of $32 \times 32$.

Some deficiencies of the method have already been mentioned in the introduction. Progress has been made recently by Wu *et al* (1993). He includes information of neighbouring vectors in his correlation-relaxation method. A drawback of the method is still that it cannot account for strong shear motion. Büche *et al* (1991) developed a new preprocessing scheme which provides better results for images with a flat greyvalue structure. A review of various approaches is given by Singh (1991).

# 3  Functional Analytic Method (FAM)

Assuming the conservation of the greyvalue of a certain pattern within an image sequence, the displacement of the greyvalue structure can be reconstructed. For two-dimensional flow, Horn and Schunk (1981) determined the motion pattern by the minimization of a nonlinear functional

$$J(\mathbf{v}) = \int \left\{ [g_t + \mathbf{v} \cdot \nabla g]^2 + \lambda^2 \left[ |\nabla v_1|^2 + |\nabla v_2|^2 \right] \right\} d\mathbf{x} \tag{1}$$

where the second term is a regularizing constraint. Schnörr (1991) solved this variational problem (Equation 1) using the technique of finite elements. The first term expresses the individual conservation of the greyvalue $g(x, y)$, where $\mathbf{v} = (v_1, v_2)$ denotes the corresponding local displacement.

$$\frac{dg}{dt} = g_t + \mathbf{v} \cdot \nabla g = 0 \tag{2}$$

This equation is also called the brightness change constraint equation. It implicitly contains the assumption that the illumination of the feature remains constant during the displacement. Furthermore, it allows only to determine the component perpendicular to the grey value gradient. Therefore, the displaced vector can only be derived accurately at the edge of a feature. To overcome this problem a second constraint is needed which is expressed by the second term in Equation 1. It is called the general smoothness constraint and is required to derive the two vector components $\mathbf{v} = (v_1, v_2)$. In featureless areas, the smoothness constraint performs the interpolation between neighbouring vectors. In regions of strong discontinuities the general smoothness constraint smoothes these areas and the discontinuities are washed out. This is obviously a drawback of the general smoothness constraint.

$$|\nabla v_1|^2 + |\nabla v_2|^2 = 0 \tag{3}$$

Nagel (1987) suggested the smoothness constraint be weighted depending on the local greyvalue structure. This oriented smoothness constraint is only applied in areas where the greyvalue structure does not reveal enough information to estimate both components of the displacement vector field. In the direction of the greyvalue gradient sufficient information is available to determine the displacement component. Since discontinuities in the vector field occur in real situations, the smoothness constraint in the direction of the greyvalue gradient will be suppressed and forces the vector to be parallel to the greyvalue edge. The oriented smoothness constraint is given by:

$$\frac{(\nabla v_1 \cdot \nabla g')^2 + (\nabla v_2 \cdot \nabla g')^2 + \gamma(|\nabla v_1|^2 + |\nabla v_2|^2)}{2\gamma + |\nabla g|^2} = 0 \tag{4}$$

where $\nabla g' = (g_y, -g_x)^T$, so that $\nabla g' \perp \nabla g$.

As the input data consist of first order derivatives in time and space, the derivative of a three-dimensional Gaussian filter is applied to the image sequence. The data are smoothed and the omnipresent noise is, therefore, reduced. The filter width is variable

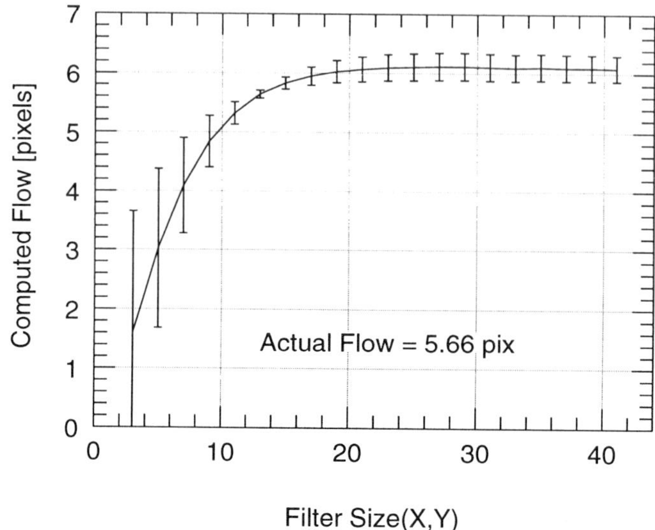



**Figure 2.** *Influence of the filter size on the estimated velocity field for a constant flow of 5.66 pixels.*

and is an important input parameter, which has a strong influence on the results. The factor chosen for the Gaussian filter depends on the scale of the structure to be observed.

The factor $\lambda$ of the second term in Equation 1 allows the smoothness constraint to be weighted. With increasing $\lambda$ the vector fields are smoothed and the absolute value of the vectors becomes smaller. From the experience achieved by processing many images it was decided to keep $\lambda = 1$ for all cases. Choosing $\lambda = 1$ treats both terms in Equation 1 equally.

Sequences of artificial images are created to study the performance of the algorithm. The advantage of simulated data is that the influence of various input parameters on the flow field can be investigated in detail and fully controlled. Furthermore, the complex verification of the results from satellite data is drastically reduced and the errors can be estimated by applying simple statistics.

Since we only simulated situations without strong flow discontinuities the results gained for the two smoothness constraints are identical.

The effect of the derivative filter is presented in Figure 2. Here the mean computed velocity is plotted versus different filter sizes. The best agreement between the calculated and actual flow is gained for the filter size twice the simulated displacement ($13 \times 13$). For this ideal case the standard deviation has its minimum. For a larger filter size the flow is slightly overestimated. A filter size smaller than the simulated velocity leads to unrealistic displacements.

The first term in Equation 1 implies no brightness change within the image sequence. However, brightness changes in consecutive satellite images can be caused by variations in viewing geometry and solar elevation, by skin effects or by natural turbidity vari-

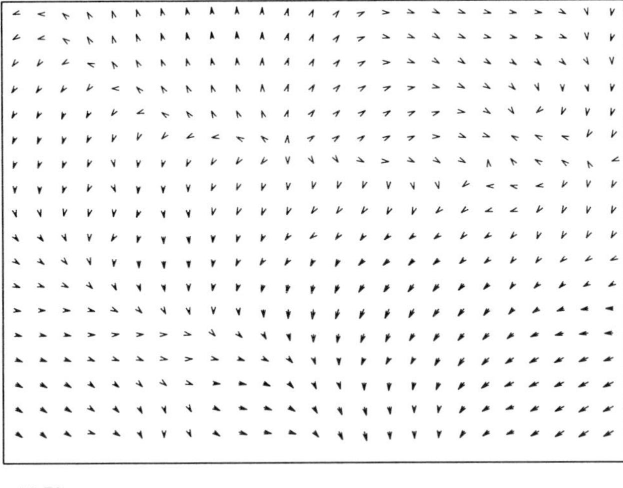



**Figure 3.** *Influence of illumination changes by an offset corresponding to 1 K temperature change.*

ations. Thus, motion can be faked by images that differ only by a constant offset of the greyvalue. A constant offset results in displacements even if the greyvalue gradient of the images is conserved. Under realistic conditions a temperature accuracy of 1K in the infrared region is expected (Schlüssel *et al.* , 1990). The result of a change in brightness by 1K is illustrated in Figure 3. The maximum displacement found is 0.79 pixels for the given input parameters of a filter size of 13. This value is small and has only to be considered if the motion is rather slow.

Two data sets, one of pure translational and one of pure rotational motion, were created. For the translation test a sequence was created by shifting a picture by 4 pixels in $x$ and $y$ direction. This implies a displacement of 5.66 pixels with 305° flow direction. Figure 4 visualizes how good the algorithm reconstructs the simulated flow field. An angular standard deviation of 0.88° and 0.14 pixels for the standard deviation in absolute velocity was found for this example.

Another experiment was carried out in order to study how well eddies, which are often found in oceanographic and meteorological images, will be reconstructed by FAM. Figure 5 shows a displacement vector field computed from a sequence of images representing a clockwise rotation of 5°. A first visual impression indicates a linear increase of velocity from the centre towards the fringes of the picture, and thus, a rather realistic reproduction of the given rotation field. The minimum and maximum displacement (0.06 pixels and 8.39 pixels) are in good agreement with the actual motion in the centre (0.0 pixels) and along a circle with a radius the width of the image (8.36 pixels). The error of 20% for the average estimated motion appears, however, rather high. Evidently, it is caused mainly by highly underestimated vectors in the upper left section of the image. In this area, a cellular pattern of small convective clouds are exhibited. Obviously, FAM is less suitable for such a pattern. For the other regions, where more homogeneous patterns are observed, FAM provides good results.

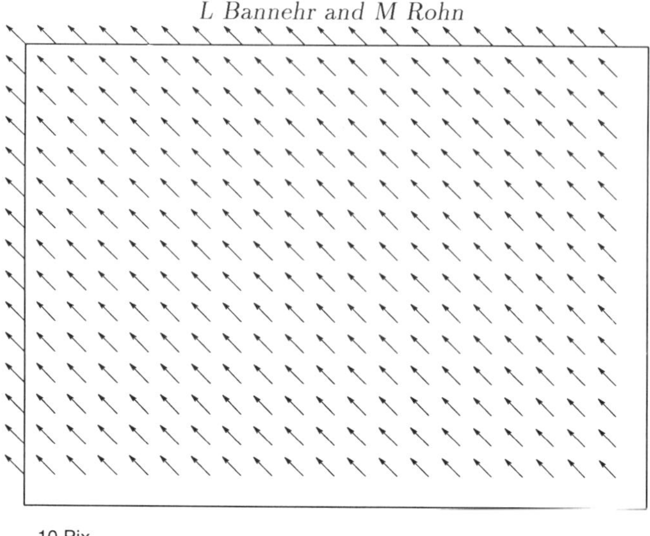


**Figure 4.** *Estimation of pure translational motion using the functional analytic method.*

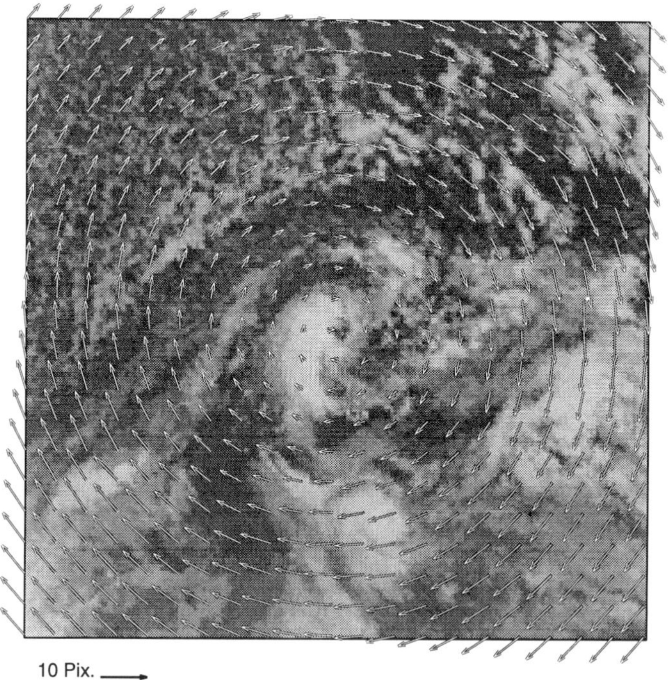


**Figure 5.** *Inferred vector field for a clockwise rotation of 5°.*

Similar results are achieved applying the MCC method to the translation and rotation test data set and, hence, they are not individually discussed. However, under real conditions both methods provide different results as shown in Section 6.

# 4 The choice of satellite

Various satellites can be used to infer the movement of different kinds of sea surface features. The choice of the satellite depends, apart from the scientific objective, mainly on the time lag between the observations and the spatial resolution of the satellite radiometer. For oceanic applications the motion varies roughly between 0 to 50cm/s. Consecutive images should not be shorter than 12 hours apart. For this reason the NOAA polar orbiting satellites and geostationary satellites such as METEOSAT, GOES and GMS are most favourable. Satellites whose global coverage takes more than 24 hours are not useful to derive the displacement vectors, since the sea surface structure may have altered in the meantime.

As we have mentioned earlier, apart from anthropogenic substances, mainly sediments, chlorophyll and yellow substance modify the ocean color. They can be best observed in the visible spectrum. The NOAA and METEOSAT satellites allow the estimation of the sediment transport from the visible channels and the sea surface velocity fields using the observed infrared radiances as a tracer.

Motion of chlorophyll and yellow substance is not observable by these satellites because it requires a high spectral resolution of 10–20nm in the visible range to derive these parameters.

For future applications the Sea Wide Field-of-View Scanner (SeaWiFS) satellite will be launched on the Seastar satellite. With eight high resolution channels this polar orbiting satellite (average height 705km) provides all spectral characteristics to retrieve the water substances.

# 5 Further problems and verification of the results

When the images are received by the ground station they are geometrically distorted. This is due to the method of scanning and the curvature and the rotation of the Earth. Therefore, before the displacement vector fields can be determined, the image sequence must be put in the same perspective. Often a stereographic or mercator map projection is chosen. With the current methods a navigation of $\pm 1$ pixel is achievable. Assuming an accuracy of $\pm 2$ pixels (1 NOAA pixel = 1.1km $\times$ 1.1km) of a geometrically corrected image sequence this results in a velocity of the sea feature movement of $\pm 2.55$ cm/s. Since the ice motion and sea surface currents are small, it is easily possible to induce an error of 20 % in the vector fields by a bad geometric correction.

Other unwanted interferences are the tides, radiance changes due to different solar-satellite geometry as well as upwelling and downwelling.

Currently IR-image sequences suit the purpose more than VIS data, because aerosols and brightness changes mask the sea surface and lead to ambiguous results. However, weather conditions also have an impact on the results since only the sea surface temperature is used as a tracer for the sea surface currents.

The verification of the retrieved vector fields is a difficult task. The verification can be performed by:

- studies on how the algorithm behaves on artificially created image sequences,

- films giving flow directions,

- *in situ* measurements,

- ocean current models.

Theoretical studies on artificial image sequences as a verification tool have already been outlined before. Such tests are simpler than real nature and can only provide the general behaviour under ideal conditions. We have the capabilities to make a movie-loop from any image sequence. It provides good information about the flow direction and to some extent about the absolute value of the displacement. Unfortunately, *in situ* measurements from ships and buoys are sparsely distributed and rare so that it remains a difficult task to verify the retrieved vector fields using observed data. Complex ocean models may assist in the verification process. For areas like the North Sea they provide the direction and absolute value of the flow pattern for the exact overpass of the satellite.

# 6    Estimated sea surface currents

In the following, two examples of the estimated sea surface currents from the North Sea and East China Sea are presented using the functional analytic method with the oriented smoothness constraint and the maximum cross correlation technique. These examples also show how different the results from both approaches can be.

The North Sea has 70% cloud cover per year - a very cloud-rich climate. In winter most parts of the North Sea are well mixed in the vertical direction by tides and winds. Figure 6 and Figure 7 show the derived sea surface flow patterns of the North Sea for the NOAA image sequence May 19 to 21, 1992, using the sea surface temperature as a tracer for the FAM and the MCC method. By choosing images 24 hours apart, the contamination of data by disturbing tidal effects in this area and possible errors due to relatively small flow velocities still exist but are minimized.

A verification of this vector field is even more difficult than one for the midtroposphere because less conventional data are available. Average summer, winter or annual flow maps do not necessarily represent the actual flow and are not always useful for verification purposes. Nevertheless, they can provide at least information about the velocity range. Since the signal is received from the very top layer of the sea surface, the motion pattern is strongly altered by variable weather conditions. Hence, weather maps must be included to analyse the flow pattern.

The concurrent weather maps show a persistent anticyclone with its centre over the North Sea. Therefore, no strong wind stress should be observed at the sea surface. It remains to explain the computed displacement vector field like the one reflecting to some extent the common northerly current generally observed in these eastern parts of the North Sea. These features cannot be observed from the MCC results. Without any further quality control (*e.g.* consistency checks) a clear statement about the flow direction appears to be impossible.

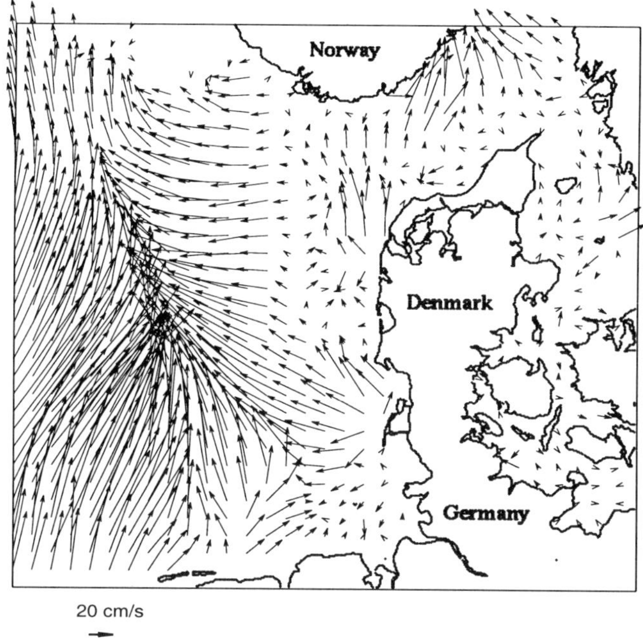


**Figure 6.** *Sea surface currents of the North Sea estimated from the functional analytic method.*

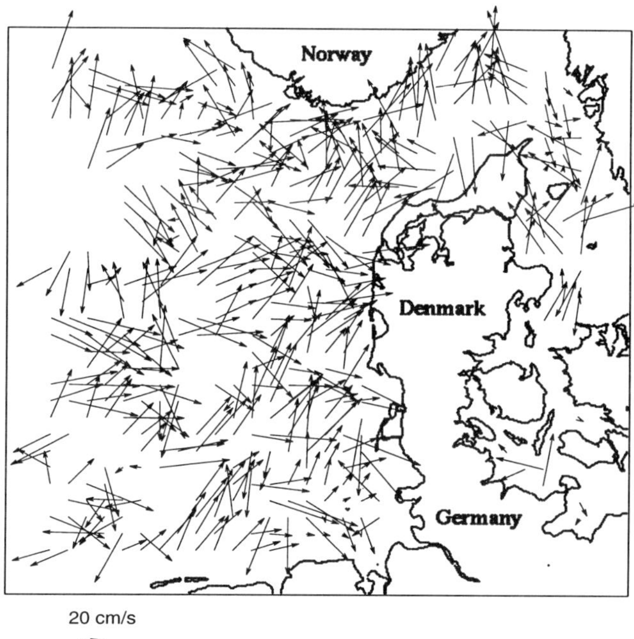


**Figure 7.** *Sea surface currents of the North Sea derived from the maximum cross correlation method.*

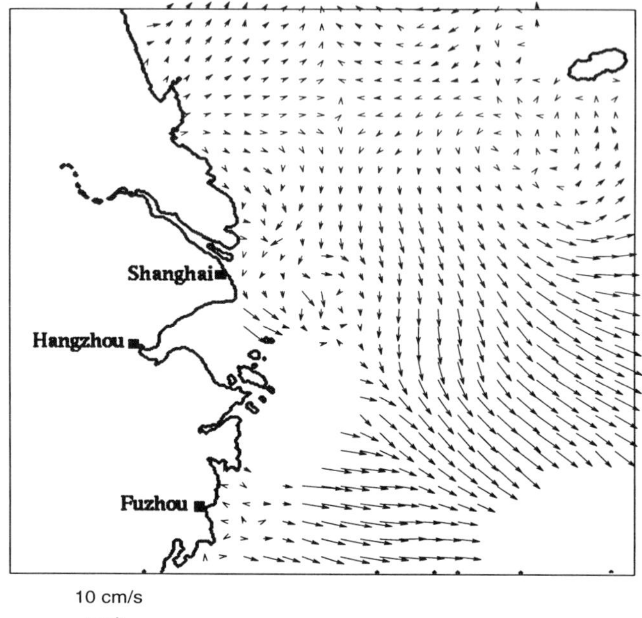


**Figure 8.** *Estimated sea surface currents of the East China Sea using the functional analytic method.*

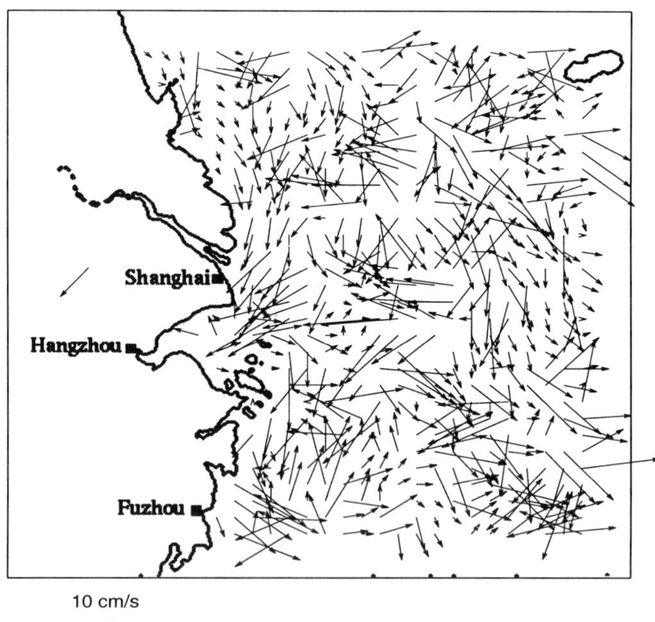


**Figure 9.** *Inferred sea surface currents of the East China Sea from the maximum cross correlation.*

Figure 8 and Figure 9 present the estimated sea surface currents from the FAM and the MCC technique for the East China Sea on December 3 to 5, 1989, 18:00 GMT. Again FAM provides a more coherent vector field than the MCC method. Chao (1991) investigated this area within a numerical study. From observations some features such as the circulation pattern below the island of Jizhou and the south-easterly flow in the lower third of Figure 8 are also reconstructed nicely. Similar to the flow pattern of the North Sea the MCC technique reveals for the East China Sea no unique velocity field (Figure 9).

# 7  Conclusions

The presented displacement vector fields indicate that the functional analytic approach can successfully be applied to oceanographic image sequences. A positive feature is that the computation is fast compared to the MCC method, so that even sequences of large image sequences can be processed in a reasonable time. Compared to the MCC technique FAM showed always realistic coherent vector fields.

Currently only the NOAA polar orbiting satellite and some geostationary satellites such as METEOSAT and GOES are available to obtain information about the sea surface motion. In the near future the SeaWiFS radiometer designed for oceanographers will open new scientific objectives and permits to infer the motion of various water substances. However, the contaminations of clouds, upwelling, downwelling and tides will make the motion analysis of the oceans further challenging.

# Acknowledgements

The authors would like to thank Professor G Warnecke for valuable stimulating discussions about the subject of motion analysis. We also like to thank Mr. Qingge Liu from Hangzhou, China, for preprocessing the presented images from the East China Sea. The project was granted by the BMFT (BEO/7103F0016A).

# References

Büche, G., Kummer, A., Ottenbacher, A. and Fischer, H., 1991, Displacement vectors from METEOSTAT-VW-images using a new extraction technique, *Proc Workshop on Wind Extraction from Operational Meteorological Satellite Data, Washington, DC*, 17 - 19th Sept. 1991 .

Chao, S.Y., 1991, Circulation of the East China Sea, A numerical Study, *Int. J. Oceanograph. Soc. of Japan*, **46**, 22

Horn, B.K.P., Schunck, B.G., 1981, Determining optical flow, *Artificial Intelligence*, **17**, 185

Kamachi, M., 1989, Advective Surface Velocities Derived from Sequential Images for Rotational Flow Field: Limitations and Applications of Maximum-Cross-Correlation Method with Rotational Registration, *J. Geophys. Res.*, **94**, No. C12, 18,277

Nagel, H.H., 1987, On the estimation of optical flow: Relations between different approaches and some new results, *Artificial Intelligence*, **33**, 299

Schlüssel, P., 1990, On the Bulk-Skin Temperature Difference and its Impact on Satellite Remote Sensing of Sea Surface Temperature, *J. Geophys. Res.*, **95**, No C8, 13,356

Schnörr, C., 1991, Determining Optical Flow for Irregular Domains by Minimizing Quadratic Functionals of a Certain Class, *Int. J. Computer Vision*, **6**, No. 1, 1991, 25

Singh, A., 1991, Optical Flow Computation, *IEEE Computer Society Press, Los Alamitos, California*

Wu, Q.X., 1993, Computing Velocity Fields From Sequential Satellite Images, *Satellite Remote Sensing of the Oceanic Environment*, eds Jones, Sugimori, Stewart, Seibutsu Kengyusha Co. Ltd

# Derivation of Wind Vector Fields From Satellite Image Sequences

M Rohn and L Bannehr

Institut für Meteorologie,
Freie Universität Berlin

## 1  Introduction

One of the most important goals of motion analysis in meteorology is the extraction of wind fields from satellite image sequences. Those wind fields are used, for example, as input data in numerical weather prediction. Beyond, the motion analysis of subsequent satellite images opens a wide field for studying various dynamic phenomena in the atmosphere (Warnecke 1987). In addition, motion analysis of satellite imagery is used in other disciplines of geoscience, *e.g.* studying sea ice movements or transport phenomena in the ocean.

For the last two decades, the Maximum Cross Correlation (MCC) method has been mainly used for the derivation of displacement vector fields from image sequences. Thus, the MCC-method is well understood and can operationally be used for the extraction of wind data. As the MCC-method shows deficiencies due to either smooth image structure as revealed in the Water-Vapour (WV) channel of METEOSTAT (Büche *et al.* 1991) or in the reconstruction of rotational and deformational motion (Kamachi 1989), the improvement of tools for motion analysis is still in progress. Various groups are working on the improved application of the MCC method for tracking of WV-features (Laurent 1990), on preprocessing schemes for the image data before applying the MCC-method (Büche *et al.* 1991) or quality control schemes which check the consistency of the resulting vector field by including the information of adjacent vectors (Wu 1993).

In Section 2 a functional analytic approach of Horn and Schunck (1981) for motion analysis is presented. It has been adopted from the computer vision field. As the input data consist of first order derivatives of the image data the design of the derivative filter is important and will be discussed in detail in Section 3.

In Section 6 a result of computer science is presented that modifies the approach of

Horn and Schunck (1981) in respect of the image structure (Nagel 1987). The modification leads to a better representation of discontinuities of the motion at greyvalue edges. In the computer vision field those approaches are developed on the basis of artificial motion scenes or rather simple real image sequences compared to the highly complex patterns of atmospheric motion. Further development of those algorithms in the computer vision field includes *a priori* knowledge because a certain approach cannot suit the variety of motion pattern. In Section 7 a modification of the approach of Horn and Schunck (1981) is discussed that considers the kinematic quantities such as vorticity, divergence and deformation.

Applications to a METEOSAT-VW image sequence are shown for all schemes. The resulting displacement vector fields are compared to radiosonde winds at 500hPa and discussed with the concurrent weather map.

## 2   The approach of Horn and Schunk

If it is supposed that the greyvalue $g(t, \mathbf{x})$ of a certain scene element remains constant as it moves in space and time, then the two-dimensional motion field $\mathbf{v}(t, \mathbf{x})$ can be reconstructed. Horn and Schunck (1981) achieved this minimising the following functional:

$$J(\mathbf{v}) = \int \left\{ [g_t + \mathbf{v}\cdot\nabla g]^2 + \lambda^2 \left[ |\nabla v_1|^2 + |\nabla v_2|^2 \right] \right\} d\mathbf{x} \tag{1}$$

The functional contains two constraints which are defined as data term

$$g_t + \mathbf{v}\cdot\nabla g \tag{2}$$

and smoothness term

$$\lambda^2 \left[ |\nabla v_1|^2 + |\nabla v_2|^2 \right]. \tag{3}$$

Figure 1(a) illustrates both terms. Assuming the greyvalue is a tracer, then an advection equation for the greyvalue is given by

$$\frac{dg}{dt} = 0 = g_t + \mathbf{v}\cdot\nabla g \tag{4}$$

which provides the data term (2) above.

By separating the displacement vector $\mathbf{v}$ we can see that (4) locally only determines the displacement parallel to the greyvalue gradient:

$$\mathbf{v}_\parallel = -\frac{g_t}{|\nabla g|} \tag{5}$$

In the field of image processing this is known as the aperture problem. It is illustrated in Figure 1(b). In Figure 1(b) the small window in the image plane indicates that for the determination of the object displacement only the displacement perpendicular to its edge can be observed. An edge marked by a strong change of greyvalue means that the motion is parallel to the greyvalue gradient. Besides, one equation (4) can only determine one component of the displacement. Therefore, a second constraint is

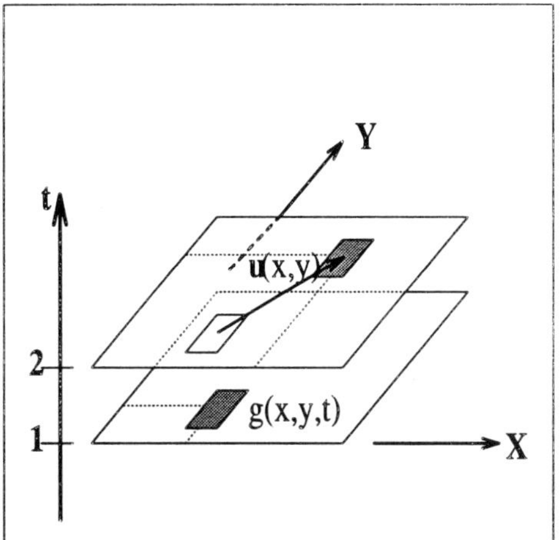


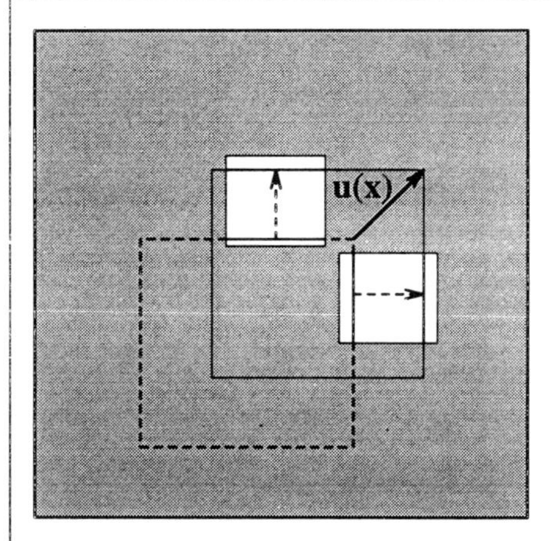


**Figure 1.** *Motion analysis of digital images. (a) Motion analysis by modeling the displacements* $\mathbf{u}(x,y)$ *of greyvalue pattern* $g(x,y,t)$ *within subsequent digital images. (b) Aperture problem. The actual displacement* $\mathbf{u}$ *is indicated at the upper right corner of the moving square while the observed motion* $\mathbf{u}\cdot(\nabla g/|\nabla g|)$ *is drawn as a dashed vector.*

needed. Horn and Schunck (1981) reconstructed the motion field by minimizing the variation between adjacent vectors. This is expressed by the gradient of both vector components.

$$|\nabla v_1|^2 + |\nabla v_2|^2 = 0 \tag{6}$$

The smoothness constraint leads to a global optimization of the displacement vectors which are locally defined by the data term (2).

As the vector field has to satisfy both constraints (4), (6) the solution is found by minimizing the deviations from (4) using (3) as a regularizing constraint. Then one obtains the quadratic functional (1) where the smoothness term is weighted by the parameter $\lambda$. Hence, a solution is sought for the variational problem:

$$J(\mathbf{u}) = \inf J(\mathbf{v}) \tag{7}$$

The numerical solution of the variational problem is achieved with the aid of functional analysis. Functional (1) together with the modified approaches discussed in later sections represents a certain class of quadratic functionals for the derivation of displacement vector fields. For such functionals Schnörr (1991) proved the existence of unique solutions. For certain properties of the given quadratic functional the variational problem (7) can be reduced to an N-dimensional set of linear equations

$$L_N \cdot u_N = b_N. \tag{8}$$

where N equals twice the number of pixel positions within the discrete image where displacements have to be calculated. The system of linear equations is compiled using the method of finite elements. The coefficients $\{L_{ij}\}, \{b_i\}, i,j = 1,\ldots,N$ are calculated by integrations over the finite elements. The way of integrating is specified by functional (1).

The vector field is estimated at the nodes of the finite element grid. By increasing the size of the elements the dimension N can be reduced in order to decrease the computation.

## 3   Estimation of first order derivatives

As the displacement vector field is reconstructed by examining the greyvalue function in temporal and spatial domains (4) the corresponding first order derivatives have to be derived from the image sequence. Both derivatives account for costs arising from the data term in (1). The derivation operator has to be approximated by a discrete derivative filter since digital images consist only of discrete data. Therefore, the computation of those derivatives is crucial and will be discussed in this section.

Derivation means an emphasis on high frequency contributions as revealed at grey-value edges. For a continuous function derivation leads to a multiplication by $(-ik)$ in Fourier space for the Fourier representation of the greyvalue function:

$$\frac{\partial}{\partial x} g(\mathbf{x}) \;=\; \frac{1}{2\pi} \int_{-\infty}^{\infty} (-ik_x)\hat{g}(\mathbf{k}) \exp\left[-i(k_x x + k_y y)\right] d\mathbf{k}$$

For discrete approximation of a derivative operator we must use a difference operator:

$$\begin{aligned}
Dg(\mathbf{x}) \;&=\; \frac{1}{\Delta x}[g(\mathbf{x}+\Delta\mathbf{x}) - g(\mathbf{x})] \\
&=\; \frac{1}{\Delta x}\frac{1}{2\pi} \int_{-\infty}^{\infty} \hat{g}(\mathbf{k}) \exp(-i\mathbf{k}{\cdot}\mathbf{x}) \left[\exp(-ik_x\Delta x) - 1\right] d\mathbf{k}
\end{aligned}$$

where for simplicity we have taken $\Delta\mathbf{x} = (\Delta x, 0, 0)$. By means of expanding the Fourier representation of the difference operator $\widehat{D}$, the deviation from the ideal derivative can be examined:

$$\begin{aligned}
\widehat{D} \;&=\; \frac{1}{\Delta x}\left[\exp(-ik_x\Delta x) - 1\right] \\
&=\; -ik_x - \frac{1}{2}k_x^2\Delta x + \frac{1}{6}ik_x^3\Delta x^2 + \cdots
\end{aligned}$$

For a given grid size $\Delta x$, the error of the derivative filter rises with increasing wave number $k$. The error caused by high frequencies can be decreased by either the expansion of the filter, using a larger neighbourhood, or by averaging the data before applying a derivative operator. The effect of an expanded filter mask is illustrated in Figure 2(a). A larger filter leads to increasing computational effort. Therefore, an expansion of the effective filter size by averaging is preferable.

The algorithm presented here uses first order derivatives of a Gaussian as a derivative filter. The discrete approximation of the Gaussian derivative filter is performed by Krawtchouk's polynomials. The mathematical description is found in Hashimoto and Sklansky (1987). Here, a more intuitive presentation is shown in Figure 2(b). Let $\{D_i\}$ be a filter mask of an odd length $N$. Applying a one dimensional Gaussian filter to the first $(N-1)$ pixel and the last $(N-1)$ pixel respectively, leads to smoothed greyvalues at positions next to the center point $\left(\frac{N-1}{2}\right)$. The difference of these smoothed values

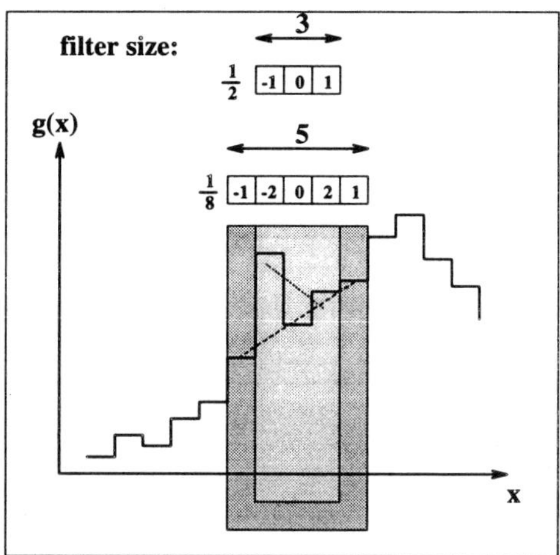
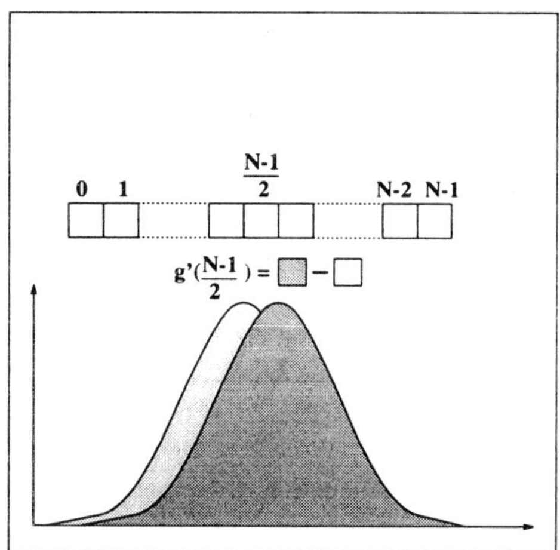



**Figure 2.** *Approximation of first order derivative filter. (a) Effect of high frequency noise on estimation of derivatives. The error can be reduced by expanding the filter width. (b) Illustration of the discrete approximation of a Gaussian derivative filter.*

results in the desired derivative at the position $i = \left(\frac{N-1}{2}\right)$. Hence, the derivative filter of a length $N$ can be obtained by subtracting two Gaussian filters of width $(N-1)$ which are displaced from each other by one pixel.

# 4 Cloud mask and parameter choice

When computing a displacement vector field with the Functional Analytic Method (FAM) the result can be influenced by the choice of three parameters and the specification of a cloud mask.

## Cloud mask

In most cases, high clouds, e.g. caused by convective phenomena, do not correspond to the large scale flow as observed in the WV-images. For the study of ocean currents, land areas as well as clouds create, however, artificial motion boundaries within the oceanic flow pattern. Since the smoothness term (3) always assimilates different motion patterns such discontinuities disturb the resulting vector field. The FAM-algorithm allows the creation of a cloud mask which accomplishes the exclusion of unwanted areas from computation.

## Size of derivative filter

First order derivatives of the image are estimated by applying the derivative of a three-dimensional Gaussian filter as described in Section 3. Because of the discrete nature of the image data, aliasing might occur. Any periodical variation in time has to be sampled twice per period. Usually the time step between subsequent satellite images $\Delta t$ is fixed. Therefore, the aliasing problem in time should be considered by its origin in the spatial domain.

In order to ensure the sampling twice per period any periodic pattern with wavelength $\lambda_0$ moving with velocity $u_0$ should be displaced by a distance less than $\lambda_0$ during the time step $\Delta t$.

$$\Delta t \leq \frac{\pi}{u_0 k_0}, \qquad k_0 = \frac{2\pi}{\lambda_0} \tag{9}$$

Obviously, the aliasing problem arises with increasing speed $u_0$ and decreasing wavelength $\lambda_0$. If $u_0$ is the maximum speed within the image sequence to be examined, (9) allows to describe the critical value for $\lambda_0$.

$$\lambda_0 \geq 2u_0 \Delta t$$

Aliasing effects can be avoided by suppressing wavenumbers larger than $k_0 = 2\pi/\lambda_0$. Therefore, the Gaussian filter is applied to the image data. For the following discussion, the representation of the Gaussian filter in the momentum space is necessary.

$$G(x) = \frac{1}{\sqrt{2\pi}\sigma} \exp\left(-\frac{x^2}{2\sigma^2}\right)$$
$$\widehat{G}(k) = \exp\left(-\frac{\sigma^2 k^2}{2}\right)$$

The suppression of wavelength $k$ depends on the width of the Gaussian filter given by $\sigma^2$. A possible criterion is to suppress wavenumbers larger than $k_0$ by a factor smaller than the value of the Gaussian at its inflexions.

$$\exp\left(-\frac{\sigma^2 k^2}{2}\right) = \exp\left(-\frac{1}{2}\right) \qquad \Longrightarrow \qquad \lambda_0 = 2\pi\sigma$$

The Gaussian is discretely approximated by the binomial distribution. Then, the standard deviation $\sigma$ can be expressed by the filter size $N$.

$$\sigma = \sqrt{\frac{N-1}{4}}$$

Considering the maximum possible displacement $\Delta x = u_0 \Delta t$, given in number of pixels, the following criterion for the suitable filter size $N$ can be derived:

$$N = 1 + \frac{4}{\pi^2} u_0^2 \Delta t^2 \tag{10}$$

**Grid size**

The numerical solution of the given variational problem (1) is performed employing the method of finite elements. This leads to a system of linear equations with dimension twice the number of the elements (8). Therefore, computation can be reduced by choosing larger elements, *i.e.* a larger grid size. For instance, choosing a mesh size of 16 compared to 4 gains a reduction in CPU time by a factor of 60. The mesh size affects the resolution so that the grid should be coarsened under consideration of the variation of the observed motion.

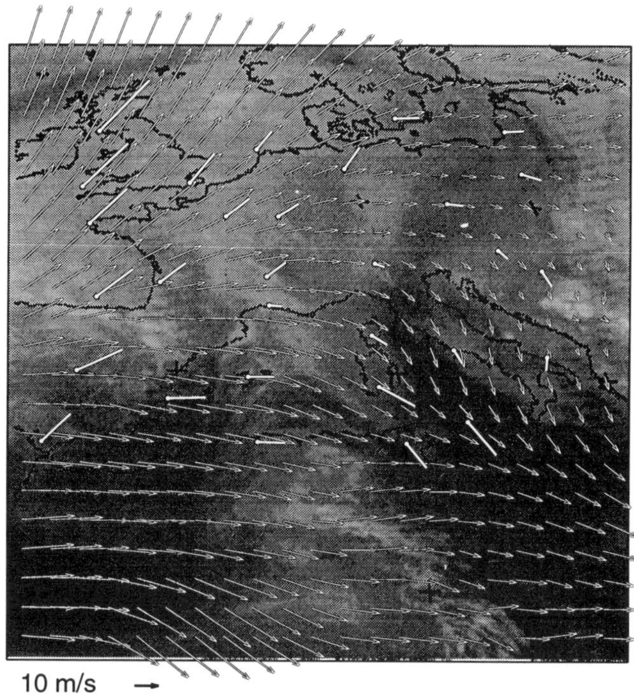



**Figure 3.** *Application of the approach of Horn and Schunck to* METEOSTAT-VW-*images. Heavy white flags indicate synchronous radiosonde winds at 500hPa.*

### Weighting factor $\lambda$

The factor $\lambda$ allows the smoothness contraint to be weighted with respect to the first term in Equation (1). With increasing $\lambda$, the variability of the flow field will be suppressed, and the derived vector field turns into the mean flow. From experience achieved by processing many image sequences, it was established that a good choice for this parameter is 1.

## 5  Applications to Meteosat WV images

FAM has been applied to METEOSAT water vapour (WV) images (May 16, 1993, 22:00, 22:30: 23:00 GMT). According to criterion (10), a size of $41 \times 41$ pixel was chosen for the derivative filter in the spatial domain. The filter size of 41 pixels corresponds to a maximum wind speed of about $45\text{ms}^{-1}$ at latitude $\phi \approx 50°$. The grid size is 24 and $\lambda = 1$.

For this parameter choice the result is shown in Figure 3. The vector field shows a coherent structure corresponding to the smooth pattern found in WV-image sequences. Maximum velocities of about $26\text{ms}^{-1}$ are found over North Africa. Radiosonde winds at 500hPa are overlaid as white lines on the determined vector field. The comparison shows good agreement in direction and absolute velocity. As the WV-signal shows structures from levels between 300 and 550hPa (5–9km) (Laurent 1990), deviations from indicated measured 500hPa winds might be caused by differences in motion at higher levels.

However, the wind speed seems to be slightly underestimated in general. This general underestimation is caused by the general smoothness constraint that shows a deficiency in reconstructing strong shear flow. A smoothing of vectors with a variable distribution in direction usually leads to a decrease in the velocity.

## 6 The oriented smoothness constraint

In the previous section it has been emphasized that because of the general smoothness constraint (3) the approach of Horn and Schunck (1981) is lacking in the reconstruction of motion discontinuities. This deficiency causes some weakness in reconstructing strong shear flow at frontal zones and in the correct estimation of high velocities as revealed in jet stream areas.

On the basis of the approach of Horn and Schunck (1), Nagel (1987) proposed a different smoothness term in order to improve the reconstruction of flow discontinuities:

$$J(\mathbf{v}) \;=\; \int \left\{ [g_t + \mathbf{v}\cdot\nabla g]^2 + \lambda^2 \left[ \frac{(\nabla v_1 \cdot \nabla g')^2 + (\nabla v_2 \cdot \nabla g')^2 + \gamma(|\nabla v_1|^2 + |\nabla v_2|^2)}{2\gamma + |\nabla g|^2} \right] \right\} d\mathbf{x}$$

where $\nabla g' = (g_y, -g_x)^T$. Clearly $\nabla g' \perp \nabla g$ and $|\nabla g'| = |\nabla g|$. This modification allows motion discontinuities at greyvalue edges. Therefore, it is called oriented smoothness constraint.

In homogeneous areas with vanishing greyvalue gradient this constraint corresponds to the general smoothness of Horn and Schunck (1981).

$$\frac{(\nabla v_1 \cdot \nabla g')^2 + (\nabla v_2 \cdot \nabla g')^2 + \gamma(|\nabla v_1|^2 + |\nabla v_2|^2)}{2\gamma + |\nabla g|^2} \approx \frac{1}{2}\left( |\nabla v_1|^2 + |\nabla v_2|^2 \right)$$

At greyvalue edges with a strong gradient the general smoothness term is suppressed by a factor $1/|\nabla g|^2$ while the vector component parallel to the edge is smoothed.

$$\frac{(\nabla v_1 \cdot \nabla g')^2 + (\nabla v_2 \cdot \nabla g')^2 + \gamma(|\nabla v_1|^2 + |\nabla v_2|^2)}{2\gamma + |\nabla g|^2} \approx \left( \nabla v_1 \cdot \frac{\nabla g'}{|\nabla g|} \right)^2 + \left( \nabla v_2 \cdot \frac{\nabla g'}{|\nabla g|} \right)^2$$

This results in a control of the general smoothness constraint parallel to the greyvalue gradient. Figure 4 shows the result for the oriented smoothness constraint. The parameter choice of filtersize, $\lambda$ and grid size is identical to the case in Figure 3, $\gamma = 0.01$.

Apparently, the oriented smoothness term leads to much more variability of the resulting vector field compared to the approach of Horn and Schunck (1981)(Figure 3). Furthermore, in high wind speed areas the maximum velocity is higher. Thus, the high wind speed of 30ms$^{-1}$ over Great Britain is in agreement with the estimated vector field in that area. This means an improvement compared to the general underestimation of velocity by the approach of Horn and Schunck.

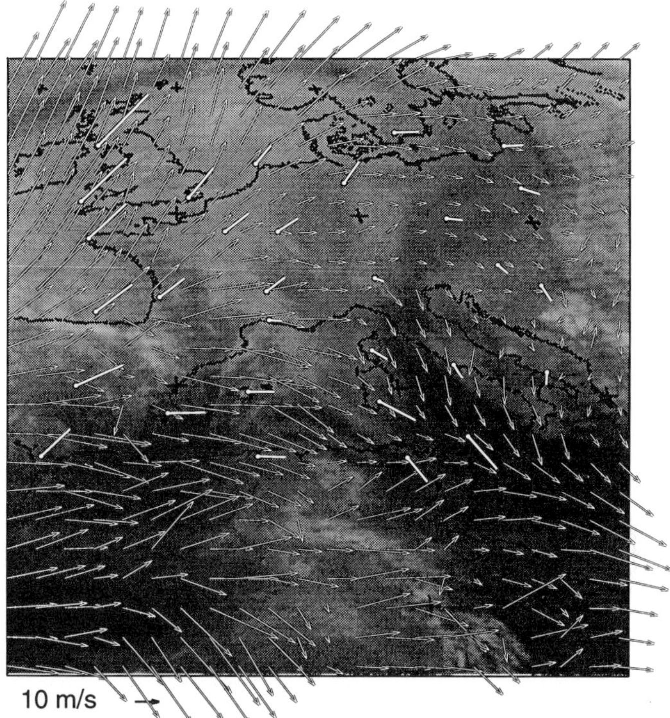


**Figure 4.** *Application of oriented smoothness constraint to METEOSAT-WV-images. Heavy white flags indicate synchronous radiosonde winds at 500hPa.*

The distortion of the vector field by convective clouds, for example, is stronger. Therefore, the masking (see Section 4) of such areas is necessary for the approach of Nagel (1987) while the general smoothness constraint appears to be much more robust.

# 7 Controlling kinematic quantities

Strong shear flow does not necessarily correspond to strong greyvalue variations. Therefore, an approach independent of the data seems to be desirable. Instead of controlling the reconstruction of motion by the image data an attempt of considering oceanic or atmospheric kinematics will be presented in the following.

By splitting the general smoothness term (3), the kinematic quantities can be weighted separately.

$$
\begin{aligned}
|\nabla v_1|^2 + |\nabla v_2|^2 &= \frac{1}{2}\left(|\mathrm{div}\,\mathbf{v}|^2 + |\mathrm{def}\,\mathbf{v}|^2 + |\mathrm{rot}\,\mathbf{v}|^2\right) \\
\mathrm{div}\,\mathbf{v} &= v_{1x} + v_{2y} \\
\mathrm{def}\,\mathbf{v} &= \sqrt{(v_{2y} - v_{1x})^2 + (v_{1y} + v_{2x})^2} \\
\mathrm{rot}\,\mathbf{v} &= v_{2x} - v_{1y}
\end{aligned}
$$

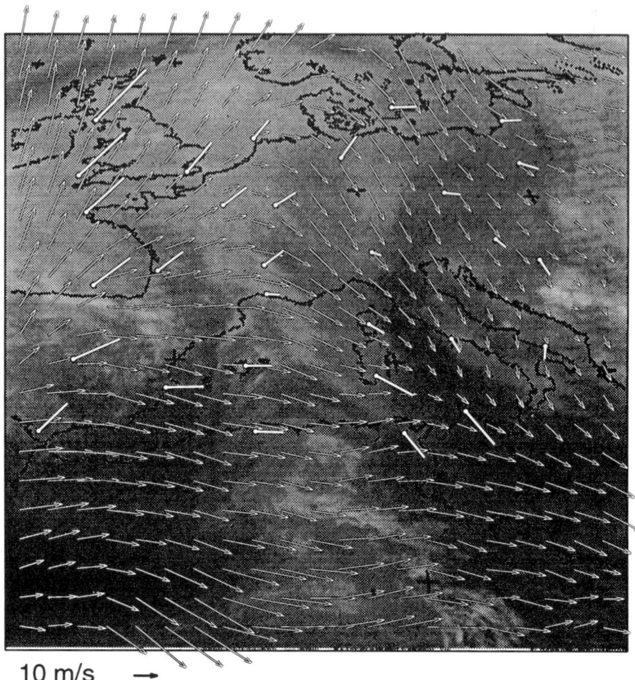



**Figure 5.** *Different weights of the kinematic quantities,* $\lambda_{\mathrm{div}}=4, \lambda_{\mathrm{def}}=\lambda_{\mathrm{rot}}=0.5$. *Heavy white flags indicate synchronous radiosonde winds at 500hPa.*

The aim of the separate treatment of the kinematic quantities is the reconstruction of motion by smoothing its divergence field and, thus, the extraction of the geostrophic component of the observed flow in the upper troposphere from METEOSAT-WV images. Figure 5 shows the result for a choice of the different weights of the kinematic quantities $\lambda_{\mathrm{div}}=4$, $\lambda_{\mathrm{def}}=\lambda_{\mathrm{rot}}=0.5$. A clear variation of the results compared to those from general and oriented smoothnesses is mainly achieved in the area over the Baltic and central Europe. The vector field in this area is turned clockwise, and therefore the eastern side of the ridge over central Europe is much more pronounced.

## 8  Conclusions

The case study of a METEOSAT-WV image sequence with the presented functional analytic approach shows the applicability of the method for motion analysis of satellite imagery. The presented modification of the approach of Horn and Schunck (1981) by the oriented smoothness (Nagel 1987) shows an improvement in respect to the correct estimation of high velocities within high wind speed areas.

High variations of the observed flow pattern can be achieved by a separate treatment of the kinematic quantities with a strong weight on the divergence compared to the weights of deformation and rotation. Nevertheless, the sparse distribution of 500hPa radiosonde winds is not sufficient for a detailed discussion. An improved verification with close wind fields at different levels is necessary.

The oriented smoothness provides a useful tool for the motion analysis of satellite scenes where flow discontinuities or strong shear flow appear in the presence of greyvalue edges. Therefore, the application to motion analysis of sea ice movements in satellite image sequences is planned for the future.

# Acknowledgements

The authors would like to thank Professor G Warnecke for many helpful and stimulating discussions. We also appreciate the explanations of many numerical details by Dr Schnörr from Universität Hamburg, Fachbereich Informatik.

The program implementation of the approach of Horn and Schunck (1981) and the oriented smoothness constraint (Nagel 1987) were performed by IITB (Fraunhofer-Institut für Informations- und Datenverarbeitung). Advices on these algorithms by Dr. Enkelmann (IITB) are gratefully acknowledged. The project was partly funded by the German BMFT (BEO/71 03F0016A).

# References

Büche G, Kummer A, Ottenbacher A, Fischer H, 1991, Displacement vectors from METEOSTAT-VW-images using a new extraction technique, *Proc Workshop on Wind Extraction from Operational Meteorological Satellite Data, Washington, DC,* 17 - 19th Sept. 1991

Hashimoto M, Sklansky J, 1987, Multiple Order Derivatives for Detecting Local Image Characteristics, *Computer Vision, Graphics and Image Processing,* **39,** 28

Horn B K P, Schunck B G, 1981, Determining optical flow, *Artificial Intelligence,* **17,** 185

Kamachi M, 1989, Advective Surface Velocities Derived from Sequential Images for Rotational Flow Field: Limitations and Applications of Maximum - Cross - Correlation Method with Rotational Registration, *J Geophys Res,* **94,** No. C12, 18,277

Laurent H, 1990, Feasibility study on water vapour wind extraction techniques, *Report of Laboratoire de Météorologie Dynamique du CRNS, Ecole Polytechnique, F-91128 Palaiseau, France*

Nagel H H, 1987, On the estimation of optical flow: Relations between different approaches and some new results, *Artificial Intelligence,* **33,** 299

Schnörr C, 1991, Determining Optical Flow for Irregular Domains by Minimizing Quadratic Functionals of a Certain Class, *Int J Computer Vision,* **6,** No. 1, 1991, 25

Warnecke G, 1987, The Visualisation of the Ceaseless Atmosphere, *Remote Sensing Applications in Meteorology and Climatology,* eds R A Vaughan, D Reidel Publ Co, Dordrecht, 245

Wu Q X, 1993, Computing Velocity Fields From Sequential Satellite Images, *Satellite Remote Sensing of the Oceanic Environment,* eds Jones, Sugimori, Stewart, Seibutsu Kengyusha Co Ltd

# References


Büche G, Kummer A, Ottenbacher A, Fischer H, 1991, Displacement vectors from METEOSTAT-vw-images using a new extraction technique, *Proc Workshop on Wind Extraction from Operational Meteorological Satellite Data, Washington, DC,* 17 - 19th Sept. 1991

Hashimoto M, Sklansky J, 1987, Multiple Order Derivatives for Detecting Local Image Characteristics, *Computer Vision, Graphics and Image Processing,* **39,** 28

Horn B K P, Schunck B G, 1981, Determining optical flow, *Artificial Intelligence,* **17,** 185

Kamachi M, 1989, Advective Surface Velocities Derived from Sequential Images for Rotational Flow Field: Limitations and Applications of Maximum - Cross - Correlation Method with Rotational Registration, *J Geophys Res,* **94,** No. C12, 18,277

Laurent H, 1990, Feasibility study on water vapour wind extraction techniques, *Report of Laboratoire de Météorologie Dynamique du CRNS, Ecole Polytechnique, F-91128 Palaiseau, France*

Nagel H H, 1987, On the estimation of optical flow: Relations between different approaches and some new results, *Artificial Intelligence,* **33,** 299

Schnörr C, 1991, Determining Optical Flow for Irregular Domains by Minimizing Quadratic Functionals of a Certain Class, *Int J Computer Vision,* **6,** No. 1, 1991, 25

Warnecke G, 1987, The Visualisation of the Ceaseless Atmosphere, *Remote Sensing Applications in Meteorology and Climatology,* eds R A Vaughan, D Reidel Publ Co, Dordrecht, 245

Wu Q X, 1993, Computing Velocity Fields From Sequential Satellite Images, *Satellite Remote Sensing of the Oceanic Environment,* eds Jones, Sugimori, Stewart, Seibutsu Kengyusha Co Ltd

# Participants


- Dr Giulia Abbate
  ENEA
  Casaccia- AMB S.P. 77
  Via Anguillarese 301
  00100 Roma, Italy

- Mr Raul Aguirre-Gomez
  University of Southampton
  Department of Oceanography
  Highfield
  Southampton SO9 5NH, UK
  EMAIL: ra1 @uk.ac.southampton

- Mr Dimitrios Alexandris
  University of Athens
  Department of Applied Physics
  Ippocratous 33
  106 80 Athens, Greece

- Mr Jose A P Almeida
  University of Dundee
  Department of APEME
  Dundee DD1 4HN, UK
  EMAIL: j.a.p.almeida@dundee.ac.uk

- Mr Ali Nadir Arslan
  Cukurova University
  Department of Electrical & Electronics
  Engineering
  01330 Adana, Turkey
  EMAIL: Alinadir@trcuniv.bitnet

- Prof J Askne
  Chalmers University of Technology
  Dept Radio and Space Science
  S-41296 Göteborg, Sweden

- Miss Jane Louise Attwell
  Linacre College
  Oxford OX1 35A, UK

- Miss Ayse Sevinc Aydinlik
  Tubitak Marmara Research Center
  PO Box 21
  Gebze, Turkey
  EMAIL: sevinc@tmmbeam.bitnet

- Dr Lutz Bannehr
  Freie Universität Berlin
  Institut fur Meteorologie
  Thielallee 50
  D-14195 Berlin, Germany

- Miss Eva Erzsebet Borbas
  Hungarian Meteorological Service
  Satellite Research Laboratory
  PO Box 39 1675
  Budapest, Hungary
  EMAIL: h10356 muh@huella.bitnet

- Ms Sophie Bouffies
  Centre d'Etude de Saclay
  LMCE- Batiment 709
  L'Orme des Merisiers
  91191 Gif-Sur-Yvette
  Cedex, France
  EMAIL: bouffie@idefix.saclay.cea.fr

- Mr Ian Anthony Brown
  University of Dundee
  Department of Geography
  Dundee DD1 4HN, UK

- Dr Matthias Carlsohn
  Am Heiddamm 36G
  D-28355 Bremen, Germany
  EMAIL: aO5a@alf.zfu.uni-bremen.de

- Mrs Eleni Charou
  Inst Info & Telecommunications
  N.C.S.R. "Demokritos"
  Ag. Paraskevi 153 10, Attiki, Greece
  EMAIL: exarou@iit.nrcps.ariadne-t.gr

- Dr Paolo Cipollini
  Dept of Information Engineering
  University of Pisa
  Via Diotisalvi 2
  56126 Pisa, Italy
  EMAIL: cipo@iet.unipi.it

- Dr Pierre Couvert
  DSM/LMCECE
  Saclay 91191
  Gif-sur-Yvette
  Cedex, France
  EMAIL: couvert@asterix.saclay.cea.fr

- Professor A P Cracknell
  University of Dundee
  Department of APEME
  Dundee DD1 4HN, UK

- Mr Terence B Davis
  66 Panoramic Drive
  SauK Ste. Marie Ontario P6B 5VL,
  Canada

- Mr A R da Silva Marcal
  8 Seafield Road
  Dundee DD1 4NS, UK
  EMAIL: amarchal@rs.dundee.ac.uk

- Miss V S de Almeida Santos
  Rua De Dona Estefania 23, 2
  1100 Lisboa, Portugal

- Mr A C de Vries
  University of Gröningen
  Department of Physical Geography
  Kerklaan 30
  9737 LG Haren
  The Netherlands
  EMAIL: a.c.de.vries@biol.rug.nl

- Miss Marta Dioszeghy
  Hungarian Meteorological Service
  Satellite Research Laboratory
  PO Box 39 1 675
  Budapest, Hungary
  EMAIL: h10356 muh@huella.bitnet

- Mr I D Downey
  Natural Resources Institute
  Central Avenue
  Chatham Maritime
  Kent ME4 4TB, UK

- Mrs Julie Englezou
  University of Pireaus
  Department of Maritime Studies
  Karaoli and Dimitriou Str. 40
  1 8532 Piraeus, Greece

- Dr Luca Facheris
  Dipartimento di Ingegneria Elettronica
  University of Florence
  Via S. Marta 3 501 39
  Firenze , Italy
  EMAIL: giuli@ingfi1.ing.unifi.it

- Dr N E Fancey
  University of Edinburgh
  Faculty of Science and Engineering
  Edinburgh EH9 3JZ, UK
  EMAIL: n.e.fancey@ed.ac.uk

- Prof A F G Fiuza
  University of Lisbon
  Oceanographic Group
  Department of Physics
  Rua da Escola Politecnica 58
  1 200 Lisbon
  Portugal

- Mr Mark Watford Freeman
  The View
  Cairnhill Terrace
  Newbottle
  Houghton-le-Spring, DH4 4SP
  Tyne and Wear, UK

- Dr I D Gardiner
  Dept Mechanical Engineering
  University of Abertay-Dundee
  Bell Street
  Dundee, UK

- Miss M A dos Santos Goncalves
  Rua Passos Manuel no. 78 - 2 Dto
  11 00 Lisboa, Portugal
  EMAIL: dgomes@sisdin9.ist.utl.pt

- Mr Michael Grech
  "Lonicera"
  St Theresa Steet
  ZBR04 Zabbar, Malta
  EMAIL: mgrec@unimt.mt

- Mr Paul Keron Guinnessy
  Queen Mary and Westfield College
  Department of Geography
  Mile End Road
  London E1 4NS, UK
  EMAIL: p.k.guinnessy@qmw.ac.uk

- Mrs Mahnaz Gumrukcuoglu
  Bogazici University
  Deptartment of Electrical
  and Electronic Engineering
  PO Box 2, 80815 Bebek,
  Istanbul, Turkey
  EMAIL: celasun@trboun.bitnet

- Dr T H Guymer
  Chilworth Research Centre
  Gamma House
  Centre for Ocean Circulation
  Chilworth
  Southampton SO1 7NS, UK

- Prof C G Helmis
  University of Athens
  Department of Applied Physics
  Laboratory of Meteorology
  Ippokratous Street 33
  10680 Athens Greece

- Mr S-I Hinopoulos
  Alamanas 36
  174.55 Alimos
  Athens, Greece

- Dr Weigen Huang
  Second Institute of Oceanography
  State Oceanic Administration
  Post Code: 310012
  PO Box 1207 Hangzhou, P.R. China

- Dr Sergey Karetnikov
  Institute for Lake Research, RAS
  Laboratory of Geography & Hydrology
  Sevastinov Str 9
  196199 St Petersburg, Russia

- Miss Judit Kerenyi
  Hungarian Meteorological Service
  Satellite Research Laboratory
  PO Box 39
  1675 Budapest, Hungary
  EMAIL: h1035kere@huella.bitnet

- Mr Sabu Kim
  University College London
  Holmbury
  St Mary
  Dorking RH5 6NT
  Surrey,UK
  EMAIL: sk@mssl ucl.ac.uk

- Ms Dimitra Kitsiou
  University of the Aegean
  Department of Environmental Studies
  Karadoni Street 17
  811 00 Mytilini, Greece

- Mr Erik Korsbakken
  Norut Informations Technology Ltd
  9005 Tromsoe, Norway
  EMAIL: erik.korsbakken@itek.norut.no

- Dr T V Kumaran
  University of Madras
  Department of Geography
  600 005 Madras, India

- Mr Tilt Kutser
  Estonian Manne Institute
  Lai 32
  EE-0001 Tallinn, Estonia
  EMAIL: mati@ak.ioc.ee

- Miss Maria Luisa Giusti Latino
  Estrada da Luz 234 -8 ESQ
  1 600 Lisboa, Portugal

- Mr Jan Leonhard Lieser
  Freie Universität
  Berlin Institut fur Meteorologie
  Thielallee 50 D-141 95
  Berlin, Germany
  EMAIL: lieser@fub46.zedat.fu-berlin.de

- Dr D Mantripp
  Mullard Space Science Lab
  UCL
  Dept Physics and Astronomy
  Holmbury St Mary
  Dorking RH5 6NT, UK

- Mr Daniel Michal Markowski
  Cylkowskiego 9A/1
  81-465 Gdynia-Redlowo, Poland

- Mr Joao Miguel Oliveira Martins
  Rua do Alecrim n: 43 4: ESQ
  1200 Lisboa, Portugal
  EMAIL: fimom@ptearn

- Dr J McGregor
  Victoria University of Wellington
  Geophysics Institute
  PO Box 600
  Wellington, New Zealand

- Mr Alex Moon
  Middlesex University
  Dept Geography & Environment
  Queensway
  Enfield Middlesex EN3 4SF, UK

- Mr Knut Müller
  Dunckerstr 73 1 0437
  Berlin, Germany
  EMAIL: knut@fub46.zedat.fu-berlin.de

- Dr Elena Palmisano
  University of Florence
  Dipartimento di Ingegneria Elettronica
  Via S Marta 3 501 39
  Firenze, Italy
  EMAIL: palmisano@ingfi1.ing.unifi.it

- Mr Dimitris Paronis
  University of the Aegean
  Department of Environmental Studies
  Karadoni Street 17
  81100 Mytilene, Greece

- Prof G E Peckham
  Heriot-Watt University
  Department of Physics
  Riccarton, Currie
  Edinburgh EH14 4AS, UK

- Miss Sofia de Oliveira Pires
  R Frei Antonio das Chaqas 13-4 ESQ
  2900 Setubal, Portugal

- Mr Desmond Power
  Memorial University
  Centre for Cold Ocean
  Resources Engineering
  St John's
  Newfoundland A1 B 3X5, Canada
  EMAIL: des@tera.engr.mun.ca

- Mr Christian Renschler
  Gabelsbergerstrasse 5
  3811 8 Braunschweig, Germany

- Mr Adrianos Retalis
  University of Athens
  Meteorology Laboratory
  Department of Applied Physics
  Ippocratous 33
  106 80 Athens, Greece

- Prof A C B Roberts
  Simon Fraser University
  Department of Geography
  V5A 1 S6 Burnaby, BC, Canada

- Dr Michael Rohn
  Freie Universität Berlin
  Institut fuer Meteorologie
  Thielallee 50
  D-141 95 Berlin, Germany
  EMAIL: rohn@fub46.zedat.fu-berlin.de

- Mr P Saravanakumar
  Kalainagar Colony
  Old Natham Road, Plot No 68
  Reserve Line Post
  625 014 Madurai, India

- Mr Boris Schroeder
  TU Braunschweig
  Institut fur Photogrammetrie
  & Bildverarbeitung
  Gausstr 22
  D-381 06 Braunschweig, Germany

- Mr Volker Schumacher
  Waetjenstr 102
  D-2821 3 Bremen, Germany
  EMAIL: vschumac@rs.dundee.ac.uk

- Dr-Ing Pierluigi Silvestrin
  European Space Agency
  ESTEC / JPP
  Keplerlaan 1
  2201 AZ Noordwijk
  The Netherlands
  EMAIL: psilvest@vmprofs.estec.esa.nl

- Dr D Sloggett
  EOS Ltd
  Broadmede
  Farnham Business Park
  Farnham Surrey GU9 8QL, UK

- Mr Constantinos Stephanidis
  University of Dundee
  Department of APEME
  Perth Road
  Dundee DD1 4HN, UK
  EMAIL: cns@rs.dundee.ac.uk

- Miss Nicola Frances Stevens
  c/o 28 Tumulus Road
  Saltdown
  Brighton BN2 8FS, UK

- Ms Catherine Lucy Stock
  EOS, Broadmede
  Famham Business Park
  Weyon Lane, Farnham
  Surrey DD1 4HN, UK
  EMAIL: catherines@eos.co.uk

- Miss Nilgun Summak
  Middle East Techical University
  Institute of Marine Sciences
  (Phys Oceanography)
  PO Box 28
  33731 Erdemli-lcel, Turkey

- Mr Mark Alexander Tadross
  Scott Polar Research Institute
  Lensfield Road
  Cambridge CB2 1 ER, UK
  EMAIL: mat1 002@cus.cam.ac.uk

- Dr A Tooke
  University of Dundee
  Department of APEME
  Dundee DD1 4HN, UK

- Mr. Sunday Okon Udo
  University of Dundee
  Department of Applied Physics
  Dundee DD1 4HN, UK

- Dr Marius Umego
  Ahmadu Bello University
  Physics Department
  Zaria, Nigeria

- Prof C A Varotsos
  University of Athens
  Laboratory of Meteorology
  Department of Applied Physics
  Ippokratous Street 33
  1 0680 Athens, Greece

- Dr R Vaughan
  University of Dundee
  Department of APEME
  Dundee DD1 4HN, UK

- Dr Paul A Volz
  1805 Jackson Avenue
  Ann Arbor Ml 48103-4039, USA

- Prof Guenter Warnecke
  Freie Universität Berlin
  Institut fur Meteorologie
  Thielallee 50
  D-141 95 Berlin, Germany

- Dr A I Watson
  University of Stirling
  Dept Environmental Science
  Stirling FK9 4LA, UK

- Mr Yong Xue
  University of Dundee
  Department of APEME
  Dundee DD1 4HN, UK

- Mr Mohammed Zawid-Naseem
  Bemampur University
  Department of Marine Sciences
  Bhanja Bihar
  760 007 Bemampur
  Orissa, India

# Index